HUMAN HEALTH AND THE
CLIMATE CRISIS

Gail L. Carlson, PhD

Assistant Professor
Environmental Studies Program
Colby College

Director
Buck Lab for Climate and Environment
Colby College

JONES & BARTLETT
LEARNING

World Headquarters
Jones & Bartlett Learning
25 Mall Road
Burlington, MA 01803
978-443-5000
info@jblearning.com
www.jblearning.com

Jones & Bartlett Learning books and products are available through most bookstores and online booksellers. To contact Jones & Bartlett Learning directly, call 800-832-0034, fax 978-443-8000, or visit our website, www.jblearning.com.

Substantial discounts on bulk quantities of Jones & Bartlett Learning publications are available to corporations, professional associations, and other qualified organizations. For details and specific discount information, contact the special sales department at Jones & Bartlett Learning via the above contact information or send an email to specialsales@jblearning.com.

42949-7

Production Credits

Vice President, Product Management: Marisa R. Urbano
Vice President, Product Operations: Christine Emerton
Director, Product Management: Matthew Kane
Product Manager: Sophie Fleck Teague
Director, Content Management: Donna Gridley
Manager, Content Strategy: Carolyn Pershouse
Content Strategist: Sara Bempkins
Director, Project Management and Content Services: Karen Scott
Project Manager: Kristen Rogers
Project Specialist: Kelly Sylvester
Digital Project Specialist: Rachel DiMaggio
Director of Marketing: Andrea DeFronzo
Senior Marketing Manager: Susanne Walker

VP, Manufacturing and Inventory Control: Therese Connell
Composition: Exela Technologies
Cover Design: Briana Yates
Media Development Editor: Faith Brosnan
Rights & Permissions Manager: John Rusk
Rights Specialist: Benjamin Roy
Cover Image (Title Page): © Africa Studio/Shutterstock, © Tatiana Grozetskaya/Shutterstock, © Warren Faidley/Getty Images, © DisobeyArt/Shutterstock, © Riccardo Mayer/Shutterstock, © Brais Seara/Shutterstock, © Monkey Business Images/Shutterstock.
Printing and Binding: LSC Communications

Library of Congress Cataloging-in-Publication Data

Names: Carlson, Gail L., author.
Title: Human health and the climate crisis / Gail L. Carlson.
Description: Burlington, MA: Jones & Bartlett Learning, [2023] | Includes bibliographical references and index. | Summary: "Written specifically for undergraduate students, Human Health and the Climate Crisis examines the direct and indirect human health impacts of climate change while uniquely exploring climate justice - the equitable protection of all people from climate impacts and the participation of all people in climate-related decision-making regardless of race/ethnicity, class, national origin, indigenous status, and gender. This comprehensive text balances appropriate technical content with sufficient contextual information about public health, epidemiology, and climate modeling for students to be able to comprehend the scientific literature on health impacts"–Provided by publisher.
Identifiers: LCCN 2021033559 | ISBN 9781284207293 (paperback)
Subjects: LCSH: Climatic changes–Health aspects. | Environmental health. | BISAC: MEDICAL / Public Health
Classification: LCC QC902.8 .C37 2023 | DDC 363.738/7452–dc23/eng/20211008
LC record available at https://lccn.loc.gov/2021033559

6048

Printed in the United States of America
25 24 23 22   10 9 8 7 6 5 4 3 2 1

© Monkey Business Images/Shutterstock.

Brief Contents

Contents

About the Author

Gail L. Carlson, Ph.D., is director of the Buck Lab for Climate and Environment at Colby College in Waterville, Maine, and is on the faculty of Colby's Environmental Studies Program. Her course Climate Change, Justice and Health formed the basis for this textbook, and she teaches other courses on environmental health, global health, environmental justice, and environmental activism. Her research focuses on the health hazards of chemical pollutants in the environment, as well as on climate change impacts and attitudes. She regularly advocates for safer chemicals policies and climate action in the media and in legislative campaigns in the state of Maine. Carlson earned a Ph.D. in biochemistry from the University of Wisconsin-Madison.

© Monkey Business Images/Shutterstock

Acknowledgments

I am constantly amazed and inspired by the passion and intellectual curiosity of my students in the Environmental Studies Program at Colby College. In particular, I would like to extend my deepest gratitude to my research assistants for this book, Megan Andersen, Hannah Richelieu, and Ingrid Sant. In addition, students who took my course Climate Change, Justice, and Health gathered global and local health data and stories from affected communities, and they offered the invaluable perspectives of young people, which I have incorporated as much as possible into this book. So a big thank you goes out to Aliza Anderson, Grace Andrews, Oakes Austin, Estelle Baldwin, Isa Berzansky, Arunika Bhatia, Cam Brooks-Miller, D'arcy Carlson, Jill Clinton, Charlotte Del Col, Julia Endicott, Owen George, Chase Goldston, Meghan Hurley, Emma Howard, Marcus Jones, Briana Killian, Ingrid Klinkenberg, Lily Lake, Melody Larson, Johanna Lawton, Bronya Lechtman, Keller Leet-Otley, Mark Leprine, Amelia Lubrano, Dominic Malia, Kathleen Mason, Anna McClean, Molly McGavick, Megan McKenna, Jocelyn Meyer, Arissa Moreno Ruiz, Ali Nislick, Meghan Parker, Sumukh Pathi, Aanavi Patodia, Sara Pipernos, Tiffany Poore, Ian Patterson, Paige Russell, Anjalee Rutah, Ingrid Sant, Lily Sethares, Jackie Seymour, Clare Stephens, Ketty Stinson, Madison Wendell, and Caroline Wren.

I am indebted to my faculty and staff colleagues in the Environmental Studies Program at Colby College for all the ways we support each other and collaborate. In particular, I would like to thank Dr. Stacy-ann Robinson, who has helped me gain a deeper understanding of climate impacts, adaptations, and justice, particularly for small island developing states.

I am grateful for the invitation to write this book that came from Sophie Teague at Jones & Bartlett Learning. It has been a pleasure to work with Sophie and Sara Bempkins, and they have been very helpful to this first-time textbook author. A big thank you also goes to the chapter reviewers, as well as to Kelly Sylvester, Faith Brosnan, and Ben Roy leading the production team.

Finally, and most importantly, I wish to thank my family for their endless support and for enduring many, many months together working and learning from home during the pandemic. They patiently tolerated my innumerable rants about climate inaction and health inequities, and kept me going when the emotional toll of a devastating pandemic, record-setting extreme weather, and ongoing racial injustices overwhelmed me. My spouse, Philip Nyhus, has always been my biggest supporter, for which I am forever and lovingly grateful, and we have been marching side by side in our quest to protect the planet and its inhabitants for over 30 years. My children, Louisa Nyhus and Soren Nyhus, are my constant inspiration to take on the climate crisis and other environmental challenges and to help create a more just, peaceful, and beautiful world. With love, I dedicate this book to them.

Introduction

I wrote this textbook while on sabbatical during a pandemic, when I rarely left my house. In some ways, it was a great time to write a book, as I could stay laser-focused on the project, but in other ways, it was extraordinarily challenging. As the world was devastated by COVID-19 and as racial injustice and political turmoil rocked the United States, a glaring spotlight was cast on the profound ways that inequity creates overwhelming and disproportionate burdens on health and well-being for much of the world's population. Layer on climate change impacts, and the situation becomes even more grim, particularly for those who have contributed very little to the anthropogenic greenhouse gas emissions driving climate change.

In 2020, the world faced record-setting heat and some of the most extreme storms, floods, droughts, and wildfires in recent history. As I write this introduction in the summer of 2021, record high temperatures, deadly heat waves, and extreme weather are again devastating communities. In April 2021, the atmospheric carbon dioxide concentration exceeded 420 ppm for the first time in recorded history. Most people are feeling the impacts of climate change today, some directly and some indirectly, some moderately and some severely, and models predict worsening impacts in the future. We have a narrowing window of time in which to enact effective and equitable strategies to mitigate greenhouse gas emissions and adapt to change. Framing climate change as a public health issue (both a *threat* and an *opportunity*), as this textbook does, gives people a familiar context in which to understand how we are being impacted, what is at stake, and how to take action.

Still, characterizing the human health impacts of climate change is quite a challenge—one that requires detecting changing trends in health outcomes and attributing at least a portion of these changes to exposure to a climate-related stressor, such as increasing temperature, air pollution, floods, or food scarcity. Many data gaps exist, especially in countries that lack the capacity to conduct robust public health surveillance, and at local scales where impacts may be apparent but are often not measured. In addition, we tend to prioritize readily quantifiable measures of health, such as mortality, case incidence, and prevalence, but many other measures and sources of information are important but not so easily quantified, including clinical anecdotes, first-person accounts, stories, and lived experiences of those impacted.

We also need to better understand how health is being affected in a wide range of populations. This requires catalyzing and amplifying research done by scholars from around the world, including from least developed countries and small island developing states. We must also listen closely to voices from diverse and particularly at-risk groups, including Indigenous peoples and communities of color, many of which have been marginalized and traumatized for centuries. These groups have their own experiences, modes of knowledge, and practices to build and sustain strength in the face of climate change and its health impacts.

This textbook is designed for undergraduate and graduate students in a range of disciplines, including environmental studies and sciences, biological sciences, public health, epidemiology, global health, sociology, and public policy, as well as those studying nursing, medicine, and other health professions. This textbook also serves as a valuable resource for professionals working on climate change and public health from within governments, businesses, and nongovernmental organizations, and across civil society at local and global scales. We hope that this textbook is also of interest to those in the general public who wish to learn more about climate justice and climate change impacts on human health and well-being.

The content and organization of textbook chapters are designed to structure a course on climate change and public health. Chapter 1 introduces basic concepts and trends in climate science, policy, and action. Chapter 2 provides an overview of public health and epidemiology and introduces a wide range of sources of health information. Chapter 3 introduces climate justice concepts and examples of people dedicated to protecting the fundamental human rights to health, well-being, and a livable planet. Chapters 4–11 each describe a specific health impact of climate change: heat-related illnesses, health impacts of extreme weather, respiratory and vector-borne diseases, water- and foodborne illnesses, malnutrition, mental health, and the health impacts of human displacement. Chapter 12 describes intersections between climate change and health professions, including clinicians' perceptions about how climate change is affecting their patients, the roles and responsibilities of health professionals to communicate and advocate for climate action, and the ways that the healthcare sector contributes to greenhouse gas emissions. In each chapter, I emphasize climate justice and include stories from a few frontline communities, with the humble acknowledgment that many other critically important stories are not being heard or told that ought to be.

As much as possible, I have incorporated up-to-date information in this textbook, including the latest epidemiological data from impacted populations, but because our knowledge is constantly advancing, no doubt some descriptions will soon be outdated. Each chapter highlights relevant sources of information, and I encourage readers to keep up with the latest information and trends that are continually being described in the scholarly literature, in the media, and in high-level reports. In 2021 and 2022, the *Sixth Assessment Report (AR6)* on climate change will be released by the Intergovernmental Panel on Climate Change, including *AR6 Climate Change 2022: Impacts, Adaptation and Vulnerability*, which summarizes the most recent evidence for human health impacts. In addition, information is available in numerous publicly available health databases that are continually updated, including those maintained by the World Health Organization and other United Nations agencies, the Institute for Health Metrics and Evaluation, the *Lancet* Countdown, the U.S. Centers for Disease Control and Prevention, and other national and local public health agencies. Readers are also encouraged to continue to listen closely to climate change stories from frontline communities.

The ways that the world is responding to climate change are constantly shifting. The book was finished after the United States announced its renewed commitment to climate action under the Paris Agreement. Who holds the reins of power makes a difference, and political transitions can be beneficial or detrimental for meaningful progress on climate action. Fortunately, many countries have pledged to work toward significant reductions in greenhouse gas emissions in the coming years, although details are scarce and commitments insufficient to limit warming to the Paris Agreement target of no more than 1.5°C. At the end of 2021, another UN climate conference (COP26 in Glasgow, Scotland) will have been convened and hopefully will have advanced global climate action even further.

In the meantime, people all over the world, notably people of color, Indigenous peoples, and youth, are engaging in inspiring activism to jump-start awareness and action on climate change. They are organizing people in their own communities and catching the attention of world leaders on the biggest stages. They provide a large measure of hope and a moral imperative to commit to taking action in time to protect the health and well-being of all people. My hope for this textbook is that it will provide critical information to help catalyze this urgent movement for change.

Gail Carlson
Waterville, Maine
July 12, 2021

CHAPTER 1

Introduction to the Climate Crisis

KEY TERMS

Climate change
Weather
Climate
Greenhouse effect
Greenhouse gases (GHGs)
Radiative forcing
Global warming potential (GWP)
Carbon sinks
El Niño–Southern Oscillation
 (ENSO)

El Niño
La Niña
Climate models
Intergovernmental Panel on
 Climate Change (IPCC)
United Nations Framework
 Convention on Climate
 Change (UNFCCC)
Kyoto Protocol
Paris Agreement

Climate mitigation
Climate adaptation
Climate finance
Nationally determined
 contributions (NDCs)
Project Drawdown

LEARNING OBJECTIVES

- List the principal greenhouse gases and their major anthropogenic sources.
- Describe how greenhouse gas emissions lead to net warming of the Earth.
- Describe the major observed changes happening to the Earth's climate.
- Understand generally how future climate changes are predicted.
- Understand generally how countries are committing to climate action under the Paris Agreement.

"**Climate change is the biggest global health threat of the 21st century**." So said a group of health experts raising the alarm in 2009.[1] In the years since, the world has witnessed time and again the impacts of extreme heat, weather disasters, air pollution, drought-fueled crop failures, and human migrations. In 2020 alone, record-setting wildfires in Australia and the western United States killed scores of people, displaced hundreds of thousands more, and cost billions of dollars. The 2020 Atlantic hurricane season saw a record 30 named storms, including Hurricanes Eta and Iota, which devastated regions of Central America after making landfall in the same part of Nicaragua just two weeks apart. Historic monsoon rainfall in

China, the most expensive natural disaster in the world in 2020, killed hundreds and displaced at least a million people.[2] Erratic rainfall and prolonged drought in parts of East Africa, coupled with a destructive locust outbreak and the COVID-19 pandemic, increased severe food insecurity for millions of people.[3,4]

The year 2020 was one of the warmest three years on record, along with 2016 and 2019.[5] In 2020, the average global temperature was 1.2°C (2.2°F) above the preindustrial level from 1850–1900. Much of the world experienced extreme heat in 2020 (dark brown regions in **Figure 1.1**), and the northern hemisphere set a record of 1.28°C (2.30°F) warmer than the 20th-century average.[6] The decade from 2011 to 2020 was the warmest on record, and 2020 was the 44th consecutive year that was warmer than

the 20th-century average.[6] Since the late 1970s, the United States has seen many more record *high* temperatures set than record *low* temperatures, and this trend is growing.[7] Since 2000, the land area in the contiguous 48 U.S. states where "unusually hot" daily high temperatures are recorded is growing, and "unusually hot" daily low temperatures are rising even faster.[8]

In 2020, the United States set a record for the most billion-dollar weather and climate disasters (22): 13 severe storms, six hurricanes, one tropical storm, one drought, and one wildfire event (**Figure 1.2**).[9] These disasters directly or indirectly affected the health, well-being, and livelihoods of millions of Americans.

Climate change is a profoundly human crisis and a health **threat magnifier**, in that

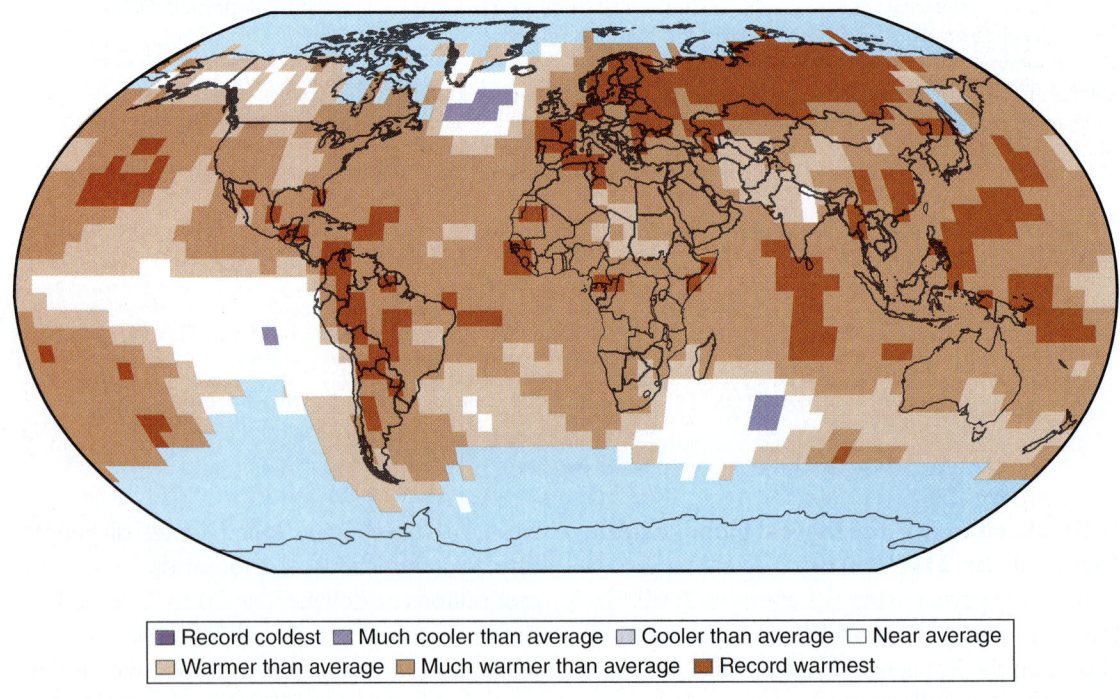

■ Record coldest ■ Much cooler than average ☐ Cooler than average ☐ Near average
■ Warmer than average ■ Much warmer than average ■ Record warmest

Figure 1.1 Map Depicting Global Average Land and Ocean Temperatures in 2020. Dark brown regions had record-warmest temperatures, and lighter brown colors indicate regions warmer or much warmer than average.

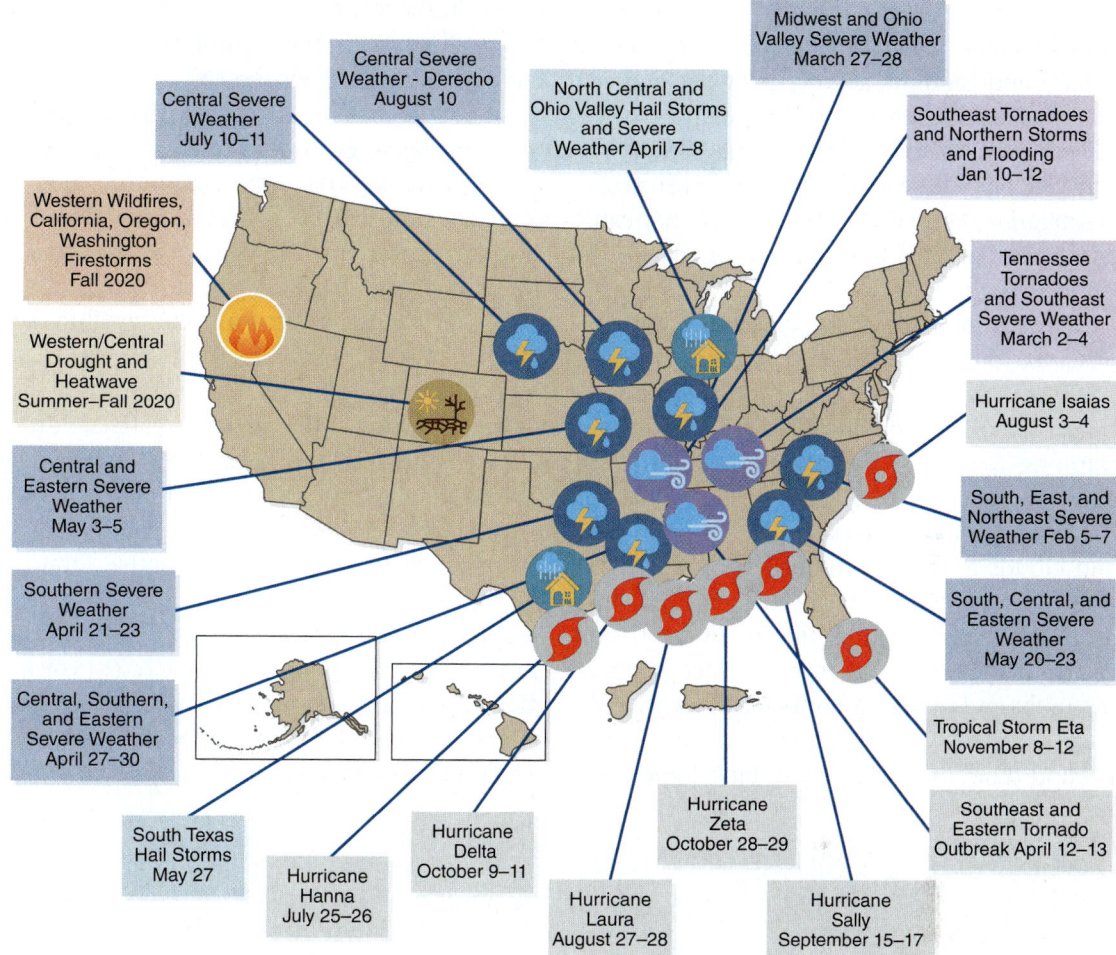

Figure 1.2 All 22 Weather and Climate Disasters That Occurred in the United States in 2020 Cost at Least $1 Billion. NOAA has tracked billion-dollar disasters in the United States since 1980.

Reproduced from National Oceanic and Atmospheric Administration National Centers for Environmental Information. Billion-dollar weather and climate disasters: overview. Accessed February 19, 2021. https://www.ncdc.noaa.gov/billions

it makes existing human health burdens and inequities worse. Understanding how, where, and why human well-being is threatened is essential to minimizing and preparing for the impacts of the climate crisis. Climate action will help to prevent these health threats and at the same time represents **"the largest public health opportunity in more than a century"** because of all the health benefits that will come with a rapid and equitable energy transition.[10]

What Is Climate Change?

Climate change refers to changes in the Earth's climate, attributed directly or indirectly to human activities, that are altering the composition of the global atmosphere and are in addition to natural climate variability observed over the same time period.[11] Climate change impacts weather patterns, but climate and weather are distinct phenomena.

Weather refers to the state of the air and atmosphere—specifically, *temperature*, *rain*, *clouds*, and *storms*—at a specific time and place. **Climate** refers to "average weather," the statistical description of the mean, extremes, and variability of features such as *temperature*, *precipitation*, and *wind* over a period that can range from months to thousands of years but is typically three decades.

The primary change in climate being observed on Earth is increased surface temperatures driven by an enhanced **greenhouse effect**, the phenomenon in which heat-trapping **greenhouse gases (GHGs)** in the atmosphere absorb and emit radiation coming from the Earth's surface, the atmosphere, and clouds after they are heated by the sun. This has the effect of warming the Earth's surface and troposphere (lower layer of the atmosphere). The greenhouse effect occurs naturally and creates a temperature range on Earth that allows for life, but fossil fuel burning and other human activities have increased

GHGs in the atmosphere such that the greenhouse effect has become amplified, leading to increased warming (**Figure 1.3**).

The magnitude of the **anthropogenic** (*human-caused*) greenhouse effect is expressed by **radiative forcing**, a measure of the energy imbalance on Earth that is calculated by subtracting the energy flowing out into space (radiating away from the Earth's surface and troposphere) from the sun's energy flowing in. This number is measured in watts per square meter of surface. A positive number indicates warming, and a negative number indicates cooling. Radiative forcing has increased since the preindustrial reference year of 1750 (**Table 1.1**).

Most of the increase in radiative forcing in recent decades is due to human activities that produce heat-trapping greenhouse gases. Natural phenomena, such as volcanic eruptions and plant decomposition, also release GHGs, but anthropogenic sources predominate today. The primary GHGs are

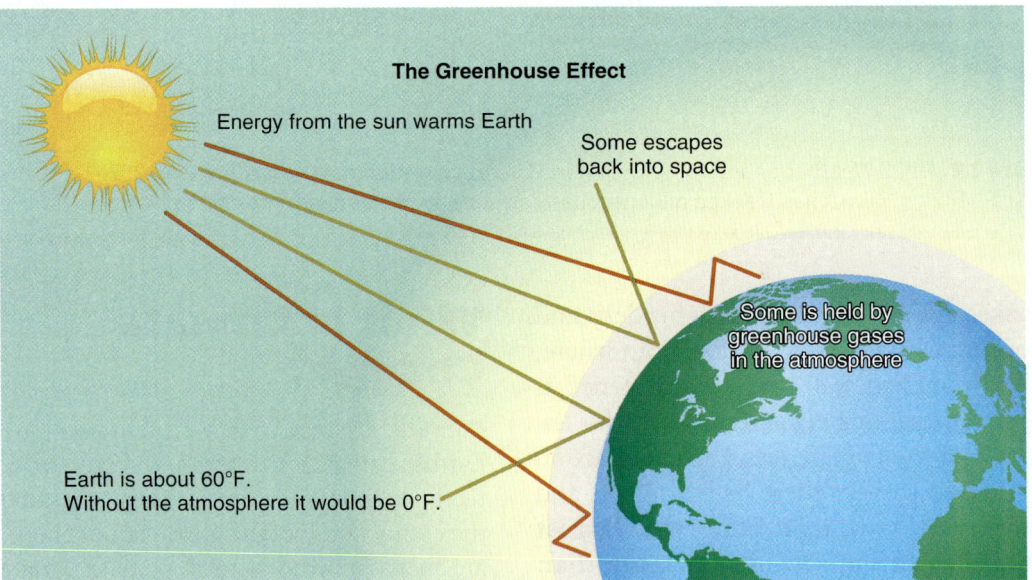

The Greenhouse Effect

Energy from the sun warms Earth

Some escapes back into space

Some is held by greenhouse gases in the atmosphere

Earth is about 60°F.
Without the atmosphere it would be 0°F.

Figure 1.3 Diagram of the Greenhouse Effect, Showing Energy from the Sun Warming the Earth, and the Atmosphere Trapping Some of This Energy as it Radiates Away from the Earth's Surface.

Table 1.1 Increasing Radiative Forcing in Recent Years Relative to 1750.

Year	Radiative Forcing (watts/m²)
1750	0
1950	0.57
1980	1.75
1990	2.17
2000	2.47
2010	2.80
2020	3.18

Data from National Oceanic and Atmospheric Administration. The NOAA Annual Greenhouse Gas Index (AGGI). Updated Spring 2021. Accessed June 16, 2021. https://www.esrl.noaa.gov/gmd/aggi/aggi.html; Intergovernmental Panel on Climate Change. Summary for Policymakers. In: Stocker TF, Qin D, Plattner G-K, et al., eds. *Climate Change 2013: The Physical Science Basis. Contribution of Working Group I to the Fifth Assessment Report of the Intergovernmental Panel on Climate Change.* Cambridge University Press; 2013. https://www.ipcc.ch/site/assets/uploads/2018/02/WG1AR5_SPM_FINAL.pdf

carbon dioxide (CO_2), methane (CH_4), nitrous oxide (N_2O), fluorinated gases such as sulfur hexafluoride (SF_6), and water vapor.[12]

Carbon dioxide (CO_2) is the principal anthropogenic GHG. **Figure 1.4** shows GHG emissions, of which CO_2 is the primary component, in the United States in 2019 and in the world in 2015.[12,13] CO_2 is an atmospheric by-product of fossil fuel combustion, land-use changes such as deforestation and vegetation and peat burning that release stored carbon, and industrial processes, including cement production, in which limestone ($CaCO_3$) is heated and releases CO_2. CO_2 is removed from the atmosphere primarily when it is absorbed by plants via photosynthesis and by oceans and soils as part of the carbon cycle.

In the 1950s, the geochemist **Dr. Charles Keeling** developed an accurate system for measuring atmospheric CO_2 concentration.

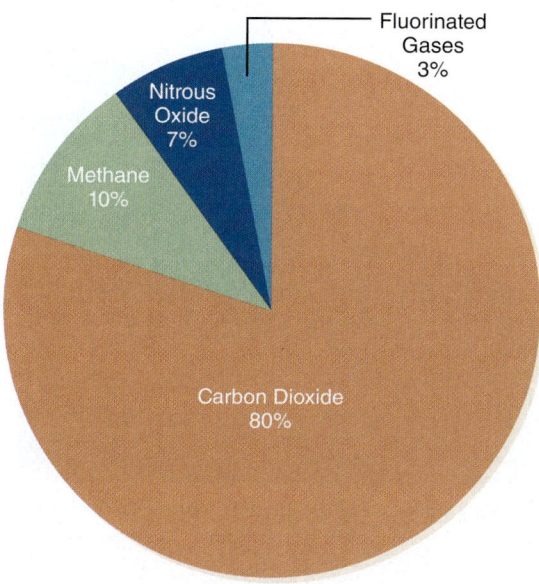

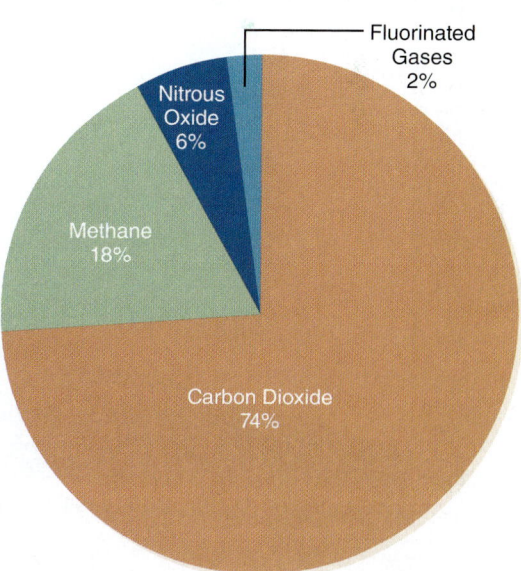

Figure 1.4 Left: U.S. Greenhouse Gas Emissions in 2019. Right: Global Greenhouse Gas Emissions in 2015.

Left: Reproduced from Environmental Protection Agency. Overview of greenhouse gases. Accessed June 7, 2021. https://www.epa.gov/ghgemissions/overview-greenhouse-gases; Right: Data from Environmental Protection Agency. Climate change indicators: global greenhouse gas emissions. Updated April 2021. Accessed June 7, 2021. https://www.epa.gov/climate-indicators/climate-change-indicators-global-greenhouse-gas-emissions

Since 1958, continuous CO_2 measurements have been made at the Mauna Loa Observatory in Hawaii and recorded on the **Keeling Curve** (**Figure 1.5**), which is updated daily and made publicly available.[14] The regular periodic oscillation in CO_2 concentration shown on the curve reflects the seasonal difference in CO_2 uptake via plant photosynthesis. Vegetative leaf cover is more abundant during summers in the northern hemisphere, where land and extratropical forested areas exceed those in the southern hemisphere. Atmospheric CO_2 concentration has increased steadily since 1958, exceeding 420 parts per million (ppm) for the first time in recorded history on April 3, 2021.[14]

Other principal GHGs include methane, nitrous oxide, and fluorinated gases. **Methane** (CH_4) is a gas emitted during the production, distribution, and use of fossil fuels; by livestock digestion and agricultural practices; and when organic waste decays, largely in solid waste landfills. Drilling, flaring, and transport of natural gas, which is composed mostly of methane, constitute a particularly significant source of methane emissions. **Nitrous oxide** (N_2O) is a gas emitted during agricultural and industrial activities, fossil fuel and solid waste combustion, and wastewater treatment. Importantly, N_2O is a by-product of microbial metabolism of nitrogen in synthetic fertilizers used in agriculture. **Fluorinated gases** are synthetic gases containing fluorine that are used in a variety of industrial applications. One such gas is **sulfur hexafluoride**, SF_6, used as an insulating chemical in electricity generation and transmission.

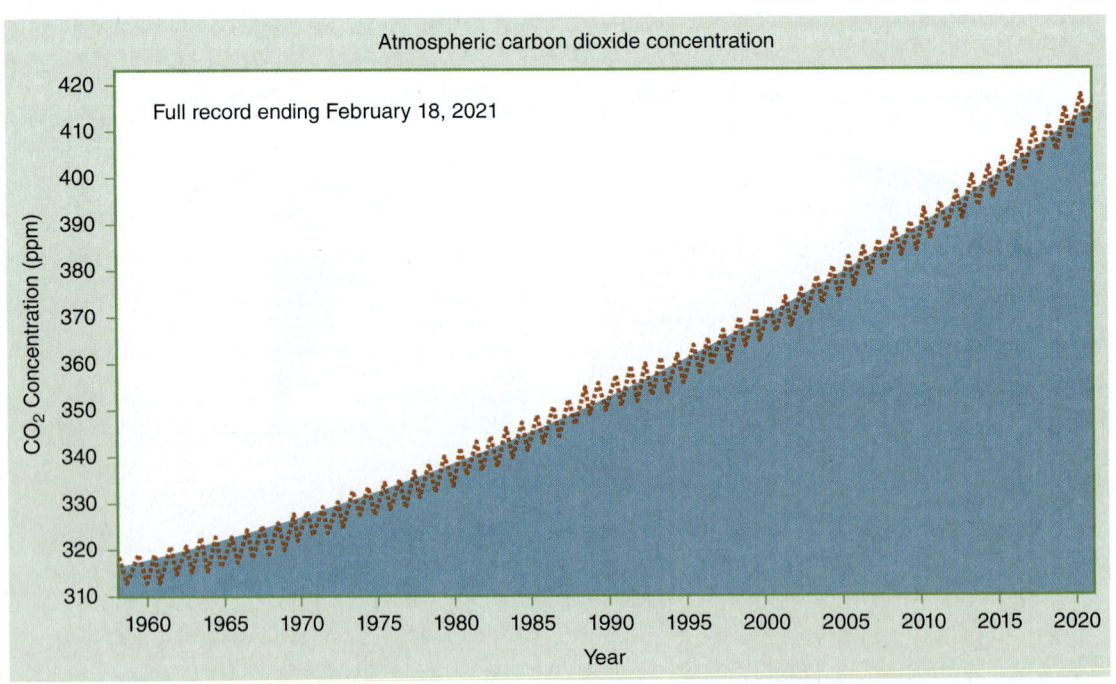

Figure 1.5 The Keeling Curve, a Plot of Global Atmospheric Carbon Dioxide Concentration Recorded at Mauna Loa Observatory in Hawaii and Maintained by Scripps Institution of Oceanography at the University of California San Diego. On February 18, 2021, the CO_2 concentration was 417 ppm.

The relative heat-trapping property of each GHG is expressed as **global warming potential (GWP)**, a measure of how much energy the emissions of one ton of a GHG will absorb over a given period—usually 100 years—relative to the emissions of one ton of CO_2 (**Table 1.2**). The larger the GWP, the more a given gas warms the Earth compared with CO_2. Even though SF_6 is a trace gas in the atmosphere, it has the highest GWP (23,500 times higher than CO_2) and thus contributes significantly to warming. CH_4 and N_2O are also present at lower concentrations than CO_2, but they have higher GWPs.

Another component of the atmosphere that affects warming is **aerosols**, tiny airborne solid or liquid particles of varying chemical composition that absorb and scatter atmospheric radiation, influence cloud formation and properties, and depending on conditions and composition, may cause warming or cooling. Warming is counteracted by **sinks**, processes that remove a GHG, an aerosol, or their precursor chemicals from the atmosphere. **Carbon sinks** remove CO_2 from the atmosphere via **sequestration**, which has the effect of increasing the content of carbon pools on land and in water. Major carbon sinks on Earth are forests, where trees and other plants take up and store CO_2 during photosynthesis, oceans, where CO_2 is dissolved and also taken up by photosynthetic marine plants, and soils.

A Brief History of Climate Science

The phenomenon of global warming was first proposed in 1856 by **Eunice Newton Foote**, an American scientist and women's rights advocate. She reported on an experiment she conducted, showing that carbon dioxide is a GHG:

> My investigations have had for their object to determine the different circumstances that affect the thermal action of the rays of light that proceed from the sun. The highest effect of the sun's rays I have found to be in carbonic acid gas [CO_2]. An atmosphere of that gas would give to our earth a high temperature.[15]

Five years later, Irish physicist **John Tyndall** reported a similar observation: that "carbonic acid diffused through the air" must "produce a change of climate."[16] In 1896, Swedish scientist **Svante Arrhenius** quantified the heat-trapping effects of water vapor and CO_2, reporting, "We now possess all the necessary data for an estimation of the effect on the earth's temperature which would be the result of a given variation of the aerial carbonic acid." Arrhenius noted that among the sources of CO_2 was "the industrial development of our time," including coal, which can be "transformed into carbonic acid."[17]

Many other scientists have made important contributions, including **Guy Callendar**, who published a 1938 paper that linked fossil fuel burning to atmospheric CO_2 and global warming. He reported that human

Table 1.2 Global Warming Potentials of Principal Greenhouse Gases.

Greenhouse Gas	Global Warming Potential (GWP)
Carbon dioxide (CO_2)	1
Methane (CH_4)	28
Nitrous oxide (N_2O)	265
Sulfur hexafluoride (SF_6)	23,500

Data from Intergovernmental Panel on Climate Change. *AR5 Climate Change 2013: The Physical Science Basis.* Published 2013. Accessed February 19, 2021. https://www.ipcc.ch/site/assets/uploads/2018/02/WG1AR5_all_final.pdf

combustion activities had contributed 150,000 tons of CO_2 to the atmosphere and that recorded global temperature measurements at the time showed 0.005°C of warming per year.[18] Two decades later, physicist **Gilbert Plass** published a paper entitled "Carbon Dioxide and the Climate," in which he concluded that the burning of fossil fuels was the major source of atmospheric CO_2 and would lead to an average temperature increase of 1.1°C per century.

A major turning point in the public's focus on climate change came in June 1988, when climate scientist **Dr. James Hansen**, director at the time of NASA's Goddard Institute for Space Studies, testified at a hearing before the U.S. Senate's Energy and Natural Resources Committee. Hansen reported that "the greenhouse effect has been detected, and it is changing our climate now."[19]

> I would like to draw three main conclusions. Number one, the earth is warmer in 1988 than at any time in the history of instrumental measurements. Number two, the global warming is now large enough that we can ascribe with a high degree of confidence a cause and effect relationship to the greenhouse effect. And number three, our computer climate simulations indicate that the greenhouse effect is already large enough to begin to affect the probability of extreme events such as summer heat waves.[19]

Today, more than three decades later, warming continues to occur, and scientists have learned a great deal about its impacts on our weather and climate systems. People all over the world are experiencing climate-change-fueled extreme events and their devastating consequences for health, well-being, and livelihoods.

Observing a Changing Climate

So how much has the Earth warmed since preindustrial times? Scientific institutions such as the U.S. National Oceanic and Atmospheric Administration (NOAA) and the National Aeronautics and Space Administration (NASA) continuously monitor climate change indicators, including atmospheric CO_2 levels, surface temperature (land, sea, or both combined), ocean heat content, sea level rise, ocean acidification (due to CO_2 dissolving in water, forming carbonic acid), glacier extent, and Arctic and Antarctic sea ice extent. **Figure 1.6** shows the change in global land and ocean surface temperature from 1880 to 2020 relative to average temperatures in the reference period 1951–1980. Some variability exists, but net warming has been occurring since the mid-20th century. In 2020, one of the three hottest years on record, the average temperature anomaly was 1.02°C (1.84°F) above the 1951–1980 average.[20]

Warming is often described globally, but it also occurs regionally or locally. The map in **Figure 1.7** illustrates how much of the world was warmer than average in 2020, particularly in northern Russia. Arctic regions are warming much faster than the world overall.

Average global sea height has risen each year since 1993, when satellite observations began (**Figure 1.8**). Sea level rise is caused by melting ice sheets and glaciers adding meltwater to the oceans, as well as thermal expansion of seawater as it absorbs heat. As of September 2020, overall sea level had risen 97 mm (nearly 4 inches) above 1993 levels, with an annual rate of change of 3.3%.[21] By 2100, even if warming is limited to 2°C, oceans are predicted to rise 1–2 feet above 1986–2005 levels.[22]

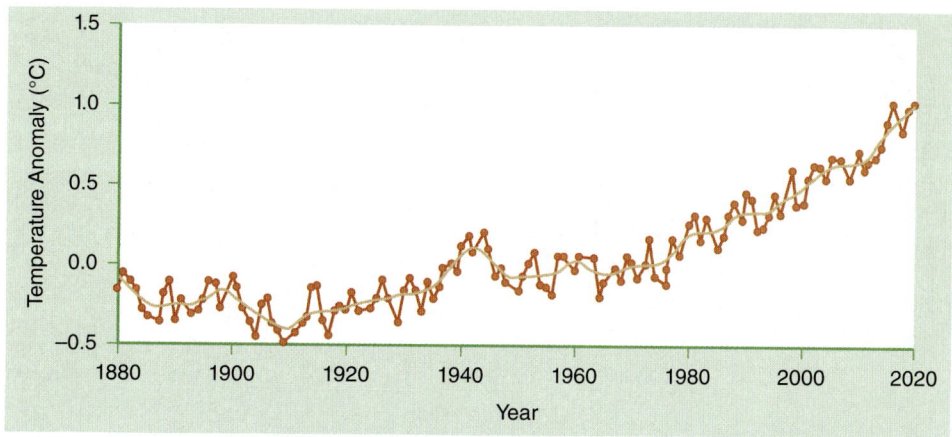

Figure 1.6 Change in Global Land and Ocean Surface Temperature from 1880 to 2020 Relative to Average Temperature in the Reference Period 1951–1980.

Reproduced from National Aeronautical and Space Administration. Facts: global temperature. Accessed February 19, 2021. https://climate.nasa.gov/vital-signs/global-temperature

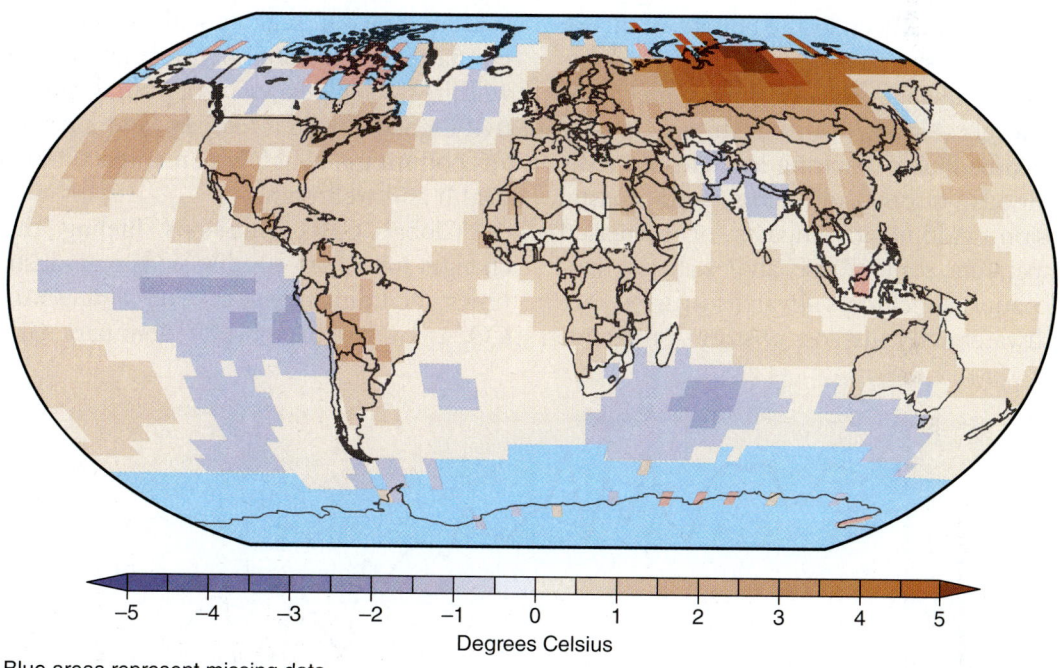

Blue areas represent missing data

Figure 1.7 Deviation of Land and Ocean Temperature During 2020 from a Baseline Average Temperature, 1980–2010. Brown and pink indicate warming.

Reproduced from National Oceanic and Atmospheric Administration National Centers for Environmental Information. Global temperature and precipitation maps. Accessed February 19, 2021. https://www.ncdc.noaa.gov/temp-and-precip/global-maps

Arctic sea ice, which reaches a yearly minimum in September, is declining at a rate of 13% per decade relative to the 1981–2010 median extent, as measured by satellites (**Figure 1.9**). The lowest sea ice extent in the satellite record was measured in 2012, and the second lowest extent was measured in 2020.[23] Land ice sheets in Antarctica and

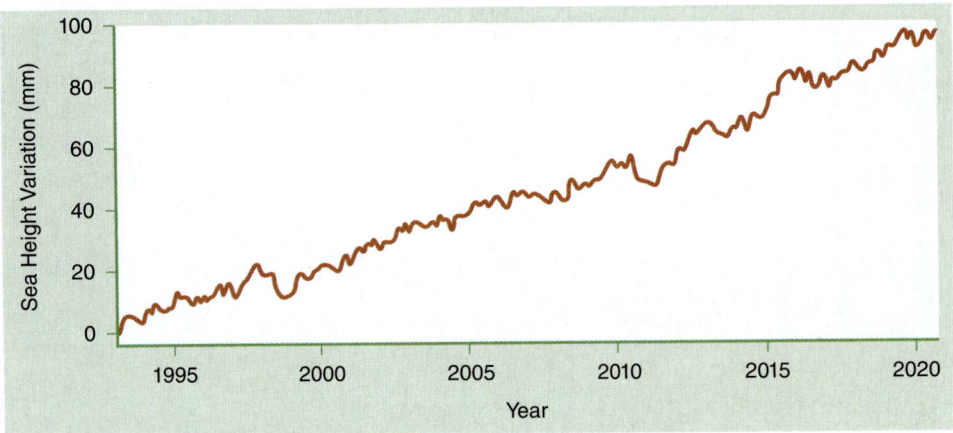

Figure 1.8 Global Sea Level Rise as Measured by Annual Sea Height Relative to the Value in 1993.
Reproduced from National Aeronautical and Space Administration. Facts: sea level. Accessed February 19, 2021. https://climate.nasa.gov/vital-signs/ocean-heat

Greenland have been losing billions of metric tons of mass since 2002.[24]

These changes in turn impact precipitation levels (**Figure 1.10**), which vary greatly across the globe, resulting in increased risks of floods, storms, droughts, and coastal erosion. Additional impacts of increased temperature, sea level rise, and/or altered rainfall patterns include saltwater intrusion into freshwater and soils, loss of storm-protective coastal sea ice in regions with cold climates, permafrost melt, more frequent and extreme wildfires, changes in agricultural productivity, greater water scarcity, and increased air pollution—all of which impact human health and well-being.

Global GHG emissions fueling these changes have risen steadily since preindustrial times, including in recent years (**Figure 1.11**).[25] CO_2 is by far the most predominant GHG,

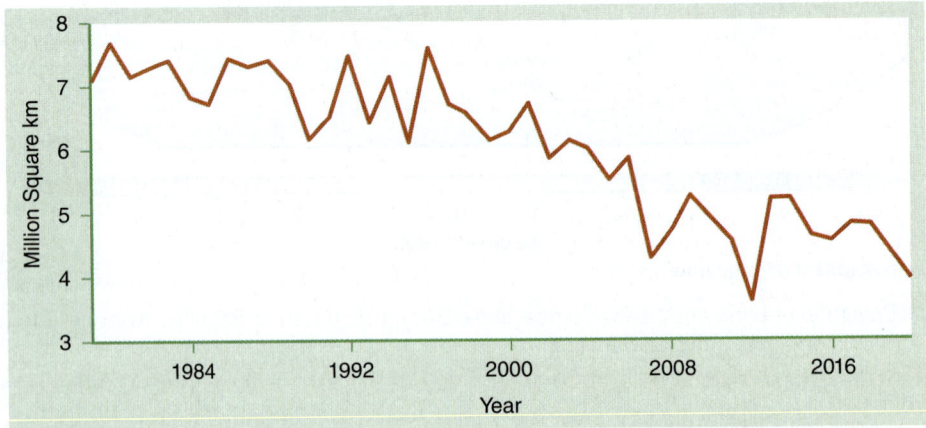

Figure 1.9 Average Monthly Arctic Sea Ice Extent Each September (When Arctic Sea Ice Reaches Its Minimum) Since 1979, Derived from Satellite Measurements. The 2012 sea ice extent is the lowest in the satellite record.
Reproduced from National Aeronautical and Space Administration. Facts: arctic sea ice minimum. Accessed February 19, 2021. https://climate.nasa.gov/vital-signs/arctic-sea-ice

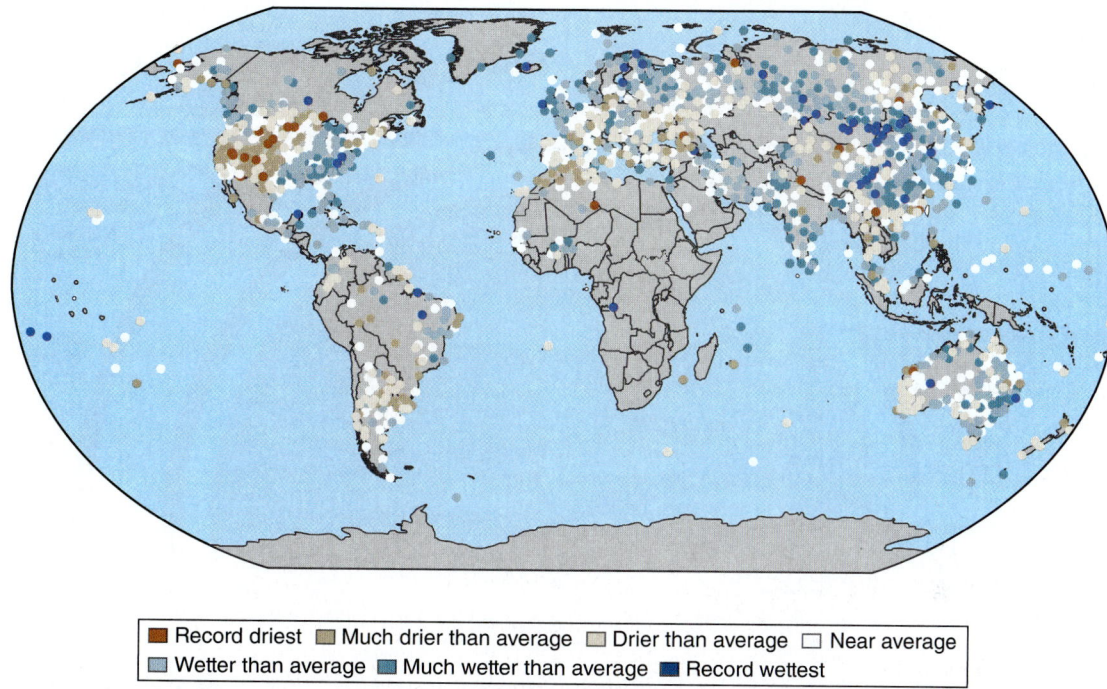

■ Record driest ■ Much drier than average □ Drier than average □ Near average
□ Wetter than average ■ Much wetter than average ■ Record wettest

Figure 1.10 Precipitation in 2020 Indicated as Deviations from Average. Blue dots indicate wetter conditions, and brown dots indicate drier conditions.

Reproduced from National Oceanic and Atmospheric Administration National Centers for Environmental Information. Global temperature and precipitation maps. Accessed February 19, 2021. https://www.ncdc.noaa.gov/temp-and-precip/global-maps

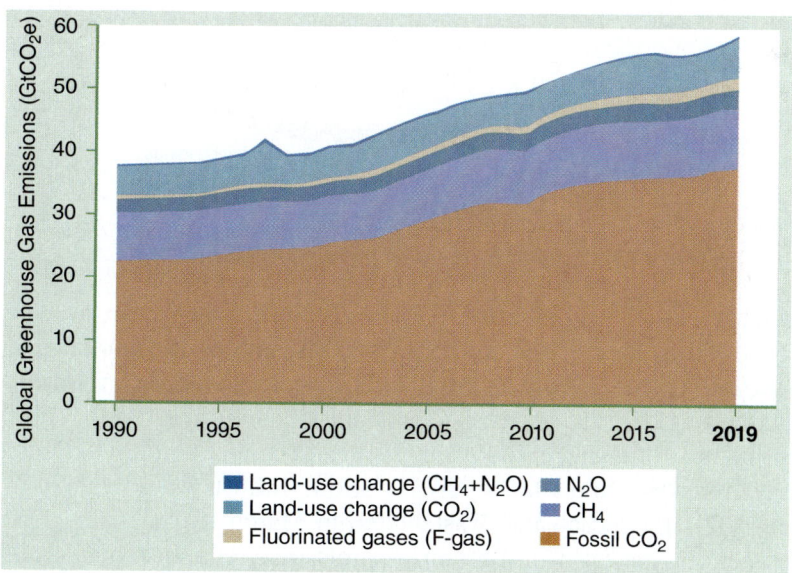

Figure 1.11 Global Increase in Emissions of the Primary Greenhouse Gases (in Gigatons of CO_2 Equivalents) from 1990 to 2019.

Reproduced from United Nations Environment Programme. *Emissions Gap Report 2020*. United Nations Environment Programme; 2020:Figure ES.1. https://www.unep.org/emissions-gap-report-2020

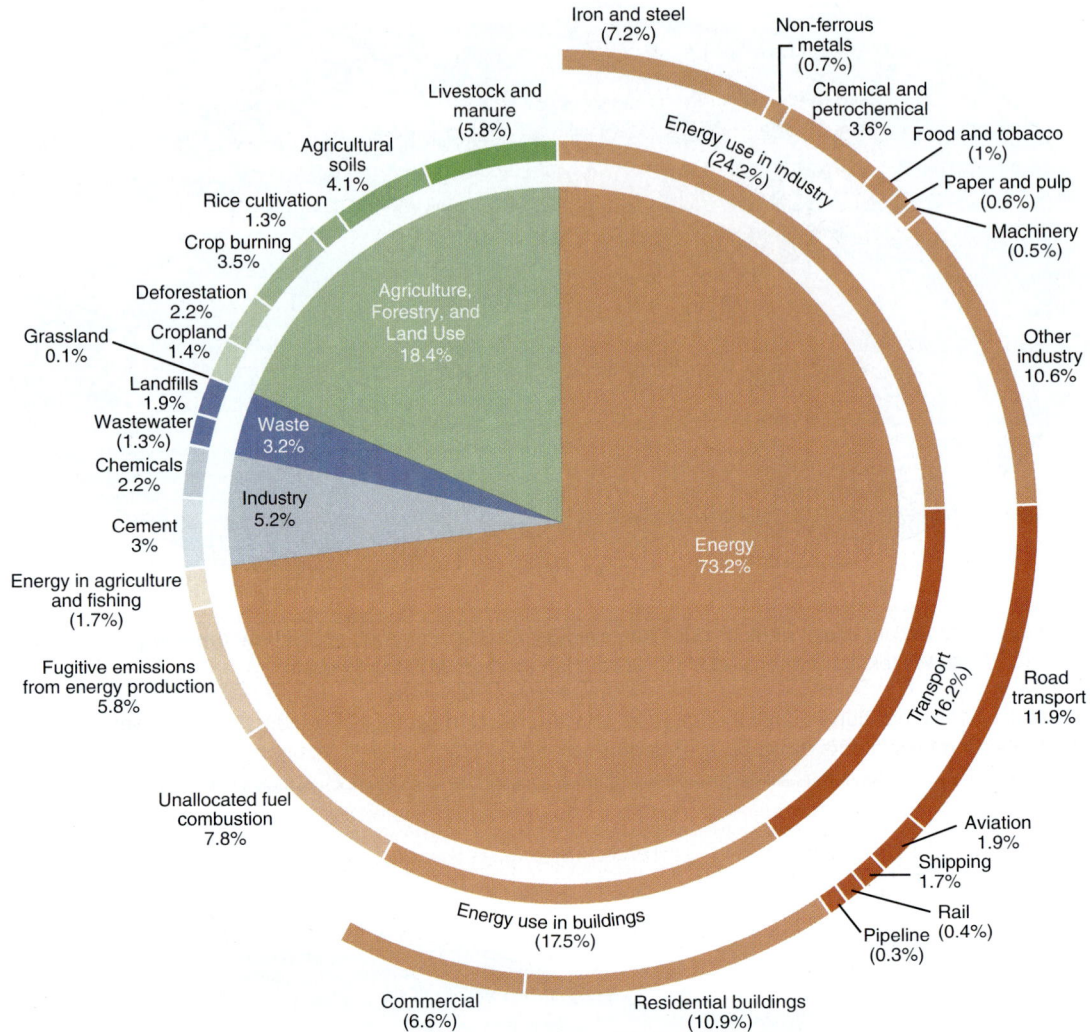

Figure 1.12 Global Greenhouse Gas Emissions by Sector in 2016.

Reproduced from Ritchie H, Roser M. Emissions by sector. Our World In Data. Accessed February 21, 2021. https://ourworldindata.org/emissions-by-sector

arising primarily from fossil fuel combustion, deforestation, and other land use changes. The energy sector, including transportation and electricity and heat generation, is the greatest source of GHGs, followed by agriculture, forestry, and land use changes (**Figure 1.12**). Industrial and waste management practices are also important sources.

In 2020, global GHG emissions decreased by 7%, due primarily to the COVID-19 pandemic and resulting lockdowns, reduced human activities, and economic slowdown (**Figure 1.13**).[25] Lower emissions from ground transport had the biggest impact, followed by power generation, industry, and aviation—particularly from March to June, the period of greatest emissions reduction in 2020.

GHG emissions vary starkly by country and by income level. Countries with the highest *total* and *per capita* GHG emissions

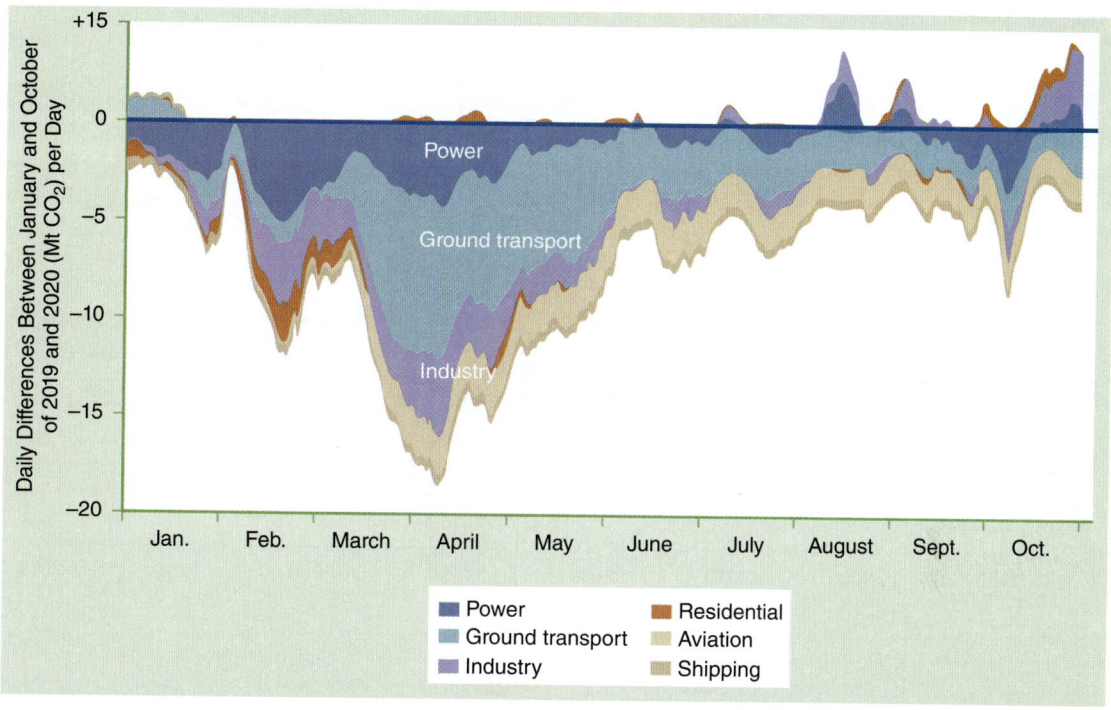

Figure 1.13 Global Reduction in CO_2 Emissions Due to COVID-19 Pandemic Shutdowns in 2020.

Reproduced from United Nations Environment Programme. *Emissions Gap Report 2020.* United Nations Environment Programme; 2020:Figure ES.3. https://www.unep.org/emissions-gap-report-2020

in 2019 are shown in **Figure 1.14**. China had the highest *total* emissions, followed by the United States, EU27+UK (27 European Union member countries plus the United Kingdom), Russia, and Japan.[25] In addition to country-level emissions, international transport is a leading global source. Globally, GHG emissions increased 43% from 1990 to 2015.[8] Since 2005, *total* U.S. GHG emissions have declined 12%.[8] The United States had the highest emissions on a *per capita* basis, followed by Russia, Japan, China, and EU27+UK.[25] Per capita emissions have declined recently in the United States, Japan, and the EU and have risen in China and India, where increased fossil fuel use has powered economic growth and development.

The world's richest 1% of income earners are responsible for more than twice the *total* GHG emissions as the bottom 50% of income earners. The richest 1% also have about 35 times higher *per capita* emissions than the global average level needed by 2030 to limit global warming to 1.5°C (**Figure 1.15**).

In addition to weather and climate variability caused by anthropogenic GHG emissions, natural phenomena also lead to shifting weather patterns, the largest of which is the **El Niño–Southern Oscillation (ENSO)**. ENSO is caused by periodic fluctuations in the temperature of surface waters in much of the tropical Pacific Ocean. Every 3–7 years, these waters warm or cool 1–3°C compared with the average and alter global air currents, which influences surface temperatures and precipitation around the world. The three phases of ENSO are **El Niño** (warmer waters), **La Niña** (cooler waters), and **Neutral** (neither warming nor

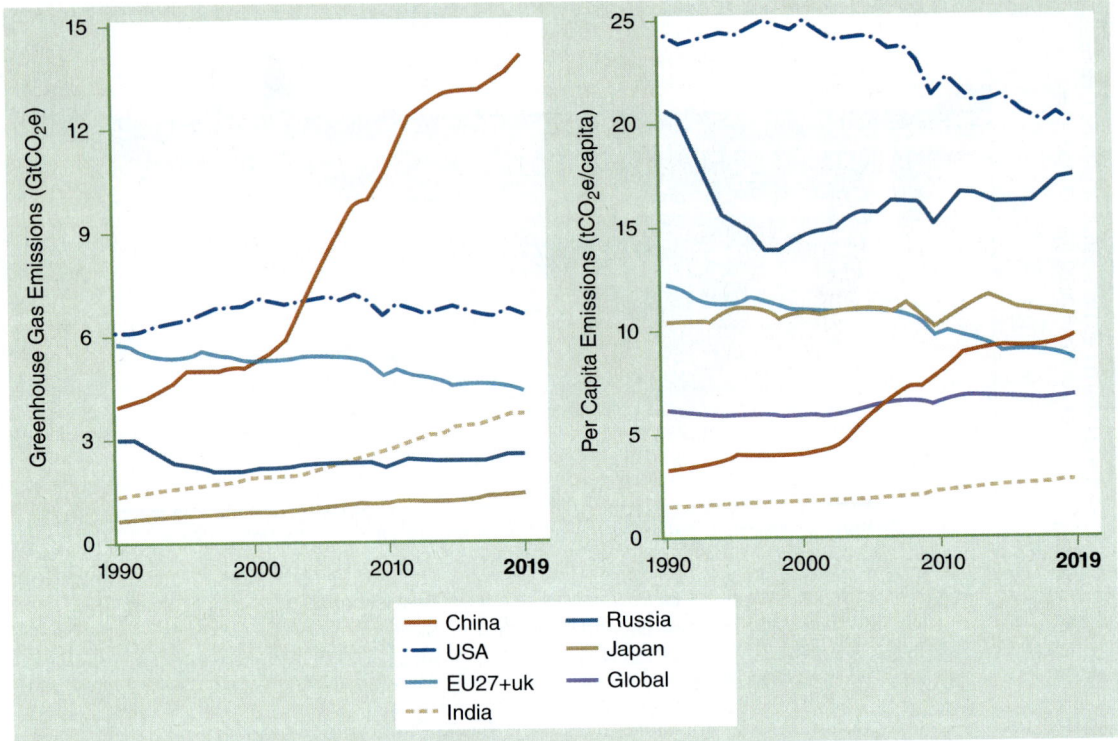

Figure 1.14 GHG Emissions by the Major Emitting Countries, 1990–2019. Left: Total emissions. Right: Per capita emissions.

Reproduced from United Nations Environment Programme. *Emissions Gap Report 2020.* United Nations Environment Programme; 2020:Figure ES.2. https://www.unep.org/emissions-gap-report-2020

cooling). El Niño and La Niña events may be categorized as *strong*, *moderate*, or *weak*.[26] The impacts of El Niño and La Niña vary regionally and may greatly impact regional temperature and precipitation (**Figure 1.16**).

In North America, El Niño tends to cause warmer temperatures and drier conditions in the northern United States and Canada, particularly in the winter months (**Figure 1.17**), and more rainfall in the southeastern United States and coast of the Gulf of Mexico. In a La Niña year, weather patterns typically grow warmer and drier across the southern United States and northern Mexico, and cooler and wetter across the north. That 2020 was one of the hottest years on record is particularly alarming given that a La Niña event

occurred in the second half of 2020, which usually results in cooler than average temperatures in many places. La Niña may also coincide with an active hurricane season in the northern Atlantic Ocean, as was the case in 2020.

Projecting Future Climate Change

Given what we know about current changes to the Earth's climate, trends in GHG emissions, and patterns of socioeconomic development, how is future climate change projected? Scientists use **climate models**, sophisticated computational representations of climate systems that incorporate a range of plausible scenarios about end-of-century

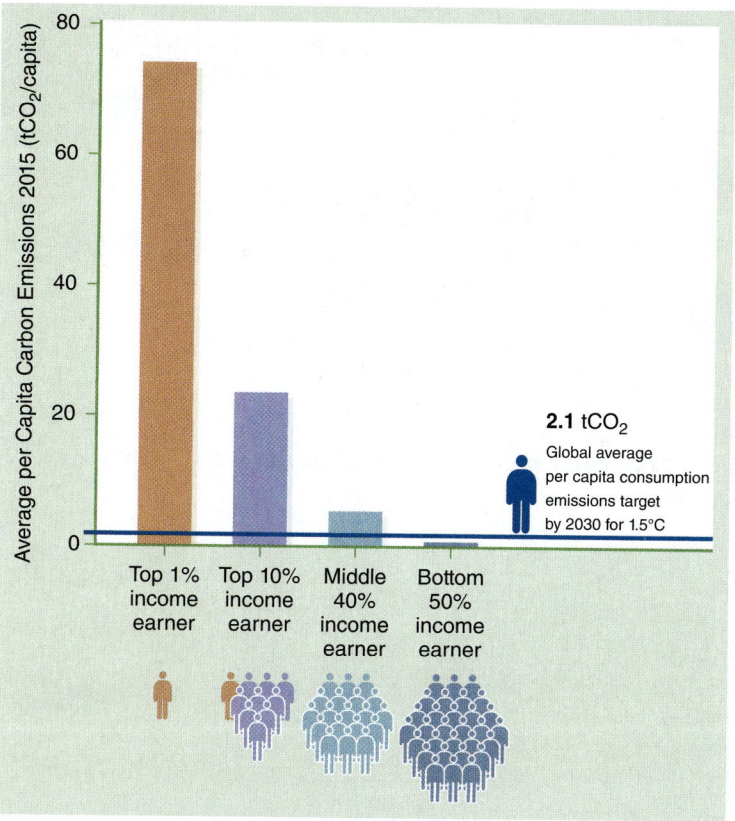

Figure 1.15 Per Capita Carbon Emissions by Global Income Level, 2015. The blue line indicates the global average per capita GHG emissions needed by 2030 to limit global warming to 1.5°C.

Reproduced from United Nations Environment Programme. *Emissions Gap Report 2020.* United Nations Environment Programme; 2020:Figure ES.8. https://www.unep.org/emissions-gap-report-2020

GHG emissions, as well as pathways of development that likely influence drivers of and responses to climate change.[27] Results from climate models are used by researchers, governments, and international and civil society organizations and are incorporated into many scientific reports, including those produced by the **Intergovernmental Panel on Climate Change (IPCC)**. Created in 1988, the IPCC is the United Nations (UN) body that assesses the state of the science of climate change, its impacts and future risks, and climate adaptation and mitigation pathways. Thousands of scientists and other experts from countries around the world contribute to the work of the IPCC,

which informs climate change decision-making at local, national, and international scales. The IPCC's latest report, the *Sixth Assessment Report* (**AR6**), was released in sections in 2021 and 2022.[28] All reports are publicly available on the IPCC website.

One example of climate model output is illustrated in **Figure 1.18**. The models used to project the changes shown for average surface temperature and precipitation in 2081–2100 are from the IPCC's 2013–2014 *Fifth Assessment Report* (**AR5**), for which scientists defined four emissions scenarios, called Representative Concentration Pathways (RCPs). Each RCP is labeled according

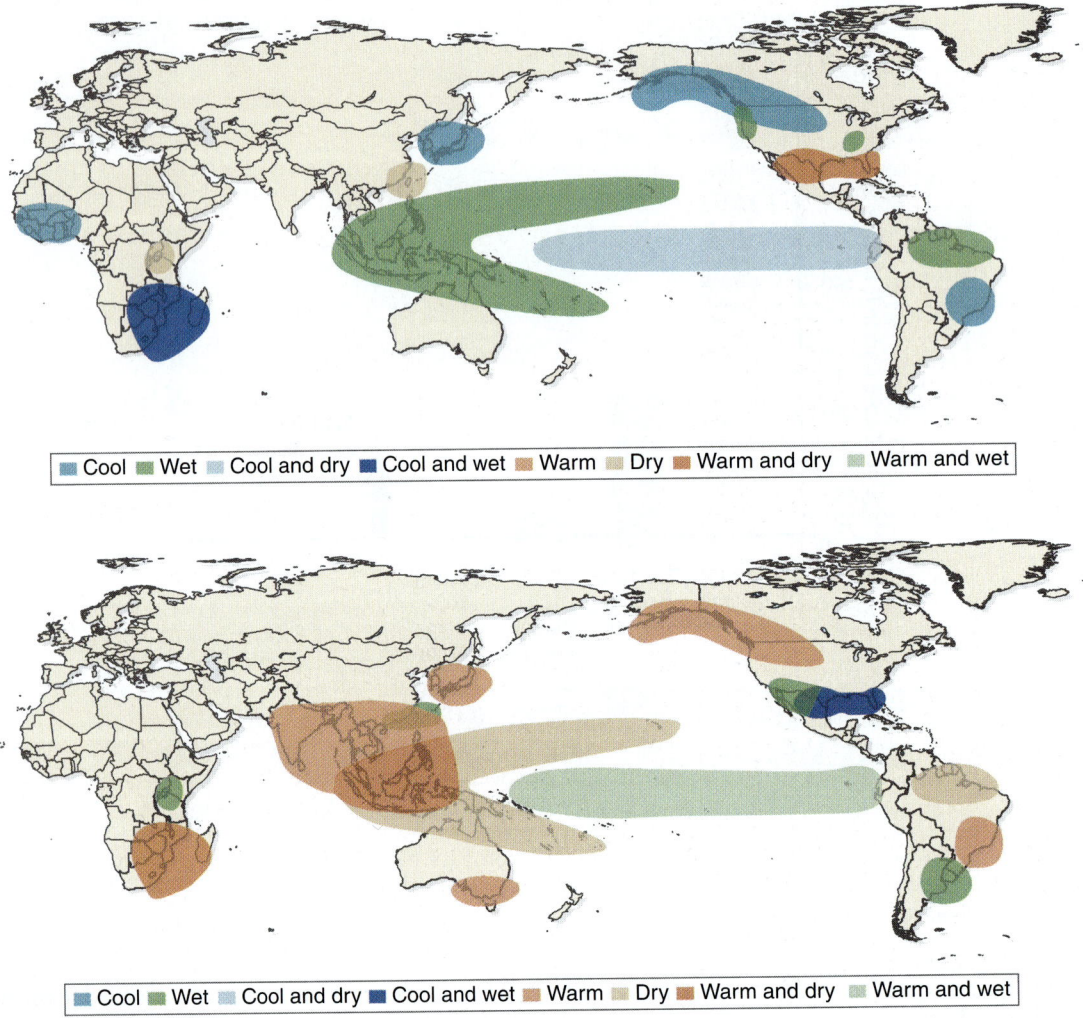

Figure 1.16 Temperature and Precipitation Variations in December, January, and February during La Niña (Top) and El Niño (Bottom) Events.

Reproduced from Lindsay R. Global impacts of El Niño and La Niña. Published February 9, 2016. Accessed February 21, 2021. https://www.climate.gov/news-features/featured-images/global-impacts-el-ni%C3%B1o-and-la-ni%C3%B1a

to its radiative forcing level in the year 2100 relative to 1750 and is linked to a specific climate policy approach (**Table 1.3**).[29]

In these RCP scenarios, concentrations of CO_2 equivalents in the atmosphere range from 475 to 1313 ppm and warming from 1.0°C to 3.7°C above 1986–2005 levels. Figure 1.18 shows projected future temperature increases and changes in precipitation under two of these emissions scenarios, RCP2.6 and RCP8.5. Significantly greater temperature increases and precipitation changes are predicted under RCP8.5, which is a very high-GHG-emissions scenario.

Climate models in the IPCC's *Sixth Assessment Report* (AR6) use a set of future scenarios called Shared Socioeconomic Pathways (SSPs), which combine assumptions about climate change, socioeconomic growth, technological advances, and human population

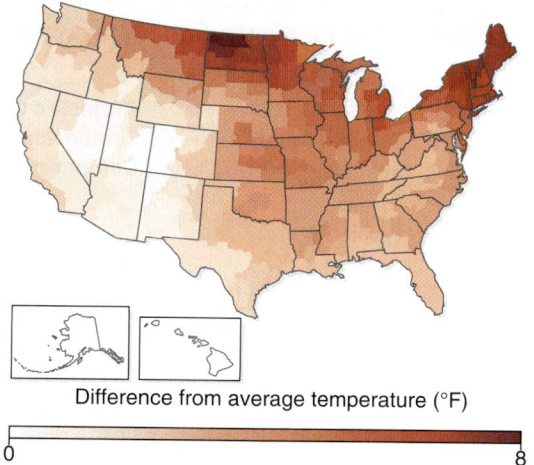

Difference from average temperature (°F)

0   8

Figure 1.17 Winter Temperature Differences from Average in the U.S. during the Strong El Niño in 2015–2016.

Reproduced from Lindsay R. U.S. winter temperatures for every El Niño since 1950. Published October 24, 2018. Accessed February 21, 2021. https://www.climate.gov/news-features/featured-images/us-winter-temperatures -every-el-ni%C3%B1o-1950

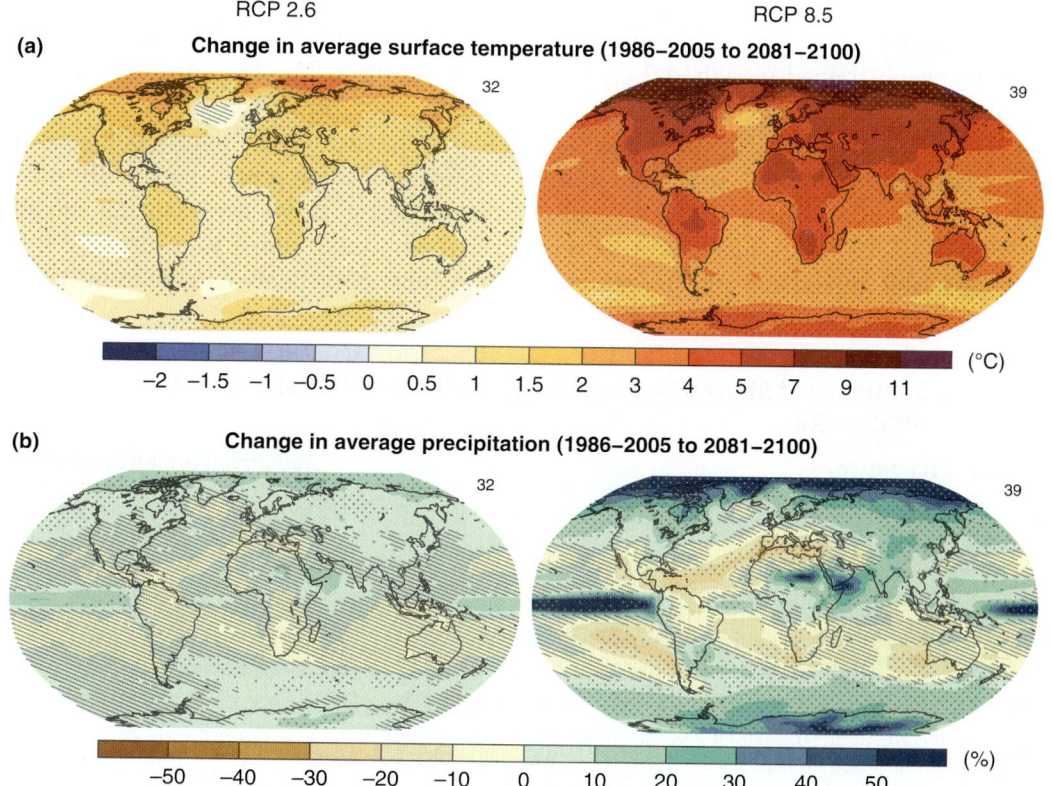

Figure 1.18 Predicted Warming and Precipitation Changes in 2080–2100 Compared with 1986–2005 Using Two Emissions Scenarios, RCP 2.6 (with GHG Emissions Mitigation) and RCP 8.5 (with No Action).

Data from Intergovernmental Panel on Climate Change. *Climate Change 2014: Synthesis Report. Contribution of Working Groups I, II and III to the Fifth Assessment Report of the Intergovernmental Panel on Climate Change.* Intergovernmental Panel on Climate Change; 2014. https://archive.ipcc.ch/pdf/assessment-report/ar5/syr/SYR_AR5_FINAL_full_wcover.pdf

Table 1.3 Summary of Representative Concentration Pathways (RCPs) Presented in the IPCC's AR5.

Scenario	Radiative Forcing Compared with 1750 (watts/m²)	Climate Policy Pathway	Projected Global Average Temperature Increase in 2100 Compared with 1986–2005 (°C)
RCP2.6	2.6	Reduce GHG emissions	1.0
RCP4.5	4.5	Stabilize GHG emissions at current levels	1.8
RCP6.0	6.0		2.2
RCP8.5	8.5	Take no action	3.7

Data from Intergovernmental Panel on Climate Change. *Climate Change 2014: Synthesis Report. Contribution of Working Groups I, II and III to the Fifth Assessment Report of the Intergovernmental Panel on Climate Change.* Intergovernmental Panel on Climate Change; 2014. https://archive.ipcc.ch/pdf /assessment-report/ar5/syr/SYR_AR5_FINAL_full_wcover.pdf

dynamics. Each SSP is labeled according to a radiative forcing level and contextual "narrative," from sustainable growth and equality to unconstrained growth and energy use (**Table 1.4**).[27] The eight SSPs expand the range of possible scenarios, including SSP1-1.9, the scenario associated with the least future warming by century's end (best estimate below 1.5°C). AR6 includes a novel publicly available interactive atlas to explore observed and projected climate data at global and regional scales.

Future climate scenarios do not forecast what is inevitable by this century's end; instead, they present a wide range of plausible outcomes, some more probable than others. In IPCC reports, the **likelihood** of a climate change or impact occurring, including a human health impact, is expressed as a probability of certainty ranging from "virtually

Table 1.4 Summary of Shared Socioeconomic Pathways (SSPs) Presented in the IPCC's AR6.

Scenario	Narrative	Radiative Forcing (watts/m²)
SSP1	"A world of sustainability-focused growth and equality"	1.9, 2.6
SSP2	"A 'middle of the road' world where trends broadly follow their historical patterns"	4.5
SSP3	"A fragmented world of 'resurgent nationalism'"	7
SSP4	"A world of ever-increasing inequality"	3.4, 6.0
SSP5	"A world of rapid and unconstrained growth in economic output and energy use"	3.4, 8.5

Data from Hausfather Z. CMIP6: the next generation of climate models explained. Published December 2, 2019. Accessed February 21, 2021. https://www .carbonbrief.org/cmip6-the-next-generation-of-climate-models-explained

Table 1.5 *Likelihood* **Language Used by the IPCC.**

Likelihood Term	Probability of the Outcome
Virtually certain	99–100%
Extremely likely	95–100%
Very likely	90–100%
Likely	66–100%
More likely than not	>50–100%
About as likely as not	33–66%
Unlikely	0–33%
Very unlikely	0–10%
Extremely unlikely	0–5%
Exceptionally unlikely	0–1%

Data from Intergovernmental Panel on Climate Change. *Climate Change 2014: Synthesis Report. Contribution of Working Groups I, II and III to the Fifth Assessment Report of the Intergovernmental Panel on Climate Change.* Intergovernmental Panel on Climate Change; 2014. https://archive.ipcc.ch/pdf/assessment-report/ar5/syr/SYR_AR5_FINAL_full_wcover.pdf

certain" to "exceptionally unlikely" (**Table 1.5**). In addition, **confidence** in a finding is reported from "very low" to "very high" based on the status of the **evidence** for this finding, categorized as *limited, medium,* or *robust,* and the level of scientific **agreement** about this finding, categorized as *low, medium,* or *high.*[29]

Taking Climate Action

Future scenarios project major climate impacts, including significant adverse effects on human health and well-being. These impacts are mostly distributed along deeply etched fault lines of inequality, burdening some people more than others. These are not inevitable outcomes, however, if swift and aggressive climate action is taken to reduce GHG emissions and prepare for these impacts.

Progress by countries to adopt climate policies and programs has been catalyzed in part by international negotiations and resulting climate change agreements and treaties. The **United Nations Framework Convention on Climate Change (UNFCCC)** is the overarching international convention on climate change. It was signed by 154 countries in 1992 at the UN Conference on Environment and Development in Rio de Janeiro, Brazil (often referred to as the "Rio Summit" or "Earth Summit") and entered into force in 1994. The main objective of the UNFCCC is

stabilization of greenhouse gas concentrations in the atmosphere at a level that would prevent dangerous anthropogenic human-induced interference with the climate system. Such a level should be achieved within a time-frame sufficient to allow ecosystems to adapt naturally to climate change, to ensure that food production is not threatened and to enable economic development to proceed in a sustainable manner.[30]

The **Conference of the Parties** (**COP**) of the UNFCCC is the decision-making body that represents all Parties to the convention. **Parties** are countries that have ratified or otherwise legally approved international agreements and treaties and thus are bound to follow them. The COP meets annually to negotiate and review the implementation of the convention and its legal instruments. Non-Party stakeholders also participate in COP meetings, including representatives of states, provinces, and cities, Indigenous peoples, civil society organizations, and businesses.

The UNFCCC itself does not set binding or enforceable GHG emissions limits, but provides the framework for international negotiations on specific treaties that do so, including the Kyoto Protocol and Paris Agreement. The **Kyoto Protocol** was adopted in 1997, and entered into force in 2005. It committed developed countries to limit their GHG emissions under the principle of "common but differentiated responsibility and respective capabilities," recognizing that developed countries are largely responsible for the high levels of atmospheric GHG emissions and thus have a greater obligation to act.[31] Currently, there are 192 Parties to the protocol. The United States is the only country that signed but did not ratify the Kyoto Protocol and thus is not bound to reduce its GHG emissions under this specific treaty. In the first phase of the Kyoto Protocol (2008–2012), 37 countries plus the European Union exceeded the pledged target of GHG emissions reductions of an average of 5% compared with 1990 levels.[31]

A second major milestone in international climate policy is the **Paris Agreement**, a treaty that compels all countries to commit to ambitious climate action—not just developed countries, as is the case under the Kyoto Protocol. It was adopted by 196 Parties at COP21, the international climate conference in Paris, France, in December 2015, and it entered into force in November 2016 (**Figure 1.19**). The main objectives of the Paris Agreement are **climate mitigation**, efforts to reduce GHG emissions to limit global warming to well below 2°C (3.6°F) and preferably below 1.5°C (2.7°F) compared with preindustrial levels, **climate adaptation** to build the capacity of countries to respond to the adverse impacts of climate change, and **climate finance** mechanisms whereby high-income countries assist low- and middle-income countries in funding low-carbon and climate-resilient development projects.[32]

The Paris Agreement sets up five-year cycles of increasingly ambitious climate action in which countries are required to prepare, communicate, and implement **nationally determined contributions (NDCs)** that outline national plans to meet emissions reductions targets and climate adaptation goals. The Paris Agreement calls for reducing GHG emissions as soon as possible, with the goal of climate neutrality (net-zero GHG emissions) by midcentury. NDCs may contain "unconditional" contributions that can be

Figure 1.19 Left: U.S. Secretary of State John Kerry Signs the Paris Agreement While Holding His Granddaughter at the United Nations in New York City, April 22, 2016. Kerry is the U.S. Special Presidential Envoy for Climate in the Biden administration. Right: Climate Activists Stage a Protest for Climate Justice at the COP21 UN climate summit in Paris, France, December 1, 2015.

implemented based on a country's own financial resources and capabilities, and "conditional" contributions that countries would undertake if international financial support is available or if "collective ambition" rises among countries to do more.[33]

The current magnitude and pace of change of GHG emissions has put the world on a dangerous warming course. As of late 2020, the world is on track for greater than 3°C of warming by century's end, and in 2021, the World Meteorological Organization predicted a 40% chance that the average annual global temperature will temporarily reach 1.5°C of warming at some point in the next five years.[34] In 2021, the UN released its first assessment of the potential for current NDCs to achieve the goal of limiting warming to below 2°C, and ideally by 1.5°C, by 2100. The report concluded that global GHG emissions must be *45% below 2010 levels by 2030*.[35] Numerous countries have pledged stricter emissions reduction targets, including net-zero emissions in some cases.

Climate change, human impacts, and actions by governments are changing rapidly. Advances in the near future, including those resulting from the COP26 climate negotiations in Glasgow, Scotland, in November 2021, are expected. The United States, a leading contributor to global GHG emissions, is playing catch-up after the Trump administration abandoned its commitments under the Paris Agreement. One of President Biden's first actions after taking office in January 2021 was to rejoin the Paris Agreement, and in April 2021, the United States announced a target to reduce its net GHG emissions by 50–52% below 2005 levels in 2030.[36] The United States has also promised to honor its financial commitments to support climate mitigation and adaptation in low- and middle-income countries[37] and to accelerate

U.S. climate action in an equitable manner by ensuring that 40% of the benefits from federal investments in clean energy, energy efficiency, improved housing, and public transport go to "disadvantaged communities."[38]

Climate action is also being taken at regional scales. For example, dozens of cities around the world have set emissions targets and developed climate adaptation plans. Reykjavik, Iceland, has already achieved 100% renewable energy use, due mostly to its plentiful geothermal energy supply, and several other cities, including Paris, San Francisco, and Canberra, have committed to the goal of 100% renewable energy.[39] In the United States, the first mandatory GHG emissions reduction program has been the successful **Regional Greenhouse Gas Initiative** (**RGGI**), an ongoing cooperative effort among nine northeastern U.S. states to cap and reduce CO_2 emissions from the power generation sector. Since 2005, CO_2 emissions from power plants have declined by more than half while economic growth increased (**Figure 1.20**).

Also in the United States, a broad coalition of nonfederal actors has committed to the goals of the Paris Agreement. Half of U.S. states; more than 500 cities and counties; hundreds of tribes; more than 250 colleges and universities; and hundreds of faith groups, cultural institutions, healthcare organizations, and businesses and investors have signaled their intention to work toward emissions targets. In 2020, this coalition represented 68% of U.S. GDP, 65% of the U.S. population, and 51% of U.S. GHG emissions.[40]

Meeting ambitious GHG emissions targets requires **decarbonizing** energy transformations that phase out carbon-intensive fossil fuels in favor of renewable energy sources. This shift must be implemented in equitable ways so that everyone benefits, especially

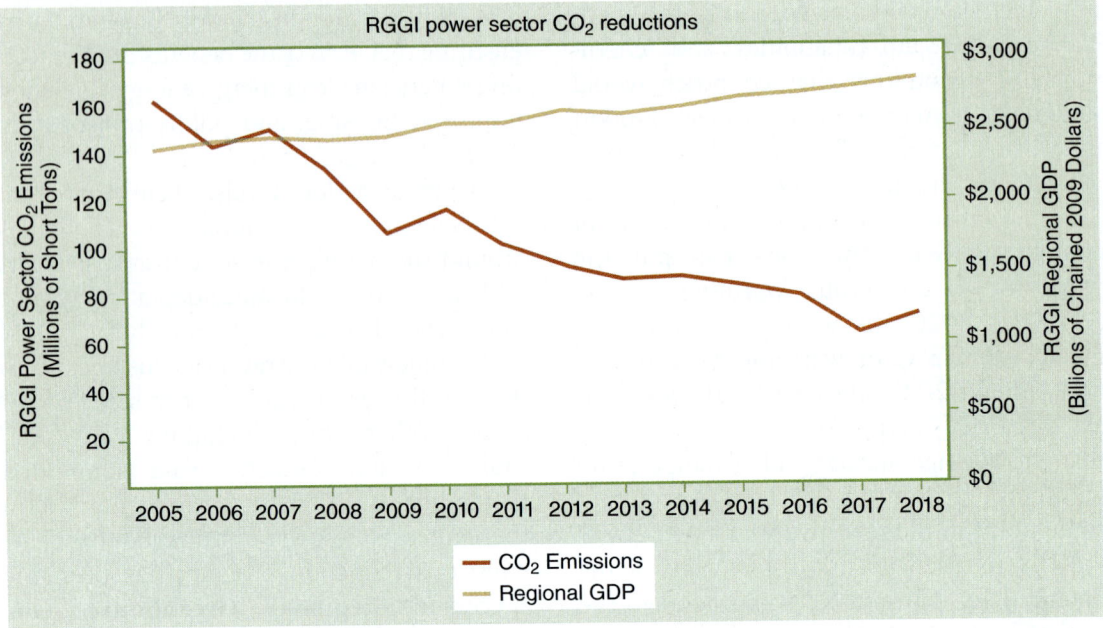

RGGI power sector CO$_2$ reductions

Figure 1.20   Reduction in CO$_2$ Emissions from the Power Sector in Nine RGGI States, and Regional GDP Increase, 2005–2018.

Reproduced from Regional Greenhouse Gas Initiative. *The Investment of RGGI Proceeds in 2018.* Regional Greenhouse Gas Initiative; 2020. https://www.rggi.org/sites/default/files/Uploads/Proceeds/RGGI_Proceeds_Report_2018.pdf

at-risk populations, vulnerable countries, and those who have contributed very little to climate change yet bear disproportionate burdens of the impacts. Ways to incentivize this transition include eliminating government subsidies paid to fossil energy companies, implementing carbon pricing on fuels that reflects their climate impacts, and enacting subsidies for clean energy from solar, wind, geothermal, and ocean sources. From 2011 to 2020, renewable energy use for U.S. electricity generation increased from 13% to 20% of all energy sources, while use of the dirtiest energy source, coal, declined from 42% to 19% (**Figure 1.21**).[41] Energy efficiency measures must be introduced simultaneously, and carbon sinks, such as forests, soils, and peatlands, must be expanded.

Alternative energy choices take into account impacts primarily on GHG emissions, but human health must also be considered.

For example, in 2020, natural gas had replaced coal as the primary energy source for U.S. electricity generation, which benefits human health by reducing CO$_2$ emissions

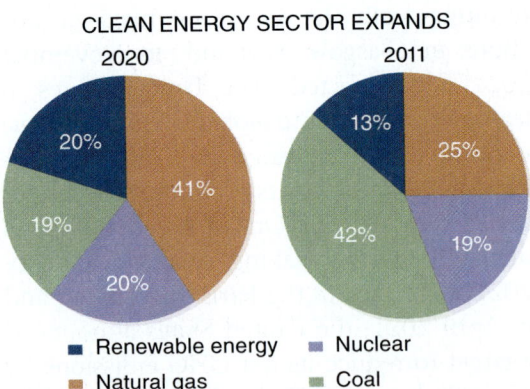

CLEAN ENERGY SECTOR EXPANDS

Figure 1.21 Energy Sources for U.S. Electricity Generation in 2011 and 2020.

Reproduced from Business Council for Sustainable Energy. *Sustainable Energy in America 2021: Factbook at-a-Glance.* Accessed June 3, 2021. https://bcse.org/wp-content/uploads/2021-Sustainability-In-America-Factbook-At-A-Glance.pdf

and eliminating hazardous by-products of coal burning, such as coal ash, airborne mercury, and particulate matter. However, the composition of natural gas is about 95% methane, a potent GHG, and fugitive methane emissions during all phases of natural gas production contribute significantly to overall GHG emissions and resulting warming. The predominant method of extracting natural gas today is hydraulic fracturing, which is associated with a broad range of health impacts, including respiratory illnesses, cancers, and adverse birth outcomes from exposure to contaminated air and water, mental illness and substance abuse among workers, and deterioration of the social fabric and economic prospects of fracking communities.[42] Generating solar energy produces no GHG emissions but requires photovoltaic cells that use toxic chemicals and create hazardous waste. Advances in alternative energy sources must prioritize less harmful means of extraction, production, use, and disposal to minimize impacts on human health.

An excellent source of information about climate solutions is **Project Drawdown**, a nonprofit organization committed to helping the world reduce GHG emissions and providing information on climate solutions for governments, businesses, institutions, activists, and scholars.[43] Project Drawdown highlights three connected areas where action is needed:

- "**Reduce Sources** [by] bringing emissions to zero"
- "**Support Sinks** [by] uplifting nature's carbon cycle"
- "**Improve Society** [by] fostering equality for all"[43]

Global climate solutions are ranked by Project Drawdown based on predicted impacts on global carbon emissions. To limit warming to 2°C, the top five most impactful solutions turn out to not relate directly to energy choices: reduce food waste, invest in education, particularly of girls and women, eat plant-rich diets, manage refrigerant chemicals that are GHGs, and restore tropical forests.[43] Successful climate action plans will incorporate a wide range of strategies like these to mitigate GHG emissions and adapt to climate change impacts.

Climate Activism

Since Dr. James Hansen publicly raised the alarm about climate change in his congressional testimony in 1988, many activists and civil society organizations have played a key role in advancing climate change awareness and action. Dr. Hansen inspired one of the most effective climate activists, writer Bill McKibben, to pen his popular book on climate change, *The End of Nature*, in 1989, in which he called for a radical repositioning of the relationship between humans and nature. Many environmental groups began focusing seriously on climate change, and in 2008, McKibben and several undergraduate students at Middlebury College launched 350, the first global grassroots organization fighting climate change. The group is named for the upper bounds of parts per million of carbon dioxide that can safely be in the atmosphere without catastrophic global warming, a level greatly exceeded at the current time. McKibben and the student leaders of 350 led the early charge to organize large-scale national and global protests, as well as local activism, to raise awareness about the need for climate action. They have supported campaigns against fossil fuel use by blocking pipeline projects and urging institutions to divest their financial holdings from fossil fuel interests. Today, many organizations large and small are fighting climate change from every place on Earth and doing impactful work on both a global scale and in their local communities.

In recent years, many youth activists have taken the lead in the climate movement and inspired a new generation to get involved, make changes in their communities, and influence world leaders on the biggest stages. Swedish activist Greta Thunberg is perhaps the most recognizable, having started the School Strike for Climate campaign and inspired young activists involved in many climate organizations, including Fridays for Future, an international youth-led coalition fighting climate change. Two youth leaders making a big difference are Xiye Bastida and Alexandria Villaseñor, co-organizers for Fridays for Future U.S. (**Figure 1.22**).

Xiye Bastida, a member of the Otomi-Toltec Nation, was a firsthand witness to catastrophic drought and flooding growing up in Mexico, and then the legacy of Hurricane Sandy after she moved with her family to New York City. Youth are using "every tool at their disposal, from traditional media to memes, to tell the world what we know and why it matters to us," she wrote recently.[44] "You don't have to know the details of the science to be part of the solution. And if you wait until you know everything, it will be too late for you to do anything. That's why we, the youth who are leading on climate, are calling this an emergency." She is focused on creating a diverse and intergenerational collaboration to work toward a "vibrant, fair and regenerative future."[44]

Alexandria Villaseñor wrote in 2020,

> I am fifteen years old and spend a lot of my time on conference calls, sending emails, speaking publicly, and going to protests. Those are probably different memories than you were making at my age, but we youth know we need to make our voices

heard now—because our generation will feel climate impacts the most.[45]

Her activism was inspired by witnessing the deadliest wildfire in California's history, the 2018 Camp Fire, which caused dangerously unhealthy air pollution that exacerbated her asthma and made her very

Figure 1.22 (Top) Alexandria Villaseñor speaks at the 2019 C40 World Mayors Summit in Copenhagen, Denmark. (Bottom) Xiye Bastida speaks at the 2019 NOVUS #WeThePlanet forum at the United Nations in New York City.

sick. "Do [I] miss just being a regular teen-ager?"[50] she asks.

> The answer is yes. I miss doing the-ater, playing volleyball, and hanging out with my friends. But the climate crisis threatens every aspect of my future. So what other choice do I have? It is a moral obligation to fight for this planet. My fight for climate action is not going to end until our planet and all its people are safe.[45]

In the United States, people of all racial and ethnic backgrounds have long been working and advocating for environmental protection, climate action, and the health of their communities, including **Heather McTeer Toney**, former U.S. EPA administra-tor for the Southeast Region and now climate justice liaison to Environmental Defense and senior advisor to Moms Clean Air Force (**Figure 1.23**). Growing up in Mississippi, Toney says she was "surrounded by the inter-weaving of nature with Black culture, pov-erty, and the rural South." She stresses that "it is the Black part of American culture that is . . . largely missing from the public con-versations about environmental and climate solutions."[46] She sees solutions in her own community, where people are experiencing climate impacts today, including flooded rivers, more heat, disease pest invasions, and problems growing food, although she laments that "no one [has been] listening to the voices of the poor, of rural folk, of south-erners."[51] She points out that women of color are not waiting to be told what to do, but rather are already climate leaders, finding ways to make change in their communities, just as they have long been on the frontlines fighting industrial polluters so prevalent in Gulf Coast states. Toney also notes that when she worked at EPA, 90% of her executive team were women of color.[46]

Figure 1.23 Heather McTeer Toney, Senior Advisor, Moms Clean Air Force.
Courtesy of Heather McTeer Toney.

Jacqueline Patterson is senior director of the Climate and Environmental Justice Program of the National Association for the Advancement of Colored People, a U.S. civil rights organization. She points out,

> It is all too clear that justice is not possible in a capitalist system pred-icated on there being winners and losers, a system rooted in racism, sexism, and xenophobia. This is the system that has put us on the path to catastrophic climate change. The only path to liberation for Black folks and all oppressed people is through revolution—total systems change.[47]

Indigenous peoples are also vital partici-pants in the environmental movement, both in the United States and around the world. Many Indigenous peoples defend their ances-tral lands, which hold much of the world's remaining biodiversity, as well as their traditional knowledge systems and practices, which provide a road map for how to respond to the climate crisis. **Sherri Mitchell**, a member of the Penobscot Nation, is an attorney, author, and founding director of the Land Peace Foundation, dedicated to the global protection of Indigenous lands, waters, and ways of life. She calls for incorporating into climate science and action Indigenous knowledge systems "based on the millennia-long study of the complex relationships that exist among all systems within creation."[48] Indigenous peoples provide the world with what she calls "living models of sustainability that are rooted in ancient wisdom and that inform us how to live in balance with all of our relations on Mother Earth." She warns, "Today prophecies are unfolding all around us. We are all observers and participants. The annihilation of Indigenous peoples is also the annihilation of humankind."[48]

Maulian Dana is the Tribal Ambassador of Penobscot Nation and cofounder of the Wabanaki Alliance in Maine, focused on protecting tribal sovereignty and improving education about Indigenous peoples. She recently helped create a comprehensive climate plan for Maine as a member of the governor-appointed Maine Climate Council.

> Frontline communities like tribal nations, new Mainers, those in poverty, people of color, and more are disproportionately affected by the climate crisis. A society is only as strong as its most vulnerable populations, and this holds true in climate work. As we make policy, we need to work from a place of inclusivity and equity to make sure our work is lasting and meaningful.[49]

Tara Houska is an attorney, member of Couchiching First Nation, and founder of the Indigenous advocacy group Giniw Collective (**Figure 1.24**). She and many other Indigenous activists are leading the charge against fossil fuel development in the United States and Canada, including protesting the Dakota Access Pipeline, which crosses the Missouri River north of the Standing Rock Sioux reservation. In 2021, Houska organized the fight against the proposed "Line 3" replacement tar sands pipeline in northern Minnesota that crosses untouched wetlands in treaty territory of Anishinaabe peoples and the headwaters of the Mississippi River. "To be humbled by the lived knowledge that our bodies cannot survive without water is to move water from the conceptual into the actual."[50]

Mitchell, Dana, and Houska each call for everyone to join them in their work, which "will require non-Indigenous people to stand with us and ensure that our lands, waters, and ways of life are not further eroded by government and industrial intrusion."[48] Houska urges Americans to "please find your bravery. Defending the land is a beautiful thing, it's a beautiful risk to take."[51]

Figure 1.24 Tara Houska, Couchiching First Nation.

Courtesy of Tara Houska.

Discussion Questions

1. How does the greenhouse effect work?
2. What are the principal greenhouse gases? What are the major anthropogenic sources of each? How can greenhouse gas emissions from these sources be reduced?
3. How are temperature, precipitation, sea level, and sea ice extent changing?
4. What are the targets for global warming in the Paris Agreement? Given current levels of climate action, how likely is it that these targets will be met? What needs to happen to achieve these goals?
5. What is the role of the IPCC?
6. How is climate activism making a difference? Describe how youth, people of color, and Indigenous peoples are contributing to the climate movement.

References

1. Costello A, Abbas M, Allen A, et al. Managing the health effects of climate change: Lancet and University College London Institute for Global Health Commission. *Lancet*. 2009;373(9676):1693–1733. doi:10.1016/S0140-6736(09)60935-1
2. Munich Re. Record hurricane season and major wildfires–The natural disaster figures for 2020. January 7, 2021. Accessed February 14, 2021. https://www.munichre.com/en/company/media-relations/media-information-and-corporate-news/media-information/2021/2020-natural-disasters-balance.html
3. Salih AAM, Baraibar M, Mwangi KK, et al. Climate change and locust outbreak in East Africa. *Nature Climate Change*. 2020;10(7):584–585. doi:10.1038/s41558-020-0835-8
4. Funk C. Ethiopia, Somalia and Kenya face devastating drought. *Nature*. 2020;586(7831):645. doi:10.1038/d41586-020-02698-3
5. World Meteorological Organization. 2020 was one of the three warmest years on record. January 15, 2021. Accessed June 3, 2021. https://public.wmo.int/en/media/press-release/2020-was-one-of-three-warmest-years-record
6. National Oceanic and Atmospheric Administration. 2020 was Earth's 2nd-hottest year, just behind 2016. Published online January 14, 2021. Accessed February 14, 2021. https://www.noaa.gov/news/2020-was-earth-s-2nd-hottest-year-just-behind-2016
7. Meehl GA, Tebaldi C, Walton G, et al. Relative increase of record high maximum temperatures compared to record low minimum temperatures in the U.S. *Geophys Res Lett*. 2009;36(23). https://doi.org/10.1029/2009GL040736
8. Environmental Protection Agency. Climate change indicators: high and low temperatures. April 14, 2021. Accessed June 3, 2021. https://www.epa.gov/climate-indicators/climate-change-indicators-high-and-low-temperatures
9. National Oceanic and Atmospheric Administration. Billion-dollar weather and climate disasters: overview. Accessed February 19, 2021. https://www.ncdc.noaa.gov/billions/
10. Patz JA. Solving the global climate crisis: the greatest health opportunity of our times? *Public Health Rev*. 2016;37(1):30. doi:10.1186/s40985-016-0047-y
11. Intergovernmental Panel on Climate Change. Glossary of terms. Accessed February 19, 2021. https://www.ipcc.ch/site/assets/uploads/2018/03/wg2TARannexB.pdf
12. Environmental Protection Agency. Overview of greenhouse gases. Accessed June 7, 2021. https://www.epa.gov/ghgemissions/overview-greenhouse-gases
13. Environmental Protection Agency. Climate change indicators: global greenhouse gas emissions. Updated April 2021. Accessed June 7, 2021. https://www.epa.gov/climate-indicators/climate-change-indicators-global-greenhouse-gas-emissions
14. Scripps Institution of Oceanography. The Keeling Curve. February 19, 2021. Accessed February 19, 2021. https://keelingcurve.ucsd.edu

15. Foote EN. Circumstances affecting the heat of the sun's rays. *Am J Sci Arts*. 1856;22(66):383–384.

16. Tyndall J. The Bakerian Lecture: on the absorption and radiation of heat by gases and vapours, and on the physical connexion of radiation, absorption, and conduction. *Philos Trans R Soc Lond*. 1861; 151:1–36.

17. Arrhenius S. On the influence of carbonic acid in the air upon the temperature of the ground. *Philos Mag J Sci*. 1896;41:237–276.

18. Callendar GS. The artificial production of carbon dioxide and its influence on temperature. *Q J R Meteorol Soc*. 1938;64(275):223–240. doi:10.1002 /qj.49706427503

19. Hansen J. Statement of Dr. James Hansen, director, NASA Goddard Institute for Space Studies, U.S. Senate, 100th Congress (1988) (Energy and Natural Resources Committee). June 23, 1988. Accessed February 19, 2021. https://babel.hathitrust.org/cgi /pt?id=uc1.b5127807&view=1up&seq=44

20. National Aeronautical and Space Administration. Facts: global temperature. Accessed February 19, 2021. https://climate.nasa.gov/vital-signs/global -temperature

21. National Aeronautical and Space Administration. Facts: sea level. Accessed February 19, 2021. https://climate.nasa.gov/vital-signs/sea-level

22. Intergovernmental Panel on Climate Change. Summary for policymakers. In: *IPCC Special Report on the Ocean and Cryosphere in a Changing Climate*. 2019. Accessed February 19, 2021. https://www .ipcc.ch/srocc/

23. National Aeronautical and Space Administration. Facts: arctic sea ice level. Accessed February 19, 2021. https://climate.nasa.gov/vital-signs/arctic-sea-ice

24. National Aeronautical and Space Administration. Facts: ice level sheets. Accessed February 19, 2021. https://climate.nasa.gov/vital-signs/ice-sheets

25. United Nations Environment Programme. *Emissions Gap Report 2020*. United Nations Environment Programme; 2020. https://www.unep .org/emissions-gap-report-2020

26. National Weather Service. What is ENSO? Accessed October 8, 2020. https://www.weather.gov/mhx /ensowhat

27. Hausfather Z. CMIP6: the next generation of climate models explained. Published December 2, 2019. Accessed February 21, 2021. https:// www.carbonbrief.org/cmip6-the-next-generation -of-climate-models-explained

28. Intergovernmental Panel on Climate Change. AR6 climate change 2021: impacts, adaptation and vulnerability. Accessed February 2, 2021. https:// www.ipcc.ch/report/sixth-assessment-report -working-group-ii

29. Intergovernmental Panel on Climate Change. *Climate Change 2014: Synthesis Report. Contribution of Working Groups I, II and III to the Fifth Assessment Report of the Intergovernmental Panel on Climate Change*. Intergovernmental Panel on Climate Change; 2014. https://archive.ipcc.ch /pdf/assessment-report/ar5/syr/SYR_AR5_FINAL _full_wcover.pdf

30. United Nations Framework Convention on Climate Change. What is the United Nations Framework Convention on Climate Change? Accessed February 15, 2021. https://unfccc.int/process-and -meetings/the-convention/what-is-the-united -nations-framework-convention-on-climate-change

31. United Nations Climate Change. What is the Kyoto Protocol? Accessed February 15, 2021. https:// unfccc.int/kyoto_protocol

32. United Nations Climate Change. The Paris Agreement. Accessed June 16, 2021. https://unfccc .int/process-and-meetings/the-paris-agreement /the-paris-agreement

33. United Nations Climate Change. Nationally determined contributions (NDCs). Accessed February 15, 2021. https://unfccc.int/process-and -meetings/the-paris-agreement/nationally -determined-contributions-ndcs/nationally -determined-contributions-ndcs

34. World Meteorological Organization. New climate predictions increase likelihood of temporarily reaching 1.5°C in next 5 years. Published May 27, 2021. Accessed June 7, 2021. https://public.wmo .int/en/media/press-release/new-climate -predictions-increase-likelihood-of-temporarily -reaching-15-%C2%B0c-next-5

35. United Nations Climate Change. Greater climate ambition urged as initial *NDC Synthesis Report* is published. Published February 26, 2021. Accessed June 7, 2021. https://unfccc.int/news/greater-climate -ambition-urged-as-initial-ndc-synthesis-report-is -published

36. The White House. Fact Sheet: President Biden sets 2030 greenhouse gas pollution reduction target aimed at creating good-paying union jobs and securing U.S. leadership on clean energy technologies. Published April 22, 2021. Accessed June 7, 2021. https://www.whitehouse.gov/briefing-room /statements-releases/2021/04/22/fact-sheet-president -biden-sets-2030-greenhouse-gas-pollution -reduction-target-aimed-at-creating-good-paying

-union-jobs-and-securing-u-s-leadership-on-clean
-energy-technologies/

37. Chemnick J. Summit will mark U.S. return to global climate talks. *E&E News*. Published January 26, 2021. Accessed February 19, 2021. https://www.eenews.net/climatewire/2021/01/26/stories/1063723513

38. Chemnick J. 40% of climate benefits? Explaining Biden's EJ promise. *E&E News*. Published February 16, 2021. Accessed February 19, 2021. https://www.eenews.net/stories/1063725151

39. Scott M. Cities are on the front line of tackling climate change—and they need to do more. *Forbes*. Published June 5, 2019. Accessed February 19, 2021. https://www.forbes.com/sites/mikescott/2019/06/05/cities-are-on-the-front-line-of-tackling-climate-change-and-they-need-to-do-more

40. America's Pledge, We Are Still In. We Are Still In to deliver on America's Pledge: A retrospective. Accessed February 19, 2021. https://assets.bbhub.io/dotorg/sites/28/2020/09/We-Are-Still-In-to-Deliver-on-Americas-Pledge_.pdf

41. Business Council for Sustainable Energy. *Sustainable Energy in America 2021: Factbook at-a-Glance*. Accessed June 3, 2021. https://bcse.org/wp-content/uploads/2021-Sustainability-In-America-Factbook-At-A-Glance.pdf

42. Concerned Health Professionals of New York, Physicians for Social Responsibility. *Compendium of Scientific, Medical, and Media Findings Demonstrating Risks and Harms of Fracking (Unconventional Gas and Oil Extraction)*. Published December 14, 2020. Accessed February 19, 2021. http://concernedhealthny.org/compendium/

43. Project Drawdown. *The Drawdown Review*. Published 2020. Accessed February 21, 2021. https://drawdown.org/drawdown-review

44. Bastida X. Calling in. In: Johnson AE, Wilkinson, KK, eds. *All We Can Save: Truth, Courage, and Solutions for the Climate Crisis*. One World; 2020:3–7.

45. Villasenor A. A Letter to adults. In: Johnson AE, Wilkinson KK, eds. *All We Can Save: Truth, Courage, and Solutions for the Climate Crisis*. One World; 2020:323–327.

46. Toney HM. Collards are just as good as kale. In: Johnson AE, Wilkinson KK, eds. *All We Can Save: Truth, Courage, and Solutions for the Climate Crisis*. One World; 2020:75–83.

47. Patterson J. At the intersections. In: Johnson AE, Wilkinson KK, eds. *All We Can Save: Truth, Courage, and Solutions for the Climate Crisis*. One World; 2020:194–202.

48. Mitchell S. Indigenous prophecy and Mother Earth. In: Johnson AE, Wilkinson KK, eds. *All We Can Save: Truth, Courage, and Solutions for the Climate Crisis*. One World; 2020:16–28.

49. Dana M. In: *Maine Won't Wait: A Four Year Plan for Climate Action*. Published December 2020. Accessed February 21, 2021. https://www.maine.gov/future/sites/maine.gov.future/files/inline-files/MaineWontWait_December2020.pdf

50. Houska T. Sacred resistance. In: Johnson AE, Wilkinson KK, eds. *All We Can Save: Truth, Courage, and Solutions for the Climate Crisis*. One World; 2020:213–219.

51. Tigue K. Urging Biden to Stop Line 3, Indigenous-led resistance camps ramp up efforts to slow construction. *Inside Climate News*. Published February 16, 2021. Accessed February 21, 2021. https://insideclimatenews.org/news/16022021/biden-line-3-minnesota-enbridge-pipeline-indigenous-resistance

CHAPTER 2

Measuring Health

KEY TERMS

Health
Millennium Development Goals (MDGs)
Sustainable Development Goals (SDGs)
Precautionary principle
Risk
Vulnerability
Public health
Epidemiology
Descriptive epidemiology
Analytic epidemiology

Incidence
Incidence rate
Prevalence
Prevalence rate
Mortality
Mortality rate
Morbidity
Disability-adjusted life year (DALY)
Odds ratio (OR)
Relative risk (RR)
Rate ratio (RR)

95% confidence interval
p-value
World Health Organization (WHO)
U.S. Global Change Research Program (USGCRP)
The Lancet Countdown
Institute for Health Metrics and Evaluation (IHME)
Global Burden of Disease (GBD) report

LEARNING OBJECTIVES

- Define health, public health, and epidemiology.
- Understand generally how climate change leads to human health impacts.
- Describe how epidemiology is used to study the relationships between climate change and human health.
- Be able to read and comprehend an epidemiological study.
- Identify sources of health information and extract data from publicly available sources.

The climate crisis affects human health in many ways—some direct and some indirect, some visible and some less so. Health effects include increasing heat-related illnesses and deaths, shifting risks of vector-borne diseases, asthma hospitalizations triggered by worsening air pollution, higher malnutrition due to crop and livestock failures in drought-stricken regions, and anxiety and posttraumatic stress after experiencing floods and storms. Between 1999 and 2018, nearly half a million people died around the world as a result of more than 12,000 extreme storms, floods, and heat waves.[1]

Haiti, Myanmar, and Puerto Rico were most affected by these weather-related health impacts because of high vulnerability to extreme weather events and a lack of robust capacity to respond and adapt. In 2020, persistent drought followed by flooding rains, heat waves, and a locust plague created severe food shortages in East Africa.[2] Hundreds perished and over half a million were displaced by hurricanes in Central America, and devastating wildfires in the western United States compounded mental health problems for survivors.[3]

Health is defined as "a state of complete physical, mental and social well-being and not merely the absence of disease or infirmity."[4] Climate change influences health primarily as a threat *magnifier*, worsening existing health burdens, risk factors, and inequities, and the health impacts of climate change are experienced differently among individuals, communities, and populations around the world. In fact, climate is best envisioned as "a mosaic of regional conditions" that impact human well-being in context-specific ways.[5] Links between climate change and human health move through complex webs of causation in which changes to the climate drive human exposures that affect health and are influenced by social and environmental factors (**Figure 2.1**).[6]

The world has moral and legal obligations to protect the fundamental human right to health for all people. Significant progress in human health has been made over the past several decades as countries raise their standards of living and improve their health systems. Scientific, medical, technological, and social innovations have extended life expectancies and created effective disease

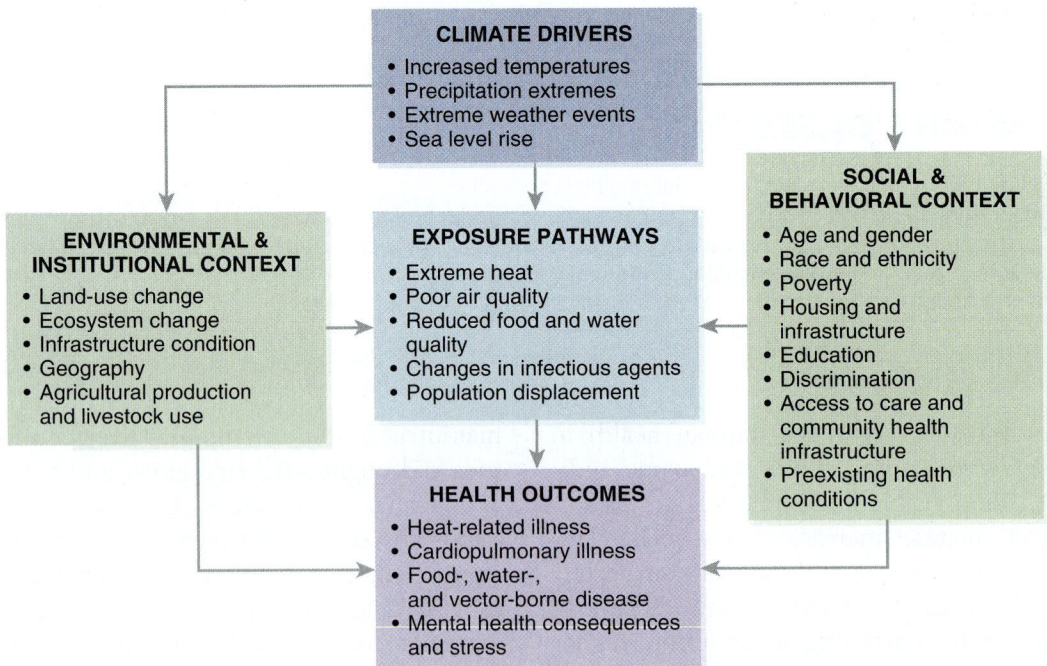

Figure 2.1 Complex Causal Pathways Linking Climate Change and Health.

Reproduced from U.S. Global Change Research Program. *The Impacts of Climate Change on Human Health in the United States: A Scientific Assessment.* U.S. Global Change Research Program; 2016. doi:10.7930/J0R49NQX

prevention and treatment strategies, and international cooperation has created political will to tackle persistent poverty, preventable disease burdens, and premature mortality. However, climate change threatens this progress, and many of these milestones are eroding and in danger of being undone.[7] The climate crisis presents an opportunity to make critical improvements to individual and community well-being through aggressive climate action.

Globally, health improvements have been driven in part by international collaborations to meet the 2000–2015 **Millennium Development Goals (MDGs)**, a set of UN goals and targets to combat extreme poverty and improve well-being, and their successors, the 2015–2030 **Sustainable Development Goals (SDGs)** (**Figure 2.2**). The SDGs are an updated set of targets created by the UN's 2030 Agenda for Sustainable Development to end "poverty and other deprivations . . . hand-in-hand with strategies that improve health and education, reduce inequality, and spur economic growth—all while tackling climate change and working to preserve our oceans and forests."[8]

The MDGs and SDGs have catalyzed national and international action to directly and indirectly protect and improve human health. Significant progress was made on the MDGs, although not all targets were achieved by 2015.[9] For example, the proportion of people in developing countries living on less than $1.25 a day dropped 70% from 1990 to 2015. In addition, global deaths in children younger than 5 declined 52%, and maternal mortality dropped 45%. Malaria cases fell 37%, due in large part to the nearly 1 billion insecticide-treated bed nets delivered to malaria-endemic countries in sub-Saharan Africa.[9]

At the same time that the international community is striving to achieve the SDGs by 2030, it is negotiating accelerated climate action to meet the goals of the Paris Agreement. These two agendas go hand in hand. Currently, progress on many of the SDGs has slowed, due in part to a lack of political will and adequate funding. The COVID-19 pandemic has further stalled this progress and eroded many global health gains, starkly revealing to the world that without good health, no society can thrive, whether the threat to health is a deadly virus or climate change.

The pandemic has also laid bare deeply etched inequalities that put certain populations at greater health risk in nonrandom ways, and these groups also tend to be disproportionately impacted by climate change. In the United States, these disparities meant that COVID-19 devastated the health of different groups in unequal ways, including higher rates of infection, hospitalization, and mortality among Indigenous, Latino, and Black people (**Table 2.1**).[10] The pandemic significantly reduced life expectancy among all groups, but particularly in Latinos, who lost more than 3 years of life expectancy at birth, and Black people, who lost more than 2 years. In contrast, whites lost just two-thirds of a year.[11] In general in the United States, health outcomes by race and ethnicity are among the most disparate in the world because of entrenched structural racism in U.S. institutions, policies and health systems, and a "matrix of oppressive forces generated by poverty and steep grades of social inequality which leads to the frustration of fundamental human needs."[12] In the United States, it tends to be "easier for a white person to live a fully healthy life than a Black person."[13]

These health disparities are not, however, intractable or inevitable. The Paris Agreement and the SDGs can make a difference and improve health equity, along

MILLENNIUM DEVELOPMENT GOALS

SUSTAINABLE DEVELOPMENT GOALS

Figure 2.2 Top: The Eight UN Millennium Development Goals (2000–2015). Bottom: The 17 UN Sustainable Development Goals (2015–2030).

Table 2.1 COVID-19 Disparities by Race and Ethnicity in the U.S. as of September 9, 2021.

COVID-19 statistic	Rates Compared with Non-Hispanic Whites			
	American Indian or Alaska Native	Hispanic or Latino	Black or African American	Asian
Cases	1.7x	1.9x	1.1x	0.7x
Hospitalizations	3.5x	2.8x	2.8x	1.0x
Deaths	2.4x	2.3x	2.0x	1.0x

Reproduced from Centers for Disease Control and Prevention. Risk for COVID-19 infection, hospitalization, and death by race/ethnicity. Accessed September 26, 2021. https://www.cdc.gov/coronavirus/2019-ncov/covid-data/investigations-discovery/hospitalization-death-by-race-ethnicity.html

with other commitments to health at global, national, and local levels. Effective climate adaptation plans can protect against current and future health impacts of climate change if public health is a strong organizing principle in these plans. Unfortunately, to date, many countries have neglected health in their nationally determined contributions (NDCs) under the Paris Agreement, including major emitter countries—such as the United States, Australia, and EU member states—which make no reference at all to health.[14] In contrast, highly vulnerable small island developing states are the most likely group of countries to mention health in their NDCs.

Many people around the world are feeling the impacts of climate change today and are dealing with significant illness, injury, and loss of property and livelihoods. Populations in countries that have contributed least to global climate change tend to suffer the most from its impacts. Pointing to deadly cyclones, erratic rainfall, and failed crops across the African continent, Ugandan youth climate activist Vanessa Nakate says, "This is my world at 1.2°C of warming. I can tell you, a 2°C hotter world is a death sentence for countries like mine."[2] Nakate stresses that African countries have "a real appetite for clean, renewable energy" even

as they face financial and technical gaps to make this transition. Gladys Habu, a climate activist from the Solomon Islands, laments, "It saddens and angers me that we face these huge daily challenges, while millions around the world are fueling the destruction we suffer, yet their own lives remain relatively unaffected."[15]

A high degree of uncertainty constrains the full characterization of the health impacts of climate change, but a precautionary approach allows for action in the face of this uncertainty to protect those experiencing these impacts today and in the future. The **precautionary principle** states that "when an activity raises threats of harm to human health or the environment, precautionary measures should be taken even if some cause and effect relationships are not fully established scientifically."[16] If stronger health adaptations can be implemented now, millions of premature deaths can be avoided, including by making climate-smart changes to our diets and food systems, lowering extreme weather and air pollution risks, and increasing forms of active transport.[17] Framing climate adaptation as a health issue creates a unique opportunity to raise awareness and political will for action because most people strongly value their health and understand how to respond to health threats.

Measuring the Health Impacts of Climate Change

To characterize and predict climate-related health impacts, many questions need to be addressed:

- Which health outcomes are changing? How are baselines and extremes shifting now, and how will they shift in the future?
- Which populations are at highest risk and why?
- Can we attribute health outcomes to climate change, and to what extent is this possible or necessary?
- Which climate mitigation and adaptation strategies offer health *co-benefits*? What are the health *co-harms* of not acting?
- What are the climate co-benefits of improving health burdens and health systems and addressing fundamental social determinants of health?
- How are health risks communicated to the public and decision makers?
- What roles should health professionals play in the climate crisis?

Understanding climate impacts on health requires characterizing risk, both quantitatively and qualitatively. **Risk** in a public health context refers to the probability that an adverse health condition will arise in an individual or a population. A more technical definition of risk is the proportion of people who were initially free of a specific condition who develop that condition during a specific time period. **Figure 2.3** shows three factors that intersect and contribute to risk: the presence of a health hazard, human exposure to the hazard, and vulnerability to the health effects of that exposure. The magnitudes of these three factors determine the level of risk,

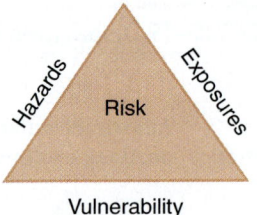

Figure 2.3 Risk Triangle Showing Risk as a Function of Hazards, Human Exposures to Hazards, and Vulnerability to the Effects of Those Exposures.

Data from Crichton D. The risk triangle. In: Ingleton J, ed. *Natural Disaster Management*. Tudor Rose; 1999:102–103.

and all three must be present simultaneously for there to be a risk.

Vulnerability is determined by three overarching factors[6]:

- *Exposure* to climate-related or other stressors that impact health, including poverty, discrimination, or hazardous occupations;
- *Susceptibility* to the adverse health impacts of that exposure, which is influenced by biological and social risk factors; and
- *Low adaptive capacity* and ability to respond, related to factors such as poverty, discrimination, low educational attainment, social norms, and lack of access to health services.

Use of the term "vulnerable" to describe at-risk populations is sometimes critiqued as labeling groups without their consent, as disempowering these groups, and for too much focus on risks and dependency rather than opportunities and self-determination.[18] Others argue that "vulnerable" may be a useful term to call attention to our moral obligation to protect at-risk people and promote equity. Using terms such as "at risk" and "underserved" highlights that systemic barriers rather than inherent characteristics of the population increase their health burdens and prevent their needs from being met.[18]

How are changes in health status detected and measured, how are future changes modeled, and how are both attributed to climate change? Answers to these questions come from numerous fields, including public health, epidemiology, biomedical sciences, and community health. **Public health** is the science and art of preventing disease and promoting health in populations, with an emphasis on controlling social and environmental determinants of health. **Epidemiology** is the study and control of the distribution and determinants of diseases and other health conditions in human populations. Traditional epidemiology allows us to detect and analyze health impacts that have already occurred, and combining this with predictive climate models allows us to forecast and quantify future health impacts. **Descriptive epidemiology** characterizes the amount and distribution of health or disease within a population (*who, what, when, where,* and *how many*). **Analytic epidemiology** examines relationships between risk factors and diseases (*how* and *why*). Many commonly used epidemiological terms and concepts are introduced here to facilitate comprehension of epidemiological studies and how they are designed to measure climate–health relationships.

Sample: A population subgroup included in an epidemiological study. A sample may be random or nonrandom, which determines whether or not study results can be extrapolated to the general population. A sample may be *intentionally* nonrandom in order—for example, to study a specific disease. If the sample is *unintentionally* nonrandom, **sampling bias** results, which limits the usefulness of the study.

Disease count: The number of cases of disease in a population.

Disease burden: The scope (*amount*) of disease in a population.

Incidence: The total number of cases of disease diagnosed in a specified population during a specified time interval (usually one year).

Incidence rate: The rate at which new cases of disease occur in a specified population during a specified time interval (usually one year); usually expressed per 100,000 people in the population.

Incidence rate = [Number of new cases during specified time interval/population at risk] × 100,000

Prevalence: The total number of cases of disease in a specified population, regardless of when the diagnosis was made.

Prevalence rate: The rate of total cases of disease in a specified population during a specified time interval (usually one year); usually expressed per 100,000 people in the population.

Prevalence rate = [Total number of cases during specified time interval/ population at risk] × 100,000

Mortality: Deaths, usually attributed to a specific disease or risk factor.

Mortality rate: The rate of total deaths in a specified population during a specified time interval (usually one year); usually indicated for a specific disease or risk factor and expressed per 100,000 people in the population.

Mortality rate = [Total number of deaths during specified time interval/ population at risk] × 100,000

Morbidity: The condition of suffering but not dying from a specified disease.

Disability-adjusted life year (DALY): One year of healthy life lost because of premature death or disability, usually indicated for a specific disease or risk factor. DALYs reflect overall suffering in a population and provide a broader characterization of health burden than solely counting deaths. DALYs are calculated by adding the number of years of life lost (YLLs) in a population due to premature deaths from a specific disease and the number of years of healthy life lost to disability (YLDs) from that disease.

$$DALYs = YLLs + YLDs$$
$$DALYs\ rate = [Total\ DALYs/population\ of\ interest] \times 100{,}000$$

A single disease may cause many DALYs in a given period (e.g., in 2019, malaria caused more than 46 million global DALYs in addition to 643,000 deaths).[19] An overarching goal of public health is to avoid as many DALYs as possible.

Crude rate: A rate that does not control for modifying variables that affect the rate (e.g., age distribution of the population).

Specific rate: A rate in a specific population subgroup (e.g., defined by age, sex, or race/ethnicity).

Adjusted rate: A rate that takes into account variance among populations being compared.

Age-adjusted (age-standardized) rate: A rate that is modified by applying a weighting process to account for differing age distributions in two or more populations being compared.

For example, **Figure 2.4** shows the crude and age-adjusted mortality rates in the United States from 1980 to 2005.[20] The trends in these two rates differ because the

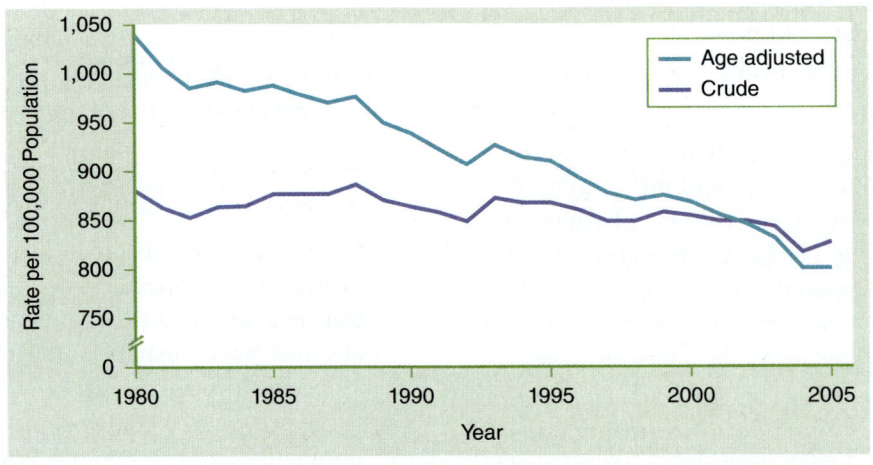

Notes: Crude death rates on an annual basis per 100,000 population. Age-adjusted rates per 100,000 U.S. standard population.

Figure 2.4 Crude vs. Age-Adjusted Death Rates in the U.S. from 1980 to 2005.

U.S. population in 2005 had a higher proportion of older people than in 1980, and older age is a risk factor for death. The crude rate did not vary significantly over the entire period, reflecting a similar number of total deaths per unit population. However, because of greater proportions of older people in the population in more recent years, the actual death rate (age-adjusted) in the population declined over time. Age-adjusted rates are also important when comparing disease burdens among countries or regions with differing age makeup of their populations.

Information on disease incidence, prevalence, morbidity, and mortality in a population may be gathered in several ways. **Data mining** involves searching existing databases for the information of interest. Increasingly, very large public health data sets are available that can be analyzed computationally to reveal patterns, trends, and associations, although data are limited by availability from local or national health surveillance and reporting systems. Where data gaps exist, estimates may be used that take into account observable disease patterns in similar populations or disease burdens and trends that are modeled rather than observed. Descriptive epidemiological studies may be used to reveal the distribution of diseases in a population and which variables (*exposures*) are associated with these diseases. For climate change, hazardous exposures include extremely high temperatures, floods, droughts, wildfires, tropical cyclones, and disease-carrying vector species.

Caution should be exercised when relying too much on quantitative data, which may obscure certain health realities. Other useful forms of information are qualitative community health surveys and testimonials about lived experiences. Many Indigenous peoples emphasize in their concepts of health not just the physical health status of individuals, but social and cultural well-being of communities, ecological health, and traditional knowledge and healing practices.[12] In addition, Indigenous peoples may claim sovereignty over their health information and self-determination with respect to their health outcomes.

Descriptive Epidemiology

Four study types in descriptive epidemiology tend to be used:

- **Case report**: A detailed description of disease occurrence an individual.
- **Case series**: A description of the characteristics of a group of subjects who all have a particular disease or condition.
- **Cross-sectional study**: An examination of the relationship between an exposure (E) and a disease (D) at a single point in time, which provides a *snapshot* of disease prevalence in a defined population.

An example of a cross-sectional study is one that characterizes the mental health impacts of exposure to extreme flooding. The study population is divided into groups that did or did not directly experience flooding and that do or do not have diagnosed or self-reported symptoms of an adverse mental health condition—say, posttraumatic stress disorder (PTSD). The prevalence rates of PTSD symptoms in the group exposed to flooding and the unexposed group are then compared. If the prevalence rate of PTSD is higher in the exposed group than in the unexposed group, flooding and PTSD are *correlated* in this study (although nothing can be concluded about *causation* from this study). This type of study is also called a **prevalence study**.

- **Ecological study**: An examination of the relationship between an exposure (E) and a disease (D) by comparing a set of populations, often organized by spatial unit (e.g., neighborhood, city, or country).

An example of an ecological study is one that characterizes the impacts of extreme heat on chronic kidney disease (CKD) by collecting data on average summertime temperatures and prevalence of CKD in cities within a country or countries around the world. Plotting the exposure variable (in this case, temperature) on the x-axis, and the disease variable (in this case, CKD prevalence) on the y-axis yields a scatterplot on which each point represents one city or country. The scatter pattern can be analyzed for an association between exposure (E) and

disease (D), with a positive slope suggesting a *positive association* between E and D, meaning that more exposure is linked to more disease. A negative slope suggests a *negative association* between E and D, meaning that exposure may protect against disease.

An ecological study reveals nothing about individual health, but it can be helpful to understand the socioeconomic or policy-related context of a health outcome. A familiar example of an ecological study is one that explores the relationship between national income and life expectancy for countries all over the world (**Figure 2.5**). As income (indicated by gross national product per capita) rises, so does life expectancy, indicating

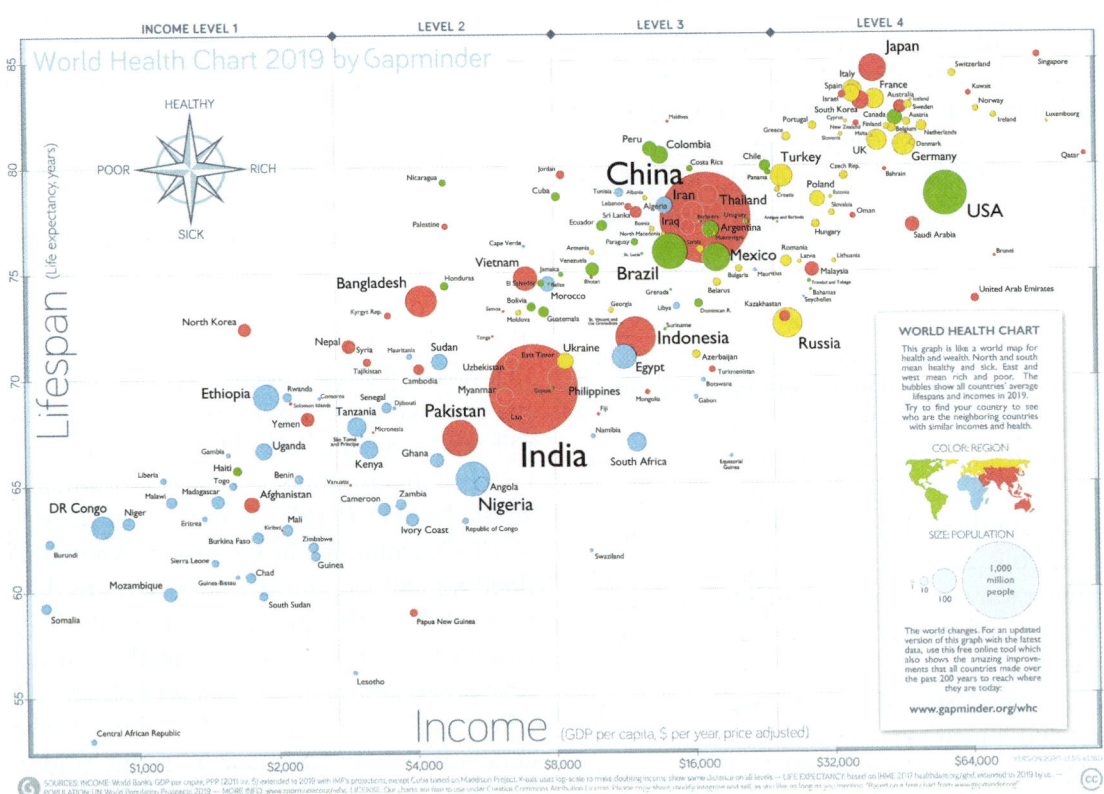

Figure 2.5 Ecological Plot of GDP Per Capita and Life Expectancy for All the Countries of the World in 2015. Each dot corresponds to one country, and the diameter of the dot reflects population size. The colors refer to the region of the world in which each country is located.

a positive association. This is not, however, a simple relationship, and critical modifying variables interact, including quality of national health systems.[21]

Analytic Epidemiology

Analytic epidemiology is used to more precisely explore associations between exposures (E) and diseases (D). Three study types are often used.

- **Case-control study**: An examination of the relationship between E and D in a population divided into *cases* (people selected because they have the disease) and *controls* (people selected with similar characteristics as cases but who do not have the disease). In this type of study, the sample is not random because participants are selected based on disease status, and the proportion of disease cases in the sample greatly exceeds what is found in the general population.

As an example, a case-control study could be used to compare a sample of babies born with a certain birth defect (cases) with babies without this birth defect (controls) for a link to maternal heat exposure during pregnancy, measured using temperature records at the mothers' places of residence. The odds of cases having been exposed is calculated and then divided by the odds that controls were exposed to give an **odds ratio (OR)**. *Odds* are calculated rather than *rates* in case-control studies because the sample is nonrandom and does not reflect the general population, so a rate (probability) in the population cannot be extrapolated. **Odds** describe how likely it is that an event will occur, compared with how likely it is that it will not occur. For example, if you have six marbles in a bag and two are blue, the *odds* of picking a blue marble is 2/4 (0.50). In contrast, the *probability* of picking a blue marble is 2/6 (0.33, 33%).

If the OR reported in a study is greater than 1, this indicates that the odds were higher that people with the disease were exposed, as compared with people without the disease. This is a *positive association* between E and D. An OR less than 1 indicates a *negative association* between E and D, suggesting that the exposure is protective and results in *less* disease. If OR equals 1, there is no association between E and D in this study.

> *If OR = 1, no association between E and D*
>
> *If OR > 1, positive association between E and D*
>
> *If OR < 1, negative association between E and D*

- **Cohort study**: An examination of the relationship between E and D by studying a large population, often with similar characteristics and often over a long period, with the goal of identifying which exposures are correlated with which diseases.

The U.S. Framingham Heart Study is an example of an ongoing cohort study. This study has followed thousands of subjects in Framingham, Massachusetts, for more than 70 years to identify exposures that are risk factors for heart disease.[22] Subjects have regular clinical checkups and respond to surveys to detect diseases and exposures. The incidence rates for a specific disease can then be compared in groups that were or were not exposed to a specific risk factor. Results from this study showed, for example, that smoking is linked to heart disease.

The results from a cohort study are expressed as a **relative risk (RR)**, calculated by dividing the rate of disease in the exposed group by the rate of disease in the unexposed group. RR is also referred to as a **rate ratio (RR)**.

If RR is greater than 1, the exposed group has a higher rate of disease than the unexposed group, indicating a *positive association* between E and D. If RR is less than 1, the exposed group has a lower rate of disease than the unexposed group, indicating a *negative association* between E and D and suggesting that the exposure is protective against the disease. If RR equals 1, no association is indicated in this study.

> *If RR = 1, no association between E and D*
>
> *If RR > 1, positive association between E and D*
>
> *If RR < 1, negative association between E and D*

- **Experimental study**: A **clinical trial** or **intervention study** in which a protective exposure (e.g., a new drug or vaccine, behavioral change, or beneficial health policy) is introduced to one group in the sample, and resulting disease rates in the exposed and unexposed groups are measured and compared. These studies are most useful to establish standards of medical or public health practice.

The relationship between E and D measured in analytic epidemiology is sometimes expressed as a **hazard ratio** (**HR**), which is analogous to RR but is used most often in *time-to-event* or *survival* analyses—for example, after a medical procedure, when researchers want to determine how long it takes for a health outcome (e.g., death) to occur in the treated versus untreated groups.

Uncertainty in the results from analytic epidemiological studies is often expressed using the statistical terms *p-value* and *confidence interval*, which may be reported along with the OR, RR, or HR values to indicate the significance of the association between E and D measured in the study.

A **95% confidence interval (95% CI)** describes a range of RR or OR values that contain the true value with 95% certainty (confidence). For an association between E and D to be considered significant, not just the RR or OR but the entire 95% CI range must be greater than 1 for a *positive association* or less than 1 for a *negative association*. If the 95% CI contains the value 1, no conclusion can be drawn about the association between E and D in this study.

An example of a cohort study reporting RR and 95% CI is one that was conducted in Spain to characterize the heat-related mortality that occurred during an extreme heat wave in 2003. The study determined that the increased risk of dying on an extreme heat day was 24% higher than on a non–extreme heat day (RR = 1.24). The 95% CI calculated for this study result was 1.19–1.30, indicating that the true value is likely to be between 19% and 30% excess risk.[23] The entire 95% CI range is greater than 1, which indicates a *positive association* between extreme heat exposure and mortality.

The term **p-value** is a statistical measure of the probability that an observed association between E and D could have occurred by chance alone. A p-value *less than 0.05* is often used as the threshold for statistical significance of the observed association. This indicates that there is a less than 5% chance that the association between E and D detected in the study occurred by chance. In other words, there is a more than 95% chance that the association is real. This 95% standard for certainty in epidemiology guards against making false claims of association between E and D, although 95% ($p < 0.05$) is an arbitrary cutoff, and results with a higher p-value may still reflect a true association.

Another important term when interpreting the results of an epidemiological study is

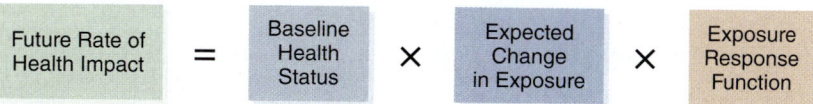

Figure 2.6 Modeling Future Health Impacts Requires Information about Baseline Health Status in the Population of Interest, the Expected Change in Exposure, and the Exposure–Response Function.

Reproduced from U.S. Global Change Research Program. *The Impacts of Climate Change on Human Health in the United States: A Scientific Assessment.* U.S. Global Change Research Program; 2016. doi:10.7930/J0R49NQX

its **power**, which is the ability of the study to demonstrate an association if one exists. Most often, power is determined by the size of the sample, with a large sample size conferring greater power to the study.

Predicting future health impacts of climate change requires using climate and health models to estimate the expected change in exposure to a change in climate, future rates of disease from measures of the baseline health status of the population of interest, and the relationship between exposure and disease (*exposure-response function*) (**Figure 2.6**). The exposure–response function quantifies how much a health outcome changes when exposure changes by a given amount—for example, how much the rate of heat stroke increases for every 1°C increase in outdoor temperature. Some epidemiological studies may use climate models that incorporate scenarios, such as representative concentration pathways (RCPs) and shared socioeconomic pathways (SSPs), to characterize a range of possible future exposures and health outcomes.

Quantifying health impacts, exposures, and their relationships may be constrained by a lack of data on the baseline incidence of disease, inability to predict future climate change exposures, and lack of understanding of the relationship between exposures and diseases. Many modifying variables also affect these relationships, making it difficult to easily identify causal relationships and predict health outcomes.[6] In addition, disease burdens, health systems, socioeconomic

conditions, and climate drivers are constantly shifting, which may complicate the study of these relationships. For example, communicable diseases, such as malaria and tuberculosis, as well as malnutrition, tend to be higher in lower income countries. As these countries develop socioeconomically and improve their health systems, they undergo an **epidemiologic transition** in which communicable diseases are increasingly controlled, and noncommunicable diseases, such as heart disease, diabetes, and cancers, cause a greater proportion of mortality and morbidity.

Sources of Health Information

Many reputable sources of information about public health and climate change are available for use by researchers, health professionals, governments, NGOs, individuals, and communities. The **World Health Organization (WHO)** is the global health agency of the United Nations (UN), and it advises countries and provides information on a wide range of global health topics. WHO maintains the **Global Health Observatory**, which has publicly available data on topics ranging from immunizations and antibiotic resistance to mental health and HIV/AIDS.[24] Climate change is a primary focus area of WHO, which helps measure and predict the health benefits of countries implementing their NDCs under the Paris Agreement.

Many other UN agencies, including the UN Children's Fund (UNICEF) and the UN Food and Agriculture Organization (FAO), conduct research and produce reports on the health impacts of climate change.

The U.S. **Centers for Disease Control and Prevention** (**CDC**) is the public health agency of the U.S. government and works to promote the health and well-being of all Americans. The CDC also works with international partners, including WHO, and runs offices in 60 countries. The CDC conducts disease surveillance, makes policy recommendations, implements government public health programs, and maintains publicly available data on its website. It also publishes the *Morbidity and Mortality Weekly Report* (*MMWR*), a periodical of timely and authoritative studies, scientific information, and recommendations for health interventions.

The CDC's **Climate and Health Program** assists state, local, tribal, and territorial public health agencies in developing and implementing climate plans, and its **Racial and Ethnic Approaches to Community Health** (**REACH**) program gives funding and technical support to state and local health departments, tribes, community organizations, and academic institutions to improve health equity.[25] REACH, founded in 1999, seeks to reduce racial and ethnic health disparities among Black, Latino, Asian, and Indigenous peoples.

The U.S. **National Institutes of Health** (**NIH**), the medical research agency of the U.S. government housed in the U.S. Department of Health and Human Services, is the largest biomedical research institution in the world. The NIH's **National Institute of Environmental Health Sciences** (**NIEHS**) conducts research within its **Climate Change and Human Health Program** and makes wide-ranging information available on its website. Some have called for funding a new branch of the NIH dedicated solely to climate change.[26]

The **U.S. Global Change Research Program (USGCRP)** is a collaboration among 13 federal government agencies to coordinate scientific and policy research on climate change. The USGCRP develops the U.S. National Climate Assessment, and two of its recent reports describe the health impacts of climate change: *The Impacts of Climate Change on Human Health in the United States: A Scientific Assessment* (2016)[6] and *Fourth National Climate Assessment Volume II: Impacts, Risks, and Adaptation in the United States* (2018). Both of these reports are publicly available on the USGCRP website.

The **Intergovernmental Panel on Climate Change** (**IPCC**), created in 1988, is the UN body that assesses the state of the science of climate change, its impacts and future risks, and climate adaptation and mitigation options. Thousands of scientists and other experts from around the world contribute to the work of the IPCC, which informs climate change decision-making at local, national, and international scales. The IPCC's latest report is the *Sixth Assessment Report* (AR6), released in 2021–2022, of which *AR6 Climate Change 2021: Impacts, Adaptation and Vulnerability* describes human health impacts.[27] All reports are publicly available on the IPCC website.

The Lancet Countdown, created in 2016, is a global collaboration among leading climate scientists, engineers, energy specialists, economists, political scientists, public health professionals, and doctors from academic institutions and UN agencies to monitor health risks, impacts, and opportunities in the climate crisis. Its work focuses on three areas: research and monitoring, policy engagement, and communications and outreach. Each year, the Lancet Countdown publishes its analysis in the medical journal *The Lancet* in advance of that year's UN climate change conference.

In the Lancet Countdown's 2020 report, researchers used 43 indicators to describe (1) climate change impacts, exposures, and vulnerabilities; (2) adaptation, planning, and resilience for health; (3) mitigation actions and health co-benefits; (4) economics and finance; and (5) public and political engagement.[7] Examples of results from their indicator tracking include the following:

- Indicator 1.1.3 showed that heat-related mortality has risen 54% in the past 20 years in people over 65 years of age, including nearly 300,000 deaths in 2018.
- Indicator 2.1.3 showed that two-thirds of global cities expect serious threats to public health and infrastructure from climate change.
- Indicator 3.5.2 showed that excess red meat consumption is increasing, particularly in the Western Pacific region, and contributed to 990,000 deaths in 2017.
- Indicator 4.1.3 showed that India and Indonesia are the worst affected countries for earnings loss due to heat-related reductions in labor, losing 4–6% of GDP annually.
- Indicator 5.1 showed that in just one year, from 2018 to 2019, media coverage of health and climate change nearly doubled worldwide.

The **Institute for Health Metrics and Evaluation (IHME)** is an independent global health research center at the University of Washington. IHME provides rigorous and comparable information on the health status and determinants of health affecting populations. This information is made freely accessible to policy makers and the public in various forms, including user-friendly data visualizations on its website.[19] IHME regularly compiles a **Global Burden of Disease (GBD)** report, a comprehensive global study of health published in *The Lancet* with data made available on IHME's website. The 2019 report analyzed 286 causes of death, 369 diseases and injuries, and 87 risk factors in 204 countries and territories. The GBD report summarizes burdens of communicable and noncommunicable diseases, as well as injuries, all of which may be impacted by various climate drivers. Although GBD does not focus explicitly on climate change, it provides important information on disease burdens and trends over time, varying by age, sex, geographic region, and socioeconomic conditions, as well as health risk factors, most of which are part of the multifactorial webs of causation for climate change impacts on health.

Other sources of health information include specialized government agencies, such as the U.S. National Weather Service, nongovernmental organizations such as the Union of Concerned Scientists, and public–private partnerships and other international collaborations, including the RBM Partnership to End Malaria. In addition, thousands of reports and research papers on climate change and health are published each year and are made available online and in peer-reviewed journals.

Some of the most useful information about the human impacts of climate change comes directly from traditional knowledge and lived experiences of affected communities, including Indigenous peoples. For example, the Alaska Native Tribal Health Consortium (ANTHC) works with communities, regional governments, and organizations to promote wellness and resilience among Indigenous groups in Alaska.[28] ANTHC conducts research and provides comprehensive health services, and it established the Alaska Native Epidemiology Center and the Center for Climate and Health, which helps communities understand the health impacts of climate change and how to respond and adapt (**Figure 2.7**).

Figure 2.7 Left: Alaska Natives Harvest Traditional Berries. Right: Remains of a House Lost to Coastal Erosion Due to Sea Ice Retreat on the Island of Shishmaref, Alaska.

© Ashley Cooper/Corbis/Getty Images.

ANTHC also supports the **Local Environmental Observer** (**LEO**) **Network**, a group of community members, journalists, and experts who serve as "the eyes, ears, and voice of our changing environment" by sharing observations online about unusual human, animal, environment, and weather events.[29] Recent examples of LEO Network reports include people falling through thin ice, power outages caused by heavy snow, underground fuel storage tanks becoming exposed as a result of thawing permafrost, shifting seasonal patterns for plants and animals, unusually high pollen counts, disease pathogens in seafood, and community impacts of coastal storms, floods, and wildfires. Today, the LEO Network continues to monitor the circumpolar region and has begun building partnerships with local observers around the world to provide valuable information about the human impacts of climate change.

Discussion Questions

1. Discuss the concept of risk. Which factors determine risk?
2. Discuss examples of climate drivers leading to human exposure pathways and health outcomes.
3. Which social and environmental risk factors influence the ways that climate change impacts human health?
4. How are incidence, prevalence, and mortality rates calculated?
5. What are DALYs, how are they calculated, and how are they a useful health metric?
6. Using examples of specific study types, discuss how descriptive and analytic epidemiology are used to study the health impacts of climate change. Find and discuss examples in peer-reviewed journals.
7. What do odds ratios, relative risks, 95% confidence intervals, and

p-values tell you about the results of an epidemiological study?

8. Describe sources of health information that are useful for understanding the relationships between climate change and human health. Explore relevant health databases to learn more about what types of information are available.

References

1. Eckstein D, Kunzel V, Schafer L, Winges M. *Global Climate Risk Index 2020*. December 4, 2019. Accessed February 8, 2021. https://www.germanwatch.org/en/17307

2. Nakate V. A 2°C hotter world is a death sentence for countries like mine. *The Independent*. February 8, 2021. Accessed March 8, 2021. https://www.independent.co.uk/climate-change/opinion/climate-change-paris-agreement-uganda-b1769562.html

3. Stern, J. A mental-health crisis is burning across the American West. *The Atlantic*. July 20, 2020. Accessed June 15, 2021. https://www.theatlantic.com/health/archive/2020/07/mental-health-aftermath-california-wildfires/608656/

4. World Health Organization. Constitution of the World Health Organization (Forty-fifth edition, Supplement). October 2006. Accessed March 8, 2021. https://www.who.int/governance/eb/who_constitution_en.pdf

5. McMichael AJ. *Climate Change and the Health of Nations: Famines, Fevers, and the Fate of Populations*. Oxford University Press; 2017.

6. U.S. Global Change Research Program. *The Impacts of Climate Change on Human Health in the United States: A Scientific Assessment*. U.S. Global Change Research Program; 2016. doi:10.7930/J0R49NQX

7. Watts N, Amann M, Arnell N, et al. The 2020 report of The Lancet Countdown on health and climate change: Responding to converging crises. *The Lancet*. 2021;397(10269):129–170. doi:10.1016/S0140-6736(20)32290-X

8. United Nations. The 17 Goals. Accessed March 8, 2021. https://sdgs.un.org/goals

9. United Nations. *The Millennium Development Goals Report 2015*. 2015. Accessed March 8, 2021. https://www.un.org/millenniumgoals/2015_MDG_Report/pdf/MDG%202015%20rev%20(July%201).pdf

10. U.S. Centers for Disease Control and Prevention. Risk for COVID-19 infection, hospitalization, and death by race/ethnicity. Accessed September 26, 2021. https://www.cdc.gov/coronavirus/2019-ncov/covid-data/investigations-discovery/hospitalization-death-by-race-ethnicity.html

11. Andrasfay T, Goldman N. Reductions in 2020 U.S. life expectancy due to COVID-19 and the disproportionate impact on the Black and Latino populations. *Proc Natl Acad Sci*. 2021;118(5):e2014746118. doi:10.1073/pnas.2014746118

12. Richardson ET. *Epidemic Illusions: On the Coloniality of Global Public Health*. MIT Press; 2020.

13. Lopez G. The Black-white life expectancy gap grew in 2020—but it can be reversed. *Vox*. February 24, 2021. Accessed March 8, 2021. https://www.vox.com/22285868/black-white-life-expectancy-gap-covid-19-health

14. Dasandi N, Graham H, Lampard P, Jankin Mikhaylov S. Engagement with health in national climate change commitments under the Paris Agreement: a global mixed-methods analysis of the nationally determined contributions. *Lancet Planet Health*. 2021;5(2):e93–e101. doi:10.1016/S2542-5196(20)30302-8

15. Habu G. *Engulfed by the Sea: The Loss and Damage from Climate Change* blog. International Institute for Environment and Development. November 18, 2020. Accessed March 8, 2021. https://www.iied.org/engulfed-sea-loss-damage-climate-change

16. Collaborative on Health and the Environment. Precautionary principle: The Wingspread Statement. Accessed June 15, 2021. https://www.healthandenvironment.org/environmental-health/social-context/history/precautionary-principle-the-wingspread-statement

17. Hamilton I, Kennard H, McGushin A, et al. The public health implications of the Paris Agreement: a modelling study. *Lancet Planet Health*. 2021;5(2):e74–e83. doi:10.1016/S2542-5196(20)30249-7

18. Clark B, Preto N. Exploring the concept of vulnerability in health care. *CMAJ*. 2018;190(11):E308–E309.

19. Institute for Health Metrics and Evaluation. GBD Compare. October 15, 2020. Accessed October 15, 2020. https://vizhub.healthdata.org/gbd-compare/

20. Kung H-C, Hoyert DL, Xu J, Murphy SL. Deaths: preliminary data for 2005. *Health E-Stats*. May 2, 2007. Accessed March 8, 2021. https://www.cdc.gov/nchs/data/hestat/prelimdeaths05/prelimdeaths05.htm

21. Gapminder. World health chart 2019. Accessed July 29, 2021. https://drive.google.com/file/d/1-WCREPFc7mu_CFrVjzLI8bJbP3kFZW0B/view. FREE TO USE! CC-BY GAPMINDER.ORG.

22. Framingham Heart Study. Three generations of research on heart disease. Accessed March 8, 2021. https://framinghamheartstudy.org/

23. Tobias A, Armstrong B, Zuza I, et al. Mortality on extreme heat days using official thresholds in Spain: a multi-city time series analysis. *BMC Public Health*. 2012;12(133):1–9. doi:10.1186/1471-2458-12-133

24. World Health Organization. The Global Health Observatory. Accessed March 8, 2021. https://www.who.int/data/gho

25. U.S. Centers for Disease Control and Prevention. Racial and ethnic approaches to community health. October 16, 2020. Accessed March 8, 2021. https://www.cdc.gov/nccdphp/dnpao/state-local-programs/reach/index.htm

26. Frumkin H, Jackson RJ. We need a National Institute of Climate Change and Health. *Sci Am*. November 22, 2020. Accessed March 8, 2021. https://www.scientificamerican.com/article/we-need-a-national-institute-of-climate-change-and-health/

27. Intergovernmental Panel on Climate Change. AR6 Climate Change 2021: Impacts, Adaptation and Vulnerability. Accessed February 2, 2021. https://www.ipcc.ch/report/sixth-assessment-report-working-group-ii/

28. Alaska Native Tribal Health Consortium. Accessed March 8, 2021. https://anthc.org/

29. Local Environmental Observer (LEO) Network. Accessed March 1, 2021. https://www.leonetwork.org/

CHAPTER 3

Climate Justice

"When it comes to the effects of climate change, there has been nothing but chronic injustice and the corrosion of human rights. To deal with climate change we must simultaneously address the underlying injustice in our world and work to eradicate poverty, exclusion and inequality. We need to create a 'people-first' platform for those on the margins suffering the worst effects of climate change, and amplify their voices to ensure them a seat at the table in any future climate change negotiations."[1]

—**Mary Robinson**, former president of Ireland and UN Special Envoy on Climate Change

KEY TERMS

Climate justice
Environmental justice (EJ)
Geographic inequity

Procedural inequity
Social inequity
Environmental racism

Bali Principles of Climate Justice
Human rights

LEARNING OBJECTIVES

- Define environmental justice and climate justice.
- Describe ways that certain populations in the United States and around the world are disproportionately impacted by climate change.
- Describe the Bali Principles of Climate Justice and apply them to examples of climate change impacts on human health.
- Describe the fundamental human right to health and how it is protected by international declarations, treaties, and climate agreements.

Bangladesh is one of the world's most at-risk countries in the climate crisis. In the summer of 2020, Bangladesh experienced such extreme precipitation that flooded rivers submerged up to one-third of the land and a million homes.[2] Climate change-fueled sea level rise, monsoon rains, and tropical cyclones are a recurrent hazard in coastal areas of Bangladesh, and people risk losing their homes, farms, fields, livestock, food supply, shops, and schools. Bangladeshis face enormous displacement pressures, and many people have been forced to settle in urban slums far from their homes. They are at risk of injuries and illnesses directly linked to storms and flooding; exacerbation of existing health burdens, including malnutrition, waterborne diseases, stress, anxiety, and depression; and loss of personal and communal identities as lands, homes, and villages are inundated and destroyed.[3] According to Sheema Sen Gupta of the United Nations Children's Fund (UNICEF), "In Bangladesh, climate change is in your face. You can't avoid it. You can see it happening. It's a given. It's now part of everyday living."[4]

The world's richest 10% of income earners contributed more than half of the global greenhouse gas (GHG) emissions fueling climate change from 1990 to 2015, whereas the poorest 50%, including most Bangladeshis, contributed just 7%.[5] In general, people in the world's least developed countries face the most significant human impacts of climate change and tend to lack adequate capacity to respond and adapt—despite having contributed little to GHG emissions and having benefited less from the intensive use of fossil energy that has driven these emissions. The risk of extreme weather impacts alone is 80 times higher in developing countries than in developed countries.[6] Many highly at-risk people live in regions where vector-borne diseases, malnutrition, and diarrheal diseases, all sensitive to climate change, are a heavy burden and where adequate and responsive healthcare services may not be accessible. Within countries, marginalized populations—including the poor, racial and ethnic minorities, women, and Indigenous peoples—often face the gravest threats.

Truly effective climate action is not possible unless those at greatest risk are protected and empowered, and this is what the climate justice movement calls for.

Climate justice is the equitable protection of all people from climate impacts, regardless of race, ethnicity, national origin, indigeneity, gender, income, or other status, and the inclusion of all people in decision-making about climate action. Climate justice is founded on the principle that no group of people should bear a disproportionate burden of climate impacts or costs, and it explicitly acknowledges that the root causes of climate change and societal inequalities—unjust economic, political, and social systems—must be corrected. **Inequality** is defined as the state of being uneven or unfair, or a disparity of distribution or opportunity. **Inequity** is the inability to attain equality because of lacking opportunities and impediments that are unevenly and unfairly distributed across populations. **Justice** is the quality of being just, impartial, and fair, and it is the moral principle that determines just conduct and righteousness.

Climate justice builds on the principles of **environmental justice (EJ)**. According to the U.S. Environmental Protection Agency (EPA), EJ is the

> fair treatment and meaningful involvement of all people regardless of race, color, national origin, or income, with respect to the development, implementation and enforcement of environmental laws, regulations and policies. It will be achieved when everyone enjoys the same degree of protection from environmental and health hazards, and equal access to the decision-making process to have a healthy environment in which to live, learn and work.[7]

Environmental injustices may occur in different ways. **Geographic inequity** is the discriminatory siting, location, and configuration of environmental hazards in proximity to certain groups or populations. **Procedural inequity** is the discriminatory application and enforcement of environmental regulations that result in disproportionate burdens on certain groups. **Social inequity** is the discriminatory impact of societal structures and practices that prevent fair participation in environmental decision-making and the balance of power. All three forms of injustice occur in the climate crisis and affect marginalized populations all over the world.

In the United States, the EJ movement emerged out of the civil rights movement and took hold in the 1980s in response to the disproportionate exposure of communities of color in the South to municipal and hazardous waste.[8] The movement identified environmental racism as an underlying cause of this unfair burden. **Environmental racism** is the imposition of disproportionate environmental impacts on groups or communities because of their racial or ethnic status, and exclusion of representatives of such groups from participation in environmental decision-making. **Racism** is the intentional or unintentional use of power to isolate, separate, and exploit others based on race that results from the combination of *racial prejudice* and *power*.

In 1991, a group of Black, Latino, Asian American, and Indigenous peoples gathered in Washington, D.C. for the First National People of Color Environmental Leadership Summit. They created a set of EJ principles that focused on addressing the unequal distribution of environmental burdens and protections, underlying structural barriers to equity, and the exclusion of certain groups from decision-making and **self-determination**, which is the right to determine one's own social, cultural, and economic development.[9] These

EJ principles have inspired and informed the climate justice movement.

The first international climate justice summit was organized by a coalition of civil society organizations to coincide with the sixth session of the Conference of the Parties (COP6) to the UN Framework Convention on Climate Change (UNFCCC) in The Hague, the Netherlands, in November 2000. This summit advanced an alternative climate change narrative not recognized in mainstream negotiations—that the people most impacted by climate change do the least to cause the problem and are being left out of negotiations. In preparation for the 2002 Earth Summit in Johannesburg, South Africa, this coalition met in Bali, Indonesia, in August 2002. Its aim was to develop a set of principles, the **Bali Principles of Climate Justice**, that put a human face on climate change (**Box 3.1**).[10] Many of the Bali Principles have been endorsed and applied by civil society groups, environmental organizations, governments, and the international community.

Several important concepts incorporated into these principles are worth highlighting. Bali Principle 4 describes the **principle of common but differentiated responsibilities**, which means that all countries have a shared obligation to address climate change but not equal responsibility for causing climate change or acting to combat it. More recently, this concept has been refined as *common but differentiated responsibilities and respective capabilities, in light of different national circumstances,* acknowledging that countries' responsibilities are not fixed and shift as socioeconomic development advances and "national circumstances" change.[11] Bali Principle 7 refers to **ecological debt**, which is the accumulated debt that wealthier polluter countries owe to poorer countries that have been plundered for resource extraction and exploitation, habitat destruction, waste

Box 3.1 Bali Principles of Climate Justice

PREAMBLE

Whereas climate change is a scientific reality whose effects are already being felt around the world;

Whereas if consumption of fossil fuels, deforestation and other ecological devastation continues at current rates, it is certain that climate change will result in increased temperatures, sea level rise, changes in agricultural patterns, increased frequency and magnitude of "natural" disasters such as floods, droughts, loss of biodiversity, intense storms and epidemics;

Whereas deforestation contributes to climate change, while having a negative impact on a broad array of local communities;

Whereas communities and the environment feel the impacts of the fossil fuel economy at every stage of its life cycle, from exploration to production to refining to distribution to consumption to disposal of waste;

Whereas climate change and its associated impacts are a global manifestation of this local chain of impacts;

Whereas fossil fuel production and consumption helps drive corporate-led globalization;

Whereas climate change is being caused primarily by industrialized nations and transnational corporations;

Whereas the multilateral development banks, transnational corporations and Northern governments, particularly the United States, have compromised the democratic nature of the United Nations as it attempts to address the problem;

Whereas the perpetration of climate change violates the Universal Declaration On Human Rights, and the United Nations Convention on Genocide;

Whereas the impacts of climate change are disproportionately felt by small island states, women, youth, coastal peoples, local communities, Indigenous Peoples, fisherfolk, poor people and the elderly;

Whereas local communities, affected people and Indigenous Peoples have been kept out of the global processes to address climate change;

Whereas market-based mechanisms and technological "fixes" currently being promoted by transnational corporations are false solutions and are exacerbating the problem;

Whereas unsustainable production and consumption practices are at the root of this and other global environmental problems;

Whereas this unsustainable consumption exists primarily in the North, but also among elites within the South;

Whereas the impacts will be most devastating to the vast majority of the people in the South, as well as the "South" within the North;

Whereas the impacts of climate change threaten food sovereignty and the security of livelihoods of natural resource-based local economies;

Whereas the impacts of climate change threaten the health of communities around the world-especially those who are vulnerable and marginalized, in particular children and elderly people;

Whereas combating climate change must entail profound shifts from unsustainable production, consumption and lifestyles, with industrialized countries taking the lead;

We, representatives of people's movements together with activist organizations working for social and environmental justice resolve to begin to build an international movement of all peoples for Climate Justice based on the following core principles:

1. Affirming the sacredness of Mother Earth, ecological unity and the interdependence of all species, Climate Justice insists that communities have the right to be free from climate change, its related impacts and other forms of ecological destruction.

2. Climate Justice affirms the need to reduce with an aim to eliminate the production of greenhouse gases and associated local pollutants.

3. Climate Justice affirms the rights of Indigenous Peoples and affected communities to represent and speak for themselves.

4. Climate Justice affirms that governments are responsible for addressing climate change in a manner that is both democratically accountable to their people and in accordance with the principle of common but differentiated responsibilities.

5. Climate Justice demands that communities, particularly affected communities play a leading role in national and international processes to address climate change.

6. Climate Justice opposes the role of transnational corporations in shaping unsustainable production and consumption patterns and lifestyles, as well as their role in unduly influencing national and international decision-making.

7. Climate Justice calls for the recognition of a principle of ecological debt that industrialized governments and transnational corporations owe the rest of the world as a result of their appropriation of the planet's capacity to absorb greenhouse gases.

8. Affirming the principle of ecological debt, Climate Justice demands that fossil fuel and extractive industries be held strictly liable for all past and current life-cycle impacts relating to the production of greenhouse gases and associated local pollutants.

9. Affirming the principle of ecological debt, Climate Justice protects the rights of victims of climate change and associated injustices to receive full compensation, restoration, and reparation for loss of land, livelihood and other damages.

10. Climate Justice calls for a moratorium on all new fossil fuel exploration and exploitation; a moratorium on the construction of new nuclear power plants; the phase out of the use of nuclear power world-wide; and a moratorium on the construction of large hydro schemes.

11. Climate Justice calls for clean, renewable, locally controlled and low-impact energy resources in the interest of a sustainable planet for all living things.

12. Climate Justice affirms the right of all people, including the poor, women, rural and Indigenous Peoples, to have access to affordable and sustainable energy.

13. Climate Justice affirms that any market-based or technological solution to climate change, such as carbon-trading and carbon sequestration, should be subject to principles of democratic accountability, ecological sustainability and social justice.

14. Climate Justice affirms the right of all workers employed in extractive, fossil fuel and other greenhouse gas producing industries to a safe and healthy work environment without being forced to choose between an unsafe livelihood based on unsustainable production and unemployment.

15. Climate Justice affirms the need for solutions to climate change that do not externalize costs to the environment and communities, and are in line with the principles of a just transition.

16. Climate Justice is committed to preventing the extinction of cultures and biodiversity due to climate change and its associated impacts.

17. Climate Justice affirms the need for socio-economic models that safeguard the fundamental rights to clean air, land, water, food and healthy ecosystems.

18. Climate Justice affirms the rights of communities dependent on natural resources for their livelihood and cultures to own and manage the

(continues)

Box 3.1 Bali Principles of Climate Justice   *(continued)*

same in a sustainable manner, and is opposed to the commodification of nature and its resources.

19. Climate Justice demands that public policy be based on mutual respect and justice for all peoples, free from any form of discrimination or bias.

20. Climate Justice recognizes the right to self-determination of Indigenous Peoples, and their right to control their lands, including sub-surface land, territories and resources and the right to the protection against any action or conduct that may result in the destruction or degradation of their territories and cultural way of life.

21. Climate Justice affirms the right of Indigenous peoples and local communities to participate effectively at every level of decision-making, including needs assessment, planning, implementation, enforcement and evaluation, the strict enforcement of principles of prior informed consent, and the right to say "No."

22. Climate Justice affirms the need for solutions that address women's rights.

23. Climate Justice affirms the right of youth as equal partners in the movement to address climate change and its associated impacts.

24. Climate Justice opposes military action, occupation, repression and exploitation of lands, water, oceans, peoples and cultures, and other life forms, especially as it relates to the fossil fuel industry's role in this respect.

25. Climate Justice calls for the education of present and future generations, emphasizes climate, energy, social and environmental issues, while basing itself on real-life experiences and an appreciation of diverse cultural perspectives.

26. Climate Justice requires that we, as individuals and communities, make personal and consumer choices to consume as little of Mother Earth's resources, conserve our need for energy; and make the conscious decision to challenge and reprioritize our lifestyles, re-thinking our ethics with relation to the environment and the Mother Earth; while utilizing clean, renewable, low-impact energy; and ensuring the health of the natural world for present and future generations.

27. Climate Justice affirms the rights of unborn generations to natural resources, a stable climate and a healthy planet.

Reproduced from International Climate Justice Network. Bali principles of climate justice. Published August 28, 2002. https://corpwatch.org/article/bali-principles-climate-justice

dumping, climate change, and resulting burdens on local people.

Bali Principle 15 describes a **just transition**, an approach to the structural changes required for climate action that will ensure workers' rights and livelihoods, provide access for all to climate-smart advances in energy, agriculture, land and resource use, and economic development, and protect communities that may bear disproportionate costs or get left behind in the transition. A just transition builds economic and political power among all groups to shift equitably to a sustainable and regenerative economy that advances self-determination, the redistribution of power, and use of alternatives to extractive and nonrenewable resources that restore human health and ecological, social, and cultural well-being.[12]

Bali Principle 17 refers to the need to safeguard fundamental human rights, recognizing that climate change undermines these rights. **Human rights** are universal entitlements (*moral rights*) that all people possess simply by virtue of being human, regardless of nationality, ethnicity, religion, sex, disability, or any other status. Protection of human rights, including the right to health, is legally recognized in numerous international declarations and treaties and also marks "the threshold at which each individual's interests generate obligations on the part of others to respect, protect, and promote those interests in various ways. The violation of an *obligation* is a moral wrong."[13]

The preamble to the Constitution of the World Health Organization (1948) states that "the enjoyment of the highest attainable standard of health is one of the fundamental rights of every human being without distinction of race, religion, political belief, economic or social condition" and that "unequal development in different countries in the promotion of health and control of disease, especially communicable disease, is a common danger."[14] It also states that "governments have a responsibility for the health of their peoples which can be fulfilled only by the provision of adequate health and social measures."[14] Climate change threatens this fundamental human right to health and the determinants necessary for the attainment of health, including a sufficient and nutritious food supply, clean water and air, and adequate livelihoods.

The primary instrument for human rights protection is the **International Bill of Human Rights**, which includes the 1948 **Universal Declaration of Human Rights** (UDHR), the 1966 **International Covenant on Economic, Social and Cultural Rights** (ICESCR), and the 1966 **International Covenant on Civil and Political Rights** (ICCPR). Each codifies the fundamental human right to health.

The UDHR was negotiated by the United Nations after World War II and adopted December 10, 1948, at the UN General

Figure 3.1 Eleanor Roosevelt, U.S. First Lady from 1933 to 1945 and Coauthor of the Universal Declaration of Human Rights.
© Everett/Shutterstock.

Assembly in Paris, France (**Figure 3.1**). Of the 58 UN member nations at the time, 48 voted to adopt the UDHR, including the United States. As a *declaration*, the UDHR is not legally binding, but it sets normative standards for its signatories. Article 25 of the UDHR states that

> everyone has the right to a standard of living adequate for the health and well-being of himself and of his family, including food, clothing, housing and medical care and necessary social services, and the right to security in the event of unemployment, sickness, disability, widowhood, old age or other lack of livelihood in circumstances beyond his control.[15]

The ICESCR and ICCPR are legally binding treaties that carry the force of law. Parties to these covenants (nations that ratify them) accept legal obligations to uphold rights and provisions set forth in the treaties. The ICCPR does not focus directly on health, but it protects the rights to life, self-determination, freedom of religion, freedom of speech, freedom of assembly, electoral rights, due process, and fair trials "without distinction of any kind, such as race, color, sex, language, religion, political or other opinion, national or social origin, property, birth or other status."[16] The ICESCR affirms "the right of everyone to the enjoyment of the highest attainable standard of physical and mental health" and calls for the "improvement of all aspects of environmental and industrial hygiene," "the prevention, treatment and control of epidemic, endemic, occupational and other diseases," and the "creation of conditions which would assure to all medical service and medical attention in the event of sickness."[17] The United States is a Party to the ICCPR but not the ICESCR.

The right to health is also affirmed in other treaties, including the International Convention on the Elimination of All Forms of Racial Discrimination (1965), the Convention on the Elimination of All Forms of Discrimination against Women (1979), the Convention on the Rights of the Child (1989), the Convention on the Rights of Persons with Disabilities (2006), and the UN Declaration on the Rights of Indigenous People (2007). Most of the world's countries are Parties to these treaties, but the United States has ratified only the Convention on the Elimination of All Forms of Racial Discrimination.

In 2019, Philip Alston, then the UN Special Rapporteur on extreme poverty and human rights, warned of looming "climate apartheid" in which wealthy people pay to avoid the impacts of climate change while the rest of the world is left to suffer. In a report to the UN Human Rights Council, Alston wrote, "Climate change threatens the full enjoyment of a wide range of rights. Rapid action and adaptation can mitigate much of this, but only if done in a way that protects people in poverty from the worst effects."[18]

He warned that fundamental human rights are threatened if countries, international and nongovernmental organizations, and the private sector do not take action now. He noted that climate change is one of many issues on the human rights agenda and ought to be prioritized because it threatens basic human rights in all aspects of society. He also stressed that the climate crisis should be seen not just as a threat, but also as an *opportunity* to catalyze action to improve lives and break down systemic causes of inequality.

Through persistent pressure from climate justice advocates and civil society organizations, the UNFCCC was amended in 2010 to state that "Parties should, in all climate change-related actions, fully respect human rights."[19] This statement stands in contrast to the early days of global climate negotiations, when there was little focus on at-risk populations, no mention of climate justice, and predominantly scientific framing of the issue, with an emphasis on mitigation over adaptation.[20]

The 2015 Paris Agreement formalized climate justice as a normative concept even though justice principles have yet to be substantially developed and enacted in practice.[20] In the preamble to the Paris Agreement, the international community acknowledged "the importance for some of the concept of 'climate justice' when taking action to address climate change" and "the specific needs and special circumstances of developing country Parties, especially those that are particularly vulnerable to the adverse effects of climate change." Article 7 of the Paris Agreement states that all nations should follow a "country-driven, gender-responsive, participatory and fully transparent approach, taking into consideration vulnerable groups, communities and ecosystems." Articles 9 and 11 prioritize assistance to those most at risk, including least developed countries and small island states.[21]

The international community's commitment to climate justice is also formalized in the 2015 **Sustainable Development Goals (SDGs)**, a set of targets at the heart of the UN 2030 Agenda for Sustainable Development to end poverty and improve health, education, equity, and economic development while preserving the environment and tackling climate change.[21] In particular, SDG 13 is a call to "take urgent action to combat climate change and its impacts." Its five targets explicitly endorse key climate justice tenets:

- Target 13.1: Strengthen resilience and adaptive capacity to climate-related hazards and natural disasters in all countries
- Target 13.2: Integrate climate change measures into national policies, strategies and planning
- Target 13.3: Improve education, awareness-raising and human and institutional capacity on climate change mitigation, adaptation, impact reduction and early warning
- Target 13.A: Implement the commitment undertaken by developed-country parties to the UNFCCC to a goal of mobilizing jointly $100 billion annually by 2020 from all sources to address the needs of developing countries. . .
- Target 13.B: Promote mechanisms for raising capacity for effective climate change-related planning and management in least developed countries and small island developing States, including focusing on women, youth and local and marginalized communities[23]

Bali Principles 3, 20, and 21 affirm the rights of Indigenous peoples, who face disproportionate environmental impacts in many places around the world. Indigenous peoples make up about 5% of the world's population but constitute more than 15% of the world's poor.[24] As a direct consequence of colonial rule, land displacement, and centuries without self-determination, Indigenous peoples have some of the worst health burdens of any population and often lack access to adequate health services. Their ancestral lands and waters support their traditional and cultural practices and contain 80% of the world's biodiversity.[25] At the same time, these lands and waters have economic value for resource extraction and are vulnerable to pollution and environmental degradation.[24]

Indigenous peoples have generated very low GHG emissions but are disproportionately impacted by climate change as a result of their geographic location, historic and current discriminatory practices, and lack of opportunity to exercise agency and participate in decision-making. Extractive fossil fuel activities that create GHGs contaminate air, water, and food sources, and Indigenous and other communities in close proximity face high rates of physical and mental illnesses. In addition, traditional practices and spiritual well-being among Indigenous peoples may erode when culturally significant lands, waters, and sites are damaged by industry, pollution, and climate change.[24]

Despite disproportionate impacts, Indigenous peoples are effective advocates for their own protection and protection of the planet from climate change and other environmental threats. Examples include Cree, Dene, and Métis peoples of northern Alberta resisting dirty tar sands oil extraction (**Figure 3.2**); Gwich'in people in Alaska campaigning against oil drilling in traditional caribou habitats; Standing Rock Sioux and other tribes protesting the Dakota Access Pipeline; and Anishinaabe and other peoples fighting

Figure 3.2 Left: First Nations Have Led the Resistance in Opposition to the Development of Tar Sands in Alberta, Canada. Right: Tar Sands Mining in Fort McMurray, Alberta.

Left: Courtesy of Eugene Viceconte; Right: © Dan_prat/iStock/Getty Images Plus/Getty Images.

the Line 3 replacement tar sands pipeline in northern Minnesota.

In the U.S. state of Maine, the Penobscot Nation has lived along the Penobscot River for thousands of years and relies on it for sustenance and traditional and spiritual practices. Climate change and environmental pollution threaten the fragility of the river ecosystem, especially fish species that Penobscot people rely on. According to John Banks, director of Penobscot Nation's Department of Natural Resources, "Many tribal people feel that we have a reciprocal agreement with the river to protect, preserve, restore and bring back the ecological integrity to where it should be."[28] The Penobscot Nation is actively focused on climate adaptation, affirming that

> tribes play a vital role due to the nature of Tribal Ecological Knowledge, for we have been on this continent for thousands of years. We flourished through the last ice age, and we will face this new era with the same resiliency that has served us through all of our challenges.[29]

The climate justice movement demands that local, national, and international climate action be informed by Indigenous peoples'

knowledge and experiences, address the risks they face, and develop adaptive measures to promote Indigenous peoples' health and well-being, including self-determination, territorial sovereignty, and freedom from the vestiges of colonial rule.[26] In general, all impacted communities need to be able to fully participate in climate-related decision-making and be heard. As writer Arundhati Roy has said, "There's really no such thing as the 'voiceless.' There are only the deliberately silenced, or the preferably unheard."[27]

Bali Principle 22 affirms the rights of women, recognizing the many gender-based disparities in climate change impacts. For example, because of traditional gender roles within households, women face increased stress collecting water and fuel, growing crops, providing food for their families, and managing their own and their children's health. In several regions, women are more likely than men to forgo food in the household following severe drought conditions, and in South Africa, unmarried women suffer the most from climate-induced food shortages.[30] In numerous heatwave events, including in France and Pakistan, more women than men died, and women were 50% more likely to have perished in Typhoon Haiyan, one of the strongest tropical cyclones ever to hit the Philippines.[30] In general, women and girls are more likely than men and boys to report mental health problems after experiencing extreme storms and flooding. In addition, women's reproductive health and the health of their offspring may be at risk because of heat waves and air pollution linked to preterm birth and adverse pregnancy outcomes, as well as infectious diseases shifting with climate change.[30] The right of future generations to a healthy planet with a stable climate is also affirmed in the Bali Principles.

One of the most effective climate solutions is investing in the education of women and girls, which improves individual, household, and community health and empowers women and girls to engage in sustainable jobs and decision-making. In addition, societal shifts needed to emphasize girls' education and empowerment are part of the systemic changes required to address root causes of climate change. Yet, to date, girls' education has not been formally included in countries' climate action plans. In fact, it is mentioned in only one of 160 nationally determined contributions (NDCs) analyzed in 2019, that of Zambia.[31] Climate change also threatens girls' education, with an estimated 4 million girls prevented from completing schooling in 2021 because of climate-related events, including drought-fueled food and water scarcity and displacement. This number could triple by 2025 if current trends persist.[32] Girls and women are also a leading force as activists in the climate movement and, as such, need to be recognized for their power and efficacy and supported at each step.

Within countries, some groups are at higher risk than others. In the United States, ample evidence points to disproportionate climate impacts on Black, Latino, and other people of color. Heather McTeer Toney, former U.S. EPA Southeast Regional Administrator, noted,

> The impacts to communities of color are coming faster and stronger than ever before. Climate change is no longer something that is future. It is now, it is existing. And we have to really talk about how it encompasses absolutely every aspect of life, particularly for black and brown communities.[33]

In 2004, the Congressional Black Caucus released a report entitled *African Americans and Climate Change: An Unequal Burden.* The report described "a stark disparity in the United States between those who benefit from the causes of climate change and those who bear the costs of climate change."[34] The report noted that Black people are less responsible for climate change than other Americans and yet are already disproportionately burdened by the health effects of climate change, including deaths related to heat and air pollution, and they will be most affected by unemployment and economic hardships associated with climate change. The report recommended that U.S. climate policies be structured to generate large health and economic benefits, not costs, for Black people and to improve equity among all Americans.

In 2005, when Hurricane Katrina hit the U.S. Gulf Coast, many communities were devastated. Among them was New Orleans, where the storm breached a shoddy system of levees and flooded approximately 80% of the city.[35] Low-income and predominantly Black neighborhoods were particularly hard hit, and many residents who had not evacuated became stranded without access to rescue, emergency health services, or shelter. The highest death rate in New Orleans was in the Lower Ninth Ward, where more than 90% of residents at the time were Black.[36] Coastal

Figure 3.3 A Capsized Shrimp Boat Lies in the Water after the U.S. Gulf Coast was Hit by Hurricane Katrina, September 2005.
© Chuck Wagner/Shutterstock.

Figure 3.4 Reverend Lennox Yearwood, Jr., Founder of the Hip Hop Caucus, Speaks about Climate Justice to Students at Colby College, November 9, 2019.
Courtesy of Dennis Griggs for Colby College.

communities in neighboring Mississippi were also devastated by the storm. Livelihoods were destroyed for many people of color and low-income people, including thousands of Vietnamese shrimp fisherfolk who had relocated to the Gulf Coast after the Vietnam War (**Figure 3.3**).[37]

During storm recovery and rebuilding, poor communities of color were left behind or denied assistance at much higher rates than white people.[38] In a national poll conducted a week after Hurricane Katrina hit, two-thirds of Black people but only 17% of white people agreed that "the government's response to the situation would have been faster if most of the victims had been white."[39] A decade later, another national poll revealed that 70% of white people, but only 44% of Black people, felt that New Orleans had mostly recovered from Katrina.[40]

The links between climate change and racial injustice were laid bare by this devastating storm. Climate justice activist Reverend Lennox Yearwood, Jr., a Louisiana native, said, "It didn't take me long to connect the dots between climate change and environmental justice because so many people who'd suffered during Katrina were the disadvantaged

and poor."[41] He organized a coalition of local and national organizations to advocate for the rights of Hurricane Katrina survivors. He also founded **Hip Hop Caucus**, a national nonpartisan nonprofit organization that empowers the hip-hop community and those "impacted first and worst by injustice" to create a "just, sustainable and prosperous world for all"[42] (**Figure 3.4**).

Black, Latino, and other people of color in the United States are much more likely than other groups to live near oil and gas operations, in zip codes with high air pollution, and near hazardous waste sites.[43] In 2018, the U.S. EPA acknowledged that the burden of particulate matter pollution, much of which comes from fossil fuel combustion, was 1.35 times higher for those living in poverty and 1.54 times higher for Black people as compared with the overall population.[44] Low-income Latinos living in fence-line neighborhoods that line the Houston Ship Channel and Port of Houston in Texas—where more than 200 polluting facilities are sited, including two of the three largest oil refineries in the United States—face extremely toxic air pollution and disproportionate illness and death.[45] In many cases, people of color and the poor spend a greater portion of their household income on energy than other groups. Bali Principle 12 affirms the right of all people to access affordable clean energy.

People of color also tend to have lower quality drinking water in the United States. For example, counties with populations that are at least 25% Latino have twice the rate of EPA violations by public drinking water suppliers compared with the rest of the country. Poorer and rural counties are also at higher risk, and many of these areas face severe drought conditions and pollution from oil and gas, mining, and other industrial activities that threaten the water supply.[46]

Populations most at risk of climate change impacts often have disproportionately poor health status as well. Common root causes include discrimination, broad social inequities, institutionalized racism, and unjust policies in all sectors, including health care.[43] In the United States and other high-income countries that are major GHG emitters, power structures and institutions have mostly protected the economic and political desires of those in power, and ignored the rights and perspectives of those with less power. In general, "communities are vulnerable because of bad policies and disinvestment, not because of the people who live in them."[43] Addressing public health disparities among people of color requires focusing on structural racism and discriminatory institutions and policies. Many argue that there can be no racial justice without simultaneously creating conditions for health, environmental, and climate justice, and vice versa.[33]

Many community-based nonprofit organizations are leading the way to provide access to equitable climate solutions. Examples in the United States include one of the most successful and enduring EJ organizations, **WE ACT for Environmental Justice** (formerly West

Harlem Environmental Action). WE ACT was founded in 1988 by community leaders to fight pollution and unjust siting of environmental hazards in the Harlem neighborhood of New York City. Today, WE ACT runs programs aimed at preventing extreme heat deaths and other climate impacts, and coordinates the **Environmental Justice Leadership Forum** (EJ Forum). The EJ Forum is a national coalition of 54 EJ organizations that work together to advance climate justice and protect and promote communities of color and low-income communities in the United States. The **National Black Environmental Justice Network** is another U.S. organization working on environmental and climate justice. It calls for addressing "systemic racism and its devastating impacts on Black Americans" as we create "climate solutions that eliminate greenhouse gases, create millions of high-wage American jobs, build green and accessible public transportation, reduce poverty and inequality, promote equal protection of workers, frontline communities and vulnerable populations, and provide safeguards against climate-induced health threats."[47]

Other grassroots organizations fighting for EJ include **Texas Environmental Justice Advocacy Services** (**T.E.J.A.S.**), which has worked on EJ issues for decades in Houston to protect residents of fence-line communities along the Houston Ship Channel. T.E.J.A.S. advocates for a just energy transition away from the oil and gas activities that have contributed so much pollution and human suffering in this area, and toward solar and wind energy development and worker training and support.[48] The **Indigenous Environmental Network** (**IEN**) works on environmental and economic justice, building the capacity of Indigenous peoples in North America to protect sacred lands, waters, natural resources, communities, and the "health of both our people and all living things."[49] Founded in

1990 by Indigenous peoples, IEN organizes campaigns and direct actions, builds alliances and collaborations, and raises public awareness.

Climate justice ultimately focuses on moving away from political, economic, and energy systems that create local and global ecological damage and social injustices, and toward systems that compensate for and overcome these impacts in just and equitable ways.[50] This requires capacity-building in affected communities and countries, with a focus on poverty alleviation and access to effective and culturally appropriate climate adaptations. Climate justice advocates also warn against GHG emissions mitigation policies that increase costs of carbon-based energy or promote expensive new technologies without subsidies for those who cannot afford them. Many people around the world live in **energy poverty**, so they rely on cheaper forms of energy for heating and cooking that may produce more GHG emissions. Shifting land use policies for climate mitigation may force poor and Indigenous peoples off their lands or decrease access to essential natural resources and ecosystem services. In addition, promoting nonmotorized transport in areas where roads are less safe can have negative health consequences, given that road injuries are a leading cause of death and disability worldwide.

Climate solutions that implement justice principles include an ongoing six-year project in Bangladesh to build the adaptive capacities of people in coastal communities, particularly women, to deal with the impacts on livelihoods and water security of saltwater intrusion into freshwater caused by storms and sea level rise (**Figure 3.5**).[51] Thousands of women and girls will receive training in climate-resilient livelihoods, community-managed rainwater harvesting solutions, and business development opportunities, and gain access

Figure 3.5 Women and Children in Floodwaters Surrounding Homes in Bangladesh.
© Stockbyte/Getty Images.

to markets and finance. They will also participate in early warning and monitoring activities for climate change impacts. The Green Climate Fund, a large international source of climate financing from wealthy countries for use by developing countries,[19] will contribute $25 million to the project, and the Bangladesh Ministry of Women and Children's Affairs will contribute $8 million.

Climate justice solutions in the United States include Coalfield Development, a social enterprise incubator committed to rebuilding economies in central Appalachia, where the coal industry is in decline.[52] Coalfield Development partners with community members to promote renewable energy, farming, and manufacturing, and academic, professional, and personal development for people facing barriers to employment. It also supports community-based affordable housing and revitalization projects. One collaboration is with Solar Holler, a full-service solar developer and installer in West Virginia committed to training the local workforce to bring solar energy to "the people and places who have always been left out"[52] (**Figure 3.6**). Another collaboration is with the Big Ugly Community Center in West Virginia, which

Figure 3.6 A Crew Installs Rooftop Solar Panels. In the U.S., Solar Holler, in partnership with Coalfield Development, creates solar industry jobs in Appalachian coal country.

© Sebastian Rothe/EyeEm/Getty Images.

provides a wide range of services to children from economically disadvantaged communities in the coalfields.

Solving the climate crisis is not just about creating scientific and technological fixes on a global scale. It is about empowering people, improving lives, and promoting equity. It is about bringing people together in their own communities and in collaborations across communities with shared goals, experiences, and opportunities. It is ultimately about promoting fairness, responsibility, and self-determination so that no one is disproportionately burdened by climate change impacts or left out of the creation of a more stable climate and a more just and equitable world.

Discussion Questions

1. Define climate justice. What can we learn from examples where climate injustices have occurred?
2. Discuss how the Bali Principles of Climate Justice highlight how climate change disproportionately impacts certain populations. How should these groups be protected, and who is responsible?
3. In what ways is Hurricane Katrina a climate injustice case study?
4. Discuss how Indigenous peoples and people of color are disproportionately impacted by climate change. How are they organizing for climate justice in the United States?

References

1. Robinson M, Palmer C. *Climate Justice: Hope, Resilience and the Fight for a Sustainable Future*. Bloomsbury; 2018:147.
2. Sengupta S, Manik JA. A quarter of Bangladesh is flooded. Millions have lost everything. *The New York Times*. July 30, 2020. Accessed August 11, 2020. https://nyti.ms/334VB7c
3. Ayeb-Karlsson S. "When we were children we had dreams, then we came to Dhaka to survive": urban stories connecting loss of wellbeing, displacement and (im)mobility. *Clim Dev*. 2021;13(4):348–359. doi:10.1080/17565529.2020.1777078
4. Goering L. With climate change driving child marriage risks, Bangladesh fights back. *Reuters*. July 20, 2017. Accessed August 11, 2020. https://www.reuters.com/article/us-climatechange-bangladesh-youth-idUSKBN1A51UA
5. Gore T. Confronting carbon inequality. Oxfam. September 21, 2020. Accessed June 17, 2021. https://www.oxfam.org/en/research/confronting-carbon-inequality
6. Levy BS, Patz JA. Climate change, human rights, and social justice. *Ann Global Health*. 2015;81(3):310–322. doi:10.1016/j.aogh.2015.08.008

7. Environmental Protection Agency. Environmental justice. Accessed August 11, 2020. https://www.epa.gov/environmentaljustice

8. Bullard RD, Gardezi M, Chennault C, et al. Climate change and environmental justice: a conversation with Dr. Robert Bullard. *J Crit Thought Prax.* 2016;5(2). doi:10.31274/jctp-180810-61

9. Bullard R. Principles of Environmental Justice turn 21. Updated October 27, 2012. Accessed August 11, 2020. https://drrobertbullard.com/principles-of-environmental-justice-turn-21/

10. International Climate Justice Network. Bali Principles of Climate Justice. *CorpWatch.* August 28, 2002. Accessed August 11, 2020. https://corpwatch.org/article/bali-principles-climate-justice

11. Pauw P, Mbeva K, van Asselt H. Subtle differentiation of countries' responsibilities under the Paris Agreement. *Palgrave Commun.* 2019;5(1):86. doi:10.1057/s41599-019-0298-6

12. Climate Justice Alliance. *Just Transition Principles.* November 2019. Accessed August 11, 2020. https://climatejusticealliance.org/wp-content/uploads/2019/11/CJA_JustTransition_highres.pdf

13. Tasioulas J, Vayena E. Getting human rights right in global health policy. *The Lancet.* 2015;385(9978): e42–e44. doi:10.1016/s0140-6736(14)61418-5

14. World Health Organization. Constitution of the World Health Organization. Accessed August 11, 2020. https://www.who.int/governance/eb/who_constitution_en.pdf

15. United Nations. Universal Declaration of Human Rights. Accessed August 11, 2020. https://www.un.org/en/about-us/universal-declaration-of-human-rights

16. United Nations. International Covenant on Civil and Political Rights. 1966. Accessed August 11, 2020. https://www.ohchr.org/en/professionalinterest/pages/ccpr.aspx

17. Office of the United Nations High Commissioner for Human Rights. International Covenant on Economic, Social and Cultural Rights. Published December 16, 1966. Accessed August 11, 2020. https://www.ohchr.org/en/professionalinterest/pages/cescr.aspx

18. Alston P. *Climate Change and Poverty: Report of the Special Rapporteur on Extreme Poverty and Human Rights.* UN Digital Library; 2019. Accessed August 11, 2020. https://digitallibrary.un.org/record/3810720?ln=en_

19. United Nations Climate Change. Cancun Agreements (2010). November 2010. Accessed August 11, 2020. https://unfccc.int/process/conferences/pastconferences/cancun-climate-change-conference-november-2010/statements-and-resources/Agreements

20. Gach E. Normative shifts in the global conception of climate change: the growth of climate justice. *Soc Sci.* 2019;8:24. doi:10.3390/socsci8010024

21. United Nations. Paris Agreement. Published 2015. https://unfccc.int/sites/default/files/english_paris_agreement.pdf

22. United Nations. The 17 Goals. Accessed August 11, 2020. https://sdgs.un.org/goals

23. United Nations. Goals: 13. Accessed September 26, 2021. https://sdgs.un.org/goals/goal13

24. Fernández-Llamazares Á, Garteizgogeascoa M, Basu N, et al. A state-of-the-art review of Indigenous peoples and environmental pollution. *Integr Environ Assess Manag.* 2020;16(3):324–341. doi:10.1002/ieam.4239

25. Mitchell S. Indigenous prophecy and Mother Earth. In: Johnson AE, Wilkinson KK, eds. *All We Can Save: Truth, Courage, and Solutions for the Climate Crisis.* One World; 2020:16–28.

26. Editors. Self-determination and Indigenous health. *The Lancet.* 2020;396(10248):361. doi:10.1016/S0140-6736(20)31682-2

27. Roy A. *Peace and the New Corporate Liberation Theology.* Sydney Peace Foundation; 2004. Accessed August 11, 2020. https://sydneypeacefoundation.org.au/peace-prize-recipients/2004-arundhati-roy/

28. Benson B. Penobscots' mission to preserve river threatened by impacts of climate change. *Maine Beacon.* June 22, 2020. Accessed August 11, 2020. https://mainebeacon.com/penobscots-mission-to-preserve-river-threatened-by-impacts-of-climate-change/

29. Penobscot Nation. Climate change. Accessed August 11, 2020. https://www.penobscotnation.org/departments/natural-resources/air-quality/climate-change

30. Dunne D. Mapped: how climate change disproportionately affects women's health. *Carbon Brief.* October 29, 2020. Accessed March 2, 2021. https://www.carbonbrief.org/mapped-how-climate-change-disproportionately-affects-womens-health?s=03

31. Kwauk C, Cooke J, Hara E, et al. *Girls' Education in Climate Strategies.* Brookings Institution. December 10, 2019. Accessed June 17, 2021. https://www.brookings.edu/research/girls-education-in-climate-strategies/

32. Fry L, Lei P. *A Greener, Fairer Future: Why Leaders Need to Invest in Climate and Girls' Education.* Malala

Fund. March 2021. Accessed June 17, 2021. https://assets.ctfassets.net/0oan5gk9rgbh/OFgutQPKIFoi5lfY2iwFC/6b2fffd2c893ebdebee60f93be814299/MalalaFund_GirlsEducation_ClimateReport.pdf

33. Curwood S. Race and environmental justice. Transcript. *Living on Earth*. June 12, 2020. Accessed August 11, 2020. https://www.loe.org/shows/segments.html?programID=20-P13-00024&segmentID=1

34. Congressional Black Caucus Foundation. African Americans and climate change: an unequal burden. July 21, 2004. Accessed August 11, 2020. https://23u0pr24qn4zn4d4qinlmyh8-wpengine.netdna-ssl.com/wp-content/uploads/2013/02/CBCF_REPORT_F.pdf

35. Brunkard J, Namulanda G, Ratard R. Hurricane Katrina deaths, Louisiana, 2005. *Disaster Med Public Health Prep*. 2008;2(4):215–223. doi:10.1097/DMP.0b013e31818aaf55

36. Schleifstein M. Study of Hurricane Katrina's dead show most were old, lived near levee breaches. *The Times-Picayune*. August 28, 2009. Accessed August 11, 2020. https://www.nola.com/news/weather/article_35741734-68e1-575e-86d0-29366eed38e5.html

37. Campbell AF, Whiteman M. Is this the end of the line for Louisiana's Vietnamese shrimpers? *The Atlantic*. October 30, 2014. Accessed August 11, 2020. https://www.theatlantic.com/politics/archive/2014/10/is-this-the-end-of-the-line-for-louisianas-vietnamese-shrimpers/431418/

38. Bullard RD, Wright B. *The Wrong Complexion for Protection: How the Government Response to Disaster Endangers African American Communities*. New York University Press; 2012.

39. Pew Research Center. Two-in-three critical of Bush's relief efforts. September 9, 2005. Accessed August 11, 2020. https://www.pewresearch.org/politics/2005/09/08/two-in-three-critical-of-bushs-relief-efforts/

40. Hamel L, Firth J, Brodie M. *New Orleans Ten Years After The Storm: The Kaiser Family Foundation Katrina Survey Project*. Kaiser Family Foundation. August 10, 2015. Accessed August 11, 2020. https://www.kff.org/other/report/new-orleans-ten-years-after-the-storm-the-kaiser-family-foundation-katrina-survey-project/

41. Maceachern D. Interview: Rev. Lennox Yearwood of the Hip Hop Caucus. Moms Clean Air Force. September 12, 2016. Accessed June 17, 2021. https://www.momscleanairforce.org/interview-rev-lennox-yearwood-hip-hop-caucus/

42. Hip Hop Caucus. About Us. 2021. Accessed August 11, 2020. https://hiphopcaucus.org/about-us/

43. Bassett MT, Graves JD. Uprooting institutionalized racism as public health practice. Article. *Am J Public Health*. 2018;108(4):457–458. doi:10.2105/AJPH.2018.304314

44. Mikati I, Benson AF, Luben TJ, Sacks JD, Richmond-Bryant J. Disparities in distribution of particulate matter emission sources by race and poverty status. *Am J Public Health*. 2018;108(4):480–485. doi:10.2105/AJPH.2017.304297

45. Johnson GS, Washington SC, King DW, Gomez JM. Air quality and health issues along Houston's Ship Channel: an exploratory environmental justice analysis of a vulnerable community (Pleasantville). *Race Gend Class*. 2014;21(3/4):273–303.

46. Holden E, Enders C, Kommenda N, et al. More than 25m drink from the worst U.S. water systems, with Latinos most exposed. *The Guardian*. February 26, 2021. Accessed February 26, 2021. https://www.theguardian.com/us-news/2021/feb/26/worst-us-water-systems-latinos-most-exposed

47. National Black Environmental Justice Network. The National Black Environmental Justice Network response to the House Select Committee on the Climate Crisis comprehensive plan. July 1, 2020. Accessed June 17, 2021. https://www.nbejn.com/news/the-national-black-environmental-justice-network-response-to-the-house-select-committee-on-the-climate-crisis-comprehensive-plan_

48. Texas Environmental Justice Advocacy Services. Everyone regardless of race, is entitled to live in a clean environment. Accessed August 11, 2020. https://www.tejasbarrios.org/

49. Indigenous Environmental Network. Accessed August 11, 2020. https://www.ienearth.org/about/

50. Schlosberg D, Collins L. From environmental to climate justice: climate change and the discourse of environmental justice. *Wiley Interdiscip Rev Clim Change*. 2014;5(3):359–374. doi:10.1002/wcc.275

51. Green Climate Fund. FP069: enhancing adaptive capacities of coastal communities, especially women, to cope with climate change induced salinity. Accessed August 11, 2020. https://www.greenclimate.fund/project/fp069#overview

52. Coalfield Development. About us. Accessed August 11, 2020. https://coalfield-development.org/about-us/

CHAPTER 4

Extreme Heat

"With this fast pace of warming of temperature extremes, we might actually pass a fundamental limit, beyond the human ability to cool itself."

—**Dr. Michael Byrne**, University of Oxford climate scientist[1]

KEY TERMS

Heat stress
Thermoregulation
Heatstroke
Urban heat island effect
Heat index (HI)

Wet-bulb globe temperature
  (WBGT)
Heat wave
Temperature mortality
  response curve

Minimum mortality
  temperature (MMT)
Heat acclimatization
Heat adaptation
Mortality displacement

LEARNING OBJECTIVES

- Define thermoregulation and heat stress.
- Describe the ways that extreme heat exposure affects the human body.
- Explain why extreme heat is worse in cities than in rural areas and why certain populations face disproportionate heat exposure.
- Understand how we characterize the human health impacts of extreme heat.
- Identify effective climate adaptations to protect people from the health effects of extreme heat.

For several days in the summer of 2017, the U.S. Southwest sweltered under extreme heat. Temperatures soared to 52.5°C (nearly 127°F) in Death Valley, California, and over 48°C (119°F) in Phoenix, Arizona, where the airport was forced to close when temperatures exceeded the operating threshold for airplanes. Numerous deaths were attributed to this heat wave, including a 57-year-old man and his 21-year-old son hiking in 100-degree-heat in New Mexico; a man who died while running a 5K race in Bakersfield, California, where temperatures hit 107°F; and two elderly residents of San Jose, California, one of whom was homeless and lived in a car.[2]

In May 2017, one of the hottest temperatures ever recorded on Earth, 53.7°C (128.7°F), hit southwestern Pakistan.[3] Two years earlier, a heat wave plagued the same area, with nearly a week of temperatures above 40°C (104°F) and a peak temperature of 49°C (120°F). At least 2,000 people died, as did more than 2,500 people in neighboring India.[4] These extreme heat events in 2017, and in the years since, follow a pattern of increasing magnitude and frequency of hot weather driven by climate change. At least a billion people around the world are at risk of extreme heat exposure annually.[5]

In 2020, one of the hottest years on record, the global average surface temperature was 1.02°C (1.84°F) above a baseline mean in the period 1950–1980, and in August of that year, perhaps the hottest temperature in the world since 1931 was recorded: 130°F in Death Valley.[6] In 2020, Phoenix experienced a record (up to that point) of at least 145 days with a high temperature ≥ 100°F and at least 53 days with a

high temperature of ≥ 110°F.[7] July 2021 was the world's hottest month ever recorded.[8] Most of the United States has warmed over the past decade, and 30-year average temperatures increased significantly as compared with the 20th-century average (**Figure 4.1**).[9]

Fueled primarily by the greenhouse effect, the world is warming rapidly and is currently predicted to exceed the internationally agreed-on target under the Paris Agreement of less than 2°C (3.6°F) warming, and ideally less than 1.5°C (2.7°F) above preindustrial levels.[10] In the future, we can expect higher temperatures and many more heat waves. In fact, some parts of the world may reach critical upper limits of survivability with respect to extreme heat exposure (**Figure 4.2**), forcing human migration on a large scale to more temperature-suitable habitats.[11] Understanding and preparing for the health impacts of extreme heat will require more research, especially in understudied parts of the world, and robust adaptation strategies to reduce

U.S. ANNUAL TEMPERATURE COMPARED TO 20th-CENTURY AVERAGE

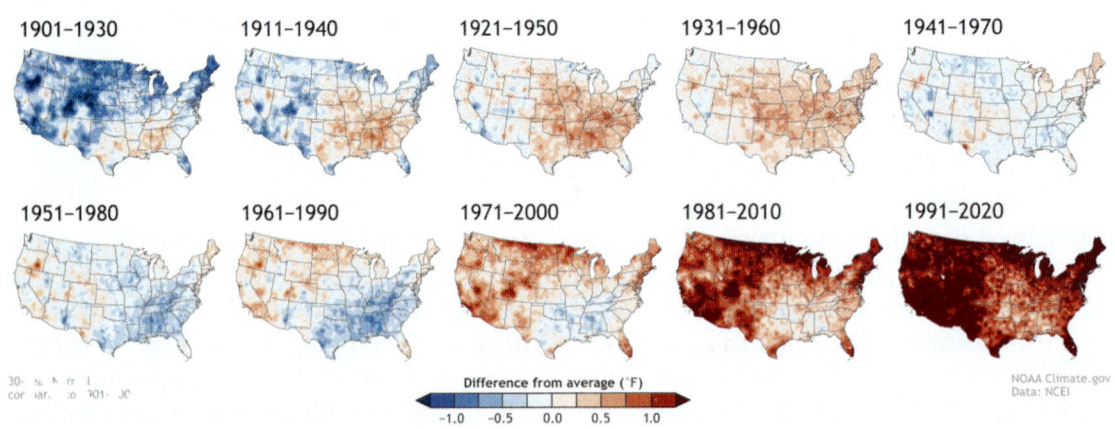

Figure 4.1 Annual U.S. Temperatures Compared with the 20th-Century Average for Each 30-Year Period from 1901–1930 (Upper Left) to 1991–2020 (Lower Right). Red colors indicate warmer temperatures, and blue colors indicate cooler temperatures.

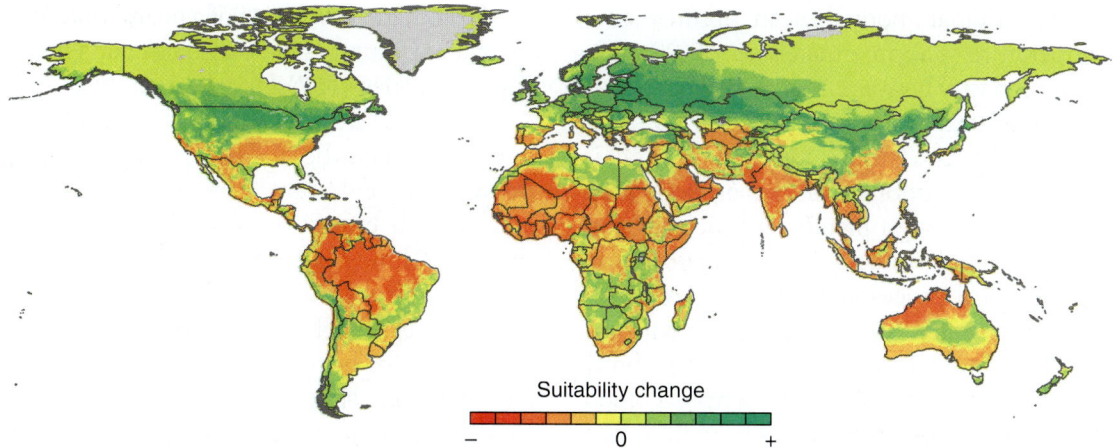

Figure 4.2 Projected Geographical Shift in the Human Temperature Niche from Now to 2070 under an Extreme Warming Scenario (RCP 8.5). Orange represents regions with less temperature suitability because of extreme warming; green areas have greater suitability.

Reproduced from Xu C, Kohler TA, Lenton TM, Svenning J-C, Scheffer M. Future of the human climate niche. *Proc Natl Acad Sci U S A.* 2020;117(21):11350–11355. doi:10.1073/pnas.1910114117

heat exposure and its associated **mortality** (deaths) and **morbidity** (illnesses and injuries), particularly among underserved populations and those living in areas most prone to extreme heat.

Heat Stress

Heat-related deaths, illnesses, and injuries result from **heat stress**, a condition in which the body cannot get rid of excess heat when exposed to high temperatures because regulation of its core temperature is compromised. Heat exposure also exacerbates existing chronic health conditions, such as cardiovascular, respiratory, and kidney diseases, diabetes, and mental illnesses. Extreme heat is not always recognized as a cause of mortality or morbidity and, as such, may not be recorded in hospital records or on death certificates. This has led to an underestimation, globally and locally, of the true burden of heat-related ill health.

The 2020 report of **The Lancet Countdown**, an international collaboration of experts monitoring the health impacts of climate change, noted that people in every region of the world are increasingly vulnerable to heat stress, particularly in Europe, the Western Pacific, Southeast Asia, and Africa. Heat-related deaths in people over 65 years of age increased by more than 50% over the past 20 years, and in 2018 totaled nearly 300,000.[12] In the United States, the **Centers for Disease Control and Prevention (CDC)** records an average of nearly 700 heat-related deaths each year, primarily in people older than 65, and an estimated 65,000 annual hospital emergency department (ED) visits for heat-related illnesses.[13] In most years, the U.S. **National Weather Service (NWS)** records more deaths from heat exposure than from other types of extreme weather, and the 30-year average annual mortality (1991–2020) is highest for heat.[14] The **U.S. Global Change Research Program** has predicted up to tens of thousands of additional heat-related premature deaths each year in the United States alone by the end of this century.[13]

The human body has an optimal core temperature of approximately 37°C (98.6°F), which can vary between 35 and 41°C depending on an individual's physical activity, food intake, hydration, sleep status, fever, and environmental conditions. The human body generates heat through metabolic processes and must constantly undergo **thermoregulation**, physiological processes to maintain an optimal temperature by exchanging heat with the environment.[15] Heat accumulated in the body is released primarily through increased blood flow to the skin and sweating. Evaporation of sweat from the surface of the skin dissipates heat and has a cooling effect. Sweating leads to loss of water and electrolytes (salts) from the body, and excessive sweating leads to dehydration, which in severe cases prevents the body from further thermoregulation. High humidity, in which ambient air is saturated with water vapor, impairs cooling because evaporation of sweat from the skin is reduced. Increased blood flow requires increased cardiac output. Without cooling interventions, **hyperthermia** (elevated core body temperature) can lead to organ failure and death.

There are many forms of heat stress, and specific health conditions linked to heat exposure are specified in the *International Classification of Diseases* (ICD), a standard diagnostic tool published by the **World Health Organization** and used by clinicians, medical examiners, epidemiologists, and public health professionals to make disease diagnoses (**Table 4.1**).[16] Causes of mortality and morbidity that result from exposure to an "external" hazard—such as "forces of nature," which include "excessive natural heat"—are classified as **external causes**. In contrast, **internal causes** refer to specific diseases that lead to mortality or morbidity. The 10th revision of the ICD (*ICD-10*) includes the code X30, which is used in medical records for *direct causal attribution* of a health outcome to heat exposure. *ICD-10* also contains *injury diagnosis* codes (T67) for "effects of heat."

Heatstroke is the deadliest form of heat stress, in which the body stops cooling itself because dehydration prevents sweating, and/or high humidity prevents sweat evaporation. Clinical symptoms that may result from heatstroke, also called **thermoplegia**, include very high body temperature, hot and dry skin, heart arrhythmias, confusion, altered mental state, slurred speech, seizures, loss of consciousness, and, if untreated, death.[15] **Pyrexia** (fever), **apoplexy** (bleeding within an internal organ), and **siriasis** (sudden collapse due to excessive heat or sun exposure) may also occur. **Exertional heatstroke** is a form of heatstroke that results from extreme physical exertion or exercise, often outdoors in very high temperatures.[15]

Other heat-related illnesses include **heat syncope**, a temporary loss of consciousness induced by prolonged standing without movement or sudden rising from a sitting or lying position when exposed to high heat. It is caused by dehydration and inadequate blood flow to the brain, causing a person to faint. **Heat cramps** are painful, involuntary muscle spasms, usually in the arms, legs, stomach, or back, caused by heavy exercise or heat exhaustion that leads to fluid and electrolyte loss. Heat cramps are often accompanied by fatigue and dizziness. **Heat exhaustion** results from a loss of fluids due to heavy sweating and includes elevated core body temperature, headache, nausea, dizziness, weakness, irritability, thirst, extreme sweating, and decreased urine output. **Heat fatigue** lacks a precise medical definition and may be used for patients who do not meet other criteria for a heat-related illness but have transient lethargy or weakness. **Heat edema** is swelling caused by excess fluid retention in organs or tissues of the body.

Table 4.1 Selected Heat-Related Classifications of Mortality and Morbidity in the *International Classification of Diseases*, 10th Revision. These codes are reported in medical records and on death certificates.

Heat-related Codes
Chapter XX: **External causes** of morbidity and mortality
Exposure to forces of nature
X30 Exposure to excessive natural heat
Chapter XIX: Injury, poisoning and certain other **consequences of external causes**
Other and unspecified effects of external causes
T67 Effects of heat and light
Heatstroke and sunstroke Heat: apoplexy pyrexia Siriasis Thermoplegia
Heat syncope
Heat cramp
Heat exhaustion (due to water or salt depletion)
Heat exhaustion
Heat fatigue
Heat edema

Data from World Health Organization. *ICD-10 Version:2010.* Accessed January 13, 2021. https://icd.who.int/browse10/2010/en

Heat rash is characterized by skin irritation caused by excessive sweating.[15]

Heat-related mortality is due primarily to heatstroke and dehydration.[15] Extreme heat exposure is toxic to internal organs, particularly the kidneys, heart, and brain. Cooling the body, which requires pumping extra blood to the skin to radiate heat off its surface, taxes the cardiovascular system and is more difficult if circulation is weakened by diseases of the heart or blood vessels. Dehydration results from inadequate fluid intake to offset fluid loss due to sweating. A resulting decrease in blood volume impairs maintenance of adequate blood pressure to supply oxygen-rich blood to the body's organs, and kidney damage and multiple organ failure can result. **Rhabdomyolysis** may also occur with heatstroke, during which muscles rapidly rupture and release electrolytes and proteins into the bloodstream, triggering irregular heart rhythms, seizures, and kidney damage.[15] Respiratory illnesses may also worsen in high heat and humidity, as well as in increased air pollution created at high temperatures. Many people who survive heatstroke often experience lingering aftereffects, such as worsened heart disease or kidney failure, which can be deadly months afterward.

At-Risk Populations

Everyone is susceptible to the adverse health impacts of extreme heat exposure. The associations between climate change, heat exposure, and human health are summarized in **Figure 4.3**. Key modifying factors that increase or decrease heat-related health risks are indicated.[13]

Certain individuals and populations are particularly at risk, including the elderly; young children; outdoor workers; those who are socially isolated, economically disadvantaged, or experiencing homelessness; those with certain chronic illnesses and disabilities; alcohol and drug users; and urban dwellers.[17] Vulnerability reflects the conditions that constrain both effective bodily thermoregulation and access to heat-protective measures, such as cooling mechanisms at home or in public spaces, social connectivity, and relevant healthcare

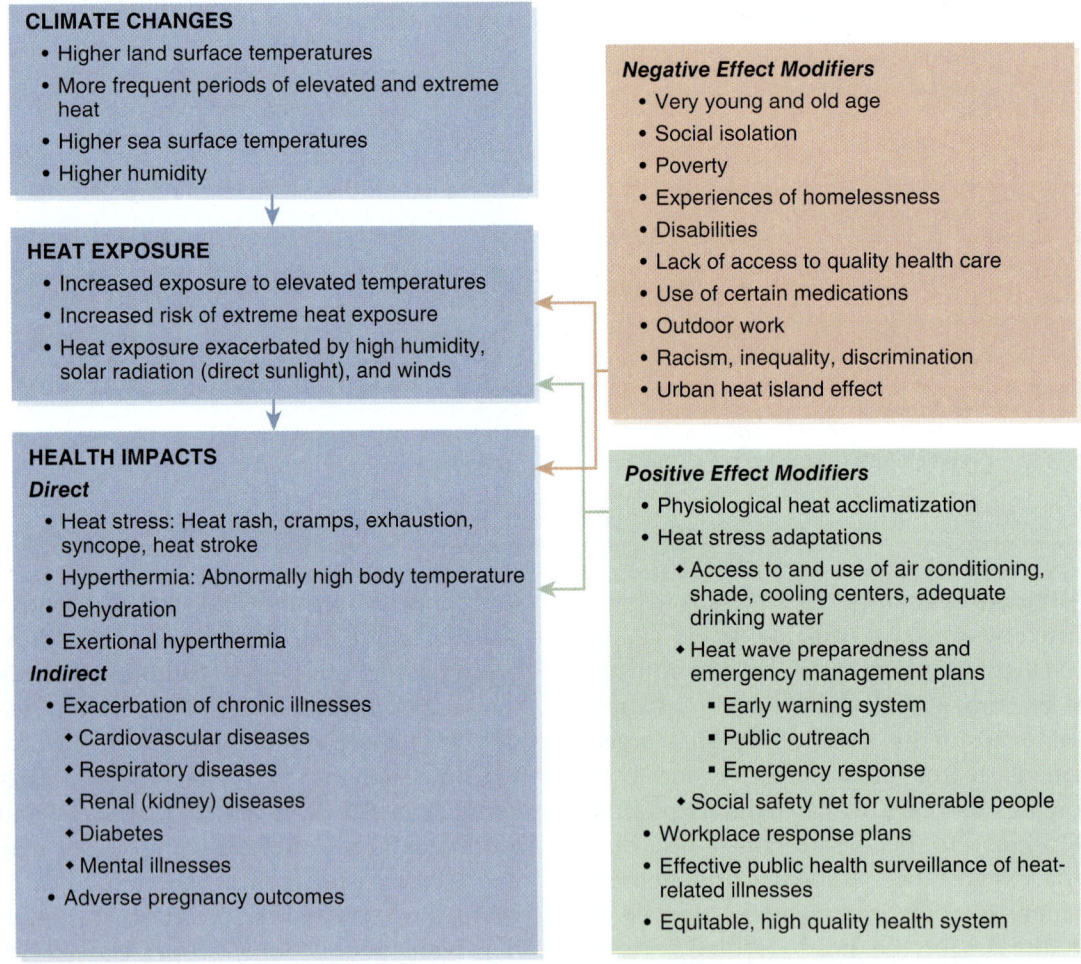

Figure 4.3 Links between Climate Change Drivers, Heat Exposure, and Human Health Impacts. Numerous modifying factors increase (pink) or decrease (green) the risks of heat exposure and resulting health impacts.

Data from U.S. Global Change Research Program. *The Impacts of Climate Change on Human Health in the United States: A Scientific Assessment.* U.S. Global Change Research Program; 2016. https://health2016.globalchange.gov

(a)

(b)

Figure 4.4 Heatstroke Is Common in the Elderly **(a)** and in Outdoor Workers, Such as Farmworkers **(b)**.
(a) © Phovoir/Shutterstock; (b) © Photo Beto/iStock Unreleased/Getty Images.

services. Understanding heat vulnerability allows health professionals to identify who is most at risk and which heat-related symptoms to look for. It is also important to raise awareness about heat impacts and access to cooling.

Older people are more susceptible to heatstroke than younger people (**Figure 4.4**). Thermoregulation in elderly people is less robust, and they tend to have higher rates of chronic health problems that increase vulnerability to heat stress, such as cardiovascular, respiratory, and kidney diseases, diabetes, and dementia.[13] In addition, elderly people have higher use of certain medications that impair thermoregulation, including beta-blockers, which slow the heart rate and are commonly prescribed for patients with high blood pressure, irregular heartbeat, and angina (chest pain); diuretics, which increase urine production and exacerbate dehydration; and psychotropic drugs, such as sedatives and some antidepressants, that lower thermal sensory input recognition.[13] They may also be more socially isolated than other groups, have more limited mobility, and live in managed care settings.[13] Populations around the world are aging, with the proportion of elderly people increasing, which will make future heat-related health burdens even more severe.

Very young children are at heightened risk because they have greater heat production during exertion compared with adults, and less effective thermoregulation during extreme heat.[18] Children also may not be able to control their environment and may rely on adults to help them access cooling measures. Youth athletes who practice outdoors in extreme heat and humidity are at high risk of heat exhaustion and heatstroke. Children are also particularly vulnerable to heat exposure when left in vehicles. From 2015 to 2019, heatstroke accounted for 26% of U.S. non-traffic-related vehicle deaths in children under 15 years old, with 214 deaths officially recorded (likely an underestimate).[19] Babies born to pregnant women who experience heat stress are at increased risk of preterm birth, low birth weight, birth defects, and infant mortality.[20]

Outdoor workers are at particular risk of exposure to extreme heat, especially if they do physically demanding work or wear heat-trapping clothing or protective equipment (Figure 4.4). Agriculture, construction, manufacturing, and mining are sectors in which occupational heat stress is frequent because of high work intensity.[21] Dehydration, physical and mental impairment, and heat-related conditions from heat rash to heatstroke may occur. Workers may lack workplace access to cooling mechanisms such as breaks during the work day, air-conditioned break spaces, shade, cold water for drinking to prevent dehydration, and health monitoring.[22] During the period 1992–2010, nearly 70,000 U.S. farmworkers were seriously injured from heat, and 783 died.[22] Lack of access to health services, language barriers, or fear of deportation may discourage migrant farmworkers from reporting adverse conditions or seeking medical attention. In general, Latino immigrants have a risk of heat-related deaths 3.6 times higher that of white people.[23] Extreme heat-related labor losses are a global problem, with India and Indonesia among the countries worst impacted, at a cost equivalent to 4–6% of their annual gross domestic product.[12]

Residents of large urban areas are also at increased risk. Cities tend to retain more heat than suburban or rural areas because of the **urban heat island effect**[13]—the increased absorption of solar radiation that results from alterations to land surfaces by roads, buildings, and other structures made of heat-trapping materials such as asphalt and concrete. In addition, the presence of fewer trees and other vegetation in cities means less cooling from shading and evapotranspiration, the natural process by which plants absorb heat from the air to evaporate water on leaf surfaces. Cities also *create* heat through high rates of human and industrial activity. Because the world's population is increasingly living in cities, the burden of urban heat-related mortality and morbidity is substantial and predicted to rise.

The intersection of urban living with discrimination only increases this vulnerability. In U.S. cities, residents of low-income neighborhoods and neighborhoods with predominantly people of color experience significantly higher outdoor temperatures under the same weather conditions—as much as 7°C (nearly 13°F)—than do residents of wealthier and predominantly white neighborhoods (**Table 4.2**).[24] In most large U.S. cities, intraurban temperature variability overlaps with historic **redlining**, the racist, discriminatory housing practice of denying mortgages, loans, and insurance to racial and ethnic minorities. Redlining had the effect of forcing people of color into less desirable neighborhoods with smaller, denser housing. Today, these areas tend to have much less tree cover and green space, and higher density of paved surfaces and large buildings made of materials like heat-retaining brick and cinder block.[25] In these neighborhoods, people tend to have fewer resources and less adaptive capacity to cope with high temperatures.

In cities, planning decisions affect urban development characteristics, including where housing, roads, highways, railways, industrial facilities, parks, and other features are sited. The needs of certain groups are invariably considered over others, and persistent systemic biases and discriminatory policies maintain these disproportionate vulnerabilities to heat (**Box 4.1**). For example, past federal housing policies that required physical barriers between white and Black neighborhoods often resulted in major roadways being built in close proximity to where Black people lived. This has implications not only for heat exposure, but also for exposure to air pollution. Another vulnerability for people of color is that, because of widespread health inequities, they tend to suffer disproportionately from chronic conditions exacerbated by extreme heat exposure, such as obesity, high

Table 4.2 Intraurban Surface Temperature Discrepancies in Selected U.S. Cities, Measured by Temperature Difference Between Historically Redlined Neighborhoods and So-Called "Best" Neighborhoods. A positive difference indicates that the redlined neighborhood was hotter. Only 7 cities out of 108 (6%) had lower temperatures in redlined neighborhoods.

U.S. City	Temperature Difference
Portland, OR	+7.09°C (+12.8°F)
Denver, CO	+6.68°C (+12.0°F)
Minneapolis, MN	+6.02°C (+10.8°F)
Baltimore, MD	+5.15°C (+9.27°F)
Miami, FL	+4.22°C (+7.60°F)
Los Angeles, CA	+4.21°C (+7.58°F)
San Francisco, CA	+4.09°C (+7.36°F)
Greater Boston, MA	+3.35°C (+6.03°F)
Chicago, IL	+3.28°C (+5.90°F)
Greater New York City, NY	+3.22°C (+5.80°F)
Dallas, TX	+2.87°C (+5.17°F)
Detroit, MI	+1.88°C (+3.38°F)
Pittsburgh, PA	+0.79°C (+1.42°F)
Seattle, WA	+0.50°C (+0.9°F)
Atlantic City, NJ	−0.81°C (−1.46°F)

Data from Hoffman J, Shandas V, Pendleton N. The effects of historical housing policies on resident exposure to intra-urban heat: a study of 108 US urban areas. *Climate.* 2020;8:12. doi:10.3390/cli8010012

Box 4.1 Racism and Heat-Related Mortality in New York City

New York City's most heat-vulnerable communities—all but one located in the boroughs of Brooklyn or the Bronx—are 79–99% people of color and have a median household income that is 2.4 times lower than that for the entire city.[27] Many residents of these neighborhoods live in public housing consisting of high-rise brick or cinder block buildings that often have higher indoor than outdoor summer temperatures. Many residents lack fans or air conditioners (AC) for cooling. The New York City Housing Authority's policies place significant barriers to AC access, including prior approval requirements, annual fees for each AC unit, and costly professional installation.[27]

When extreme heat strikes New York City, Black residents are nearly twice as likely to

(continues)

Box 4.1 Racism and Heat-Related Mortality in New York City   *(continued)*

die as white residents.[28] Given that annual heat-related deaths in New York City are projected to be well over 3,000 by the year 2080, this represents a large burden on poor people of color.[27] In the future, New York City faces a 10-fold higher heat mortality burden than cities such as Dallas in the southern United States.[29] In general, people of color are 50% more likely than other groups to live in urban heat islands.[27]

In response, New York City has implemented a number of heat adaptations, including a "Community Emergency Planning Toolkit" and the Cool Neighborhoods NYC program to increase preparedness and to build social cohesion, including creating a buddy system in which people look in on their elderly neighbors.[27] The city opens hundreds of cooling centers during extreme heat events, although there are high temperature thresholds for opening these centers, and hours of operation vary. Temperature variability around the city is also being studied closely to better refine targeting of the most vulnerable neighborhoods for cooling interventions.

Another adaptation in New York City is a network of community gardens used as heat preparedness "hubs." The benefits are numerous, including cooling shade and evapotranspiration from plants and trees, an improved healthy food supply, less storm water runoff, and a location for community gatherings, heat-related public outreach, and checking in on neighbors.[27]

Exposure to the extreme heat of 2020 overlapped with the disproportionate burden of COVID-19 in many of these vulnerable neighborhoods. Said one city council member, "The same people that are right in the line of fire for COVID are most at risk from extreme heat, and especially New York City summer heat, where you have that urban heat island effect."[30] James Grant, a retiree living in a public housing apartment with no AC, would usually go to a nearby cooling center when temperatures spiked, but in 2020, this center was closed because of the pandemic. The next nearest cooling facility was more than a mile away—too far to walk in extreme heat.[30] In general in 2020, many residents were hesitant to visit

Public Housing Complex in New York City.

cooling centers because of COVID-19 risks and restrictions, so in response, city officials distributed tens of thousands of free AC units to low-income older adults.[30] This type of targeted intervention for highly at-risk residents is one example of more equitable climate adaptation. However, because of the certainty of future warming, continued vigilance and attention to the needs of at-risk populations will be important.

Even when heat adaptation plans are rolled out, some populations may be more or less likely to adopt specific adaptation measures for various reasons. In New York City, for example, more than one-quarter of residents do not have access to or use AC.[31] The odds of not having AC were doubled for Black people compared with other racial/ethnic groups and tripled for households with annual incomes below $30,000. Only 12% of respondents to a survey about heat reported going to a public place with AC if they could not keep cool at home. This is important because heat-related health outcomes in New York City are partially explained by lack of AC access.[31] Increased use of AC also has negative climate costs because of increased electricity use, the generation of which may release more greenhouse gases (GHGs) and thus create a feedback loop and worsen warming. This problem can be solved in part if all people have equitable access to electricity from renewable energy sources that do not produce GHGs.

blood pressure, and heart disease. In addition, Black, Latino and Indigenous peoples are more likely to be uninsured than white people.[26] Some may also be linguistically isolated and/or less aware of heat-related health messaging or local climate risks.

Measuring Heat

Heat exposure is measured by proxy using ambient temperatures (outdoors) and humidity. Internal body temperatures cannot easily be measured in study populations, and indoor temperatures are too highly variable. Outdoor temperatures are recorded at weather monitoring stations, at research stations, and on ships and buoys around the world, as well as by satellites.

Both temperature and humidity affect human experiences of heat stress and are reflected in two important measures, **heat index (HI)** and **wet-bulb globe temperature (WBGT)**. HI combines temperature in the shade and relative humidity to reflect the *apparent temperature* that people actually experience (**Figure 4.5**).[32] A heat index above 103°F indicates heat stress "danger," and above 125°F indicates "extreme danger."

WBGT also indicates *apparent temperature* and incorporates measures of temperature in direct sunlight, humidity, wind speed, sun angle, and cloud cover. WBGT is used to set health guidelines for heat exposures among outdoor workers and for outdoor athletic and military training. In the United States, the NWS provides data and forecasts for WBGT and HI (**Figure 4.6**).[33] The threshold WBGT value indicating dangerous heat stress will vary depending on the level of exertion, but generally, humans are not able to effectively thermoregulate above a WBGT of 35°C (95°F).

Standard ambient temperature readings, along with HI and WBGT, may be used in epidemiological studies to determine the relationship between heat exposure and human health. Exposure to extreme daily *maximum* temperatures affects the ability to regulate core body temperature. In addition, exposure to extreme daily *minimum* temperatures that usually occur at night inhibits the body's ability to recover from

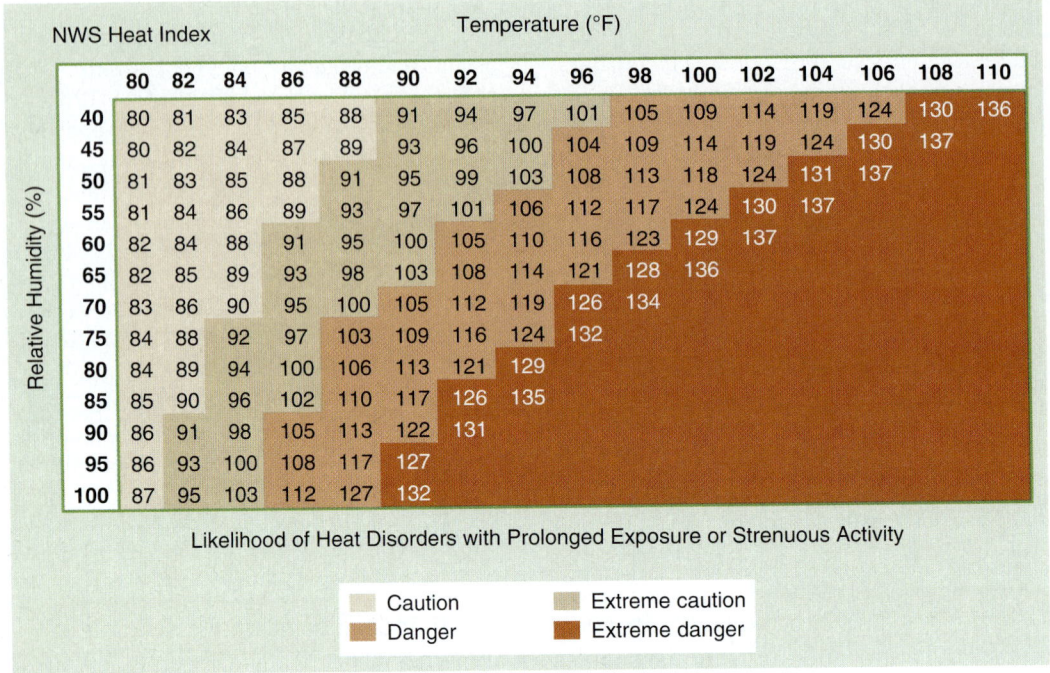

Figure 4.5 Heat Index Chart from the U.S. National Weather Service.

Reproduced from National Weather Service National Oceanic and Atmospheric Administration. Heat forecast tools. Accessed August 10, 2020. https://www.weather.gov/safety/heat-index

an excessive heat response during the day.[15] Temperature effects vary by region and season, and actual heat exposure depends on microclimates, which are often difficult to account for if high-resolution localized data are not available. Urban heat island effects, indoor and outdoor conditions, human activity levels, degrees of bodily heat adaptation, and background health status all modify heat exposure responses, such that two people exposed to the same amount of heat may not experience the same health effects.

Exposure to heat may be due to a discrete extreme weather event, like a heat wave, or general temperature increases that may be lower than during a heat wave but sustained over longer periods. A **heat wave** is defined by the **World Meteorological Organization** as an event in which the daily maximum temperature in a specific location exceeds the average maximum temperature for that location by 9°F for more than five consecutive days. When modeling future increases in heat exposure, it will likely be easier to predict general warming trends than the occurrence of heat waves, which are more sporadic and shorter in duration, yet possibly more deadly.

Measuring the Health Impacts of Heat

Epidemiologists strive to understand relationships between exposures and diseases. In the case of climate change-fueled warming, this means identifying how exposures to temperature increases or extreme heat affect human mortality and morbidity. A large body of research describes what we know about heat stress, and, as with other climate-related

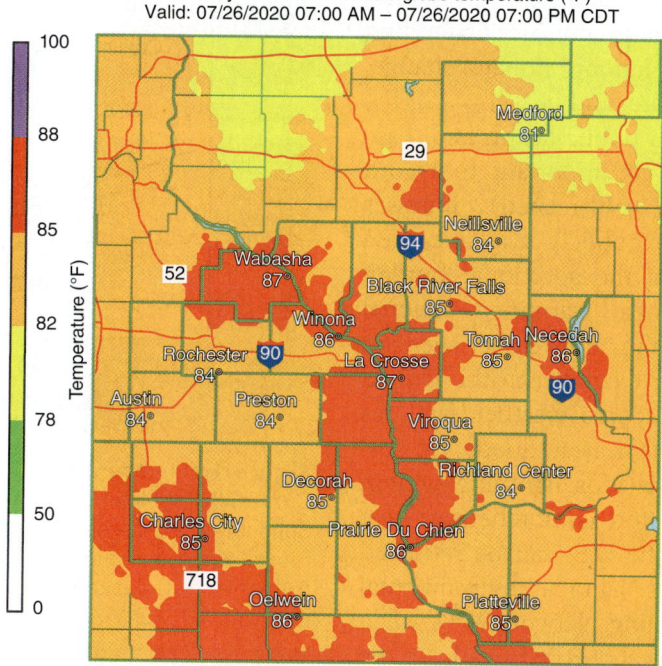

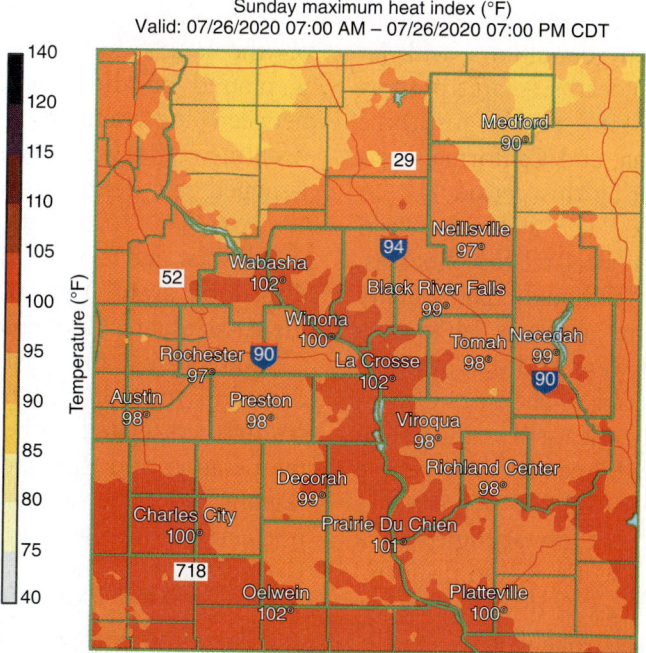

Figure 4.6 U.S. National Weather Service Forecast for Maximum Wet-Bulb Globe Temperature (Top) and Heat Index (Bottom) for the Upper Mississippi River Valley, Sunday, July 26, 2020. Different component variables in each measure result in WBGT having a lower value than HI for a given amount of heat.

health impacts, new studies are constantly being done to advance our understanding of the associations among climate change, extreme heat exposure, and health.

A critical task in public health is disease surveillance, and it is important to record heat-related (and cold-related) illnesses and make the data publicly available even without considering the additional burden brought on by global warming. National and local public health and weather agencies play a key role. For example, in the United States, the NWS records weather fatalities annually that are due to heat, floods, rip currents, hurricanes, tornados, lightning, winter, cold, and winds.[14] More comprehensive heat mortality data for the U.S. can be retrieved from the CDC via its National Environmental Public Health Tracking Network.[34] Between 1999 and 2010, a total of 8,081 deaths in the United States were attributed directly to heat or considered heat a contributing factor in people with preexisting conditions, with an annual average of 673 deaths. A large proportion of these deaths (43%) were in just three states—Texas, Arizona, and California—and among all deaths during this period, 0.44% were heat-attributable.[35] U.S. states also keep their own records on heat-related illnesses, injuries, and deaths, although data quality varies. For example, in California, 140 deaths and 16,000 ED visits were recorded during a summer 2006 heat wave.[36] The rates of hospital ED visits due to "heat stress illness" in 22 states range from 484 per 100,000 people in Florida to 120 per 100,000 in Vermont. According to the CDC, those at highest risk include men, outdoor workers, and drug and alcohol users.[15] Higher use of AC is associated with lowered heatstroke risk.

The association between temperature and mortality can be represented using a **temperature mortality response curve** (**Figure 4.7**).[37] Temperature range is plotted

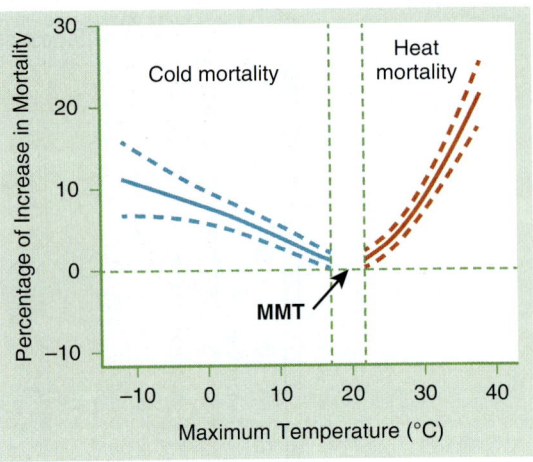

Figure 4.7 Temperature Mortality Response Curve (MMT = Minimum Mortality Temperature). Solid lines reflect central estimates of the mortality relationship as temperatures deviate from the MMT; dashed lines define 95% confidence intervals.

Data from Li T, Horton RM, Kinney P. Future projections of seasonal patterns in temperature-related deaths for Manhattan. *Nat Clim Change*. 2013;3:717–721. doi:10.1038/nclimate1902

on the x-axis and a measure of mortality is plotted on the y-axis. The relationship between the variables is defined by the resulting curve, which is usually U-shaped or J-shaped and defines a **minimum mortality temperature (MMT)**—the temperature with the lowest associated mortality.[38] In both directions away from the MMT, mortality increases as a result of cold or heat exposure.

MMTs are defined using location-specific data and thus differ by region. MMTs are strongly correlated with local annual mean temperature and thus are lower in northern regions and higher in southern regions of the northern hemisphere, including within the United States (**Table 4.3**).[38] Globally, MMTs vary from 12°C (54°F) in parts of Canada and Northern Europe to 32°C (90°F) in parts of South Africa, the Middle East, and central Amazonia.[39] MMTs not determined graphically can be closely approximated by calculating the most frequent temperature (MFT)

Table 4.3 MMT Values for Selected U.S. cities, 1973–1994.

U.S. City	MMT (°F)
Chicago	65.2
New York	66.4
Boston	69.7
Washington, D.C.	70.6
Atlanta	76.3
Miami	80.9

Data from Curriero FC, Heiner KS, Samet JM, Zeger SL, Strug L, Patz JA. Temperature and mortality in 11 cities of the eastern United States. *Am J Epidemiol.* 2002;155(1):80–87. doi:10.1093/aje/155.1.80

for a specific region. Sensitivity to heat (indicated by the slope of the curve at temperatures above the MMT) is greater in the north, and cold sensitivity is greater in the south.

Heat exposure–mortality relationships are affected by two related phenomena, **heat acclimatization** and **heat adaptation**, which are the ways in which people respond physiologically, behaviorally, technologically, and infrastructurally to increases in heat. Heat acclimatization are bodily responses in individuals that protect in cases of frequent exposure to high temperatures. They tend to set in within 7–14 days of onset of heat exposure but can be observed in as little as one day.[40] Numerous changes reduce the negative effects of continued heat stress, including lowered core temperature and heart rate, increased sweating rate, lowered temperature threshold for sweating, more blood flow from the core to the skin to release heat from the body, increased total body water volume, decreased electrolyte losses by reabsorbing sodium from sweat, and, at the molecular level, increased expression of gene products that play important roles in the heat response. To become acclimatized, people

must have adequate water intake and sun protection.

Heat adaptation includes behaviors to regulate heat exposure, including wearing light clothing and seeking cooler environments. These behaviors may be more prevalent during subsequent heat events after people have learned from a prior exposure. Adaptation also includes increasing public awareness of extreme heat risks, and greater availability of cooling measures, including AC, cooling centers, and more tree cover and green space. In contrast, certain factors may raise the risk of illness and mortality, such as power outages during heat waves.

As a result of these responses to heat exposure, heat-related mortality is higher in early summer than in late summer.[17] Another explanation for higher early season heat deaths is **mortality displacement**. In this situation, extreme heat causes a forward shift in time of excess mortality among people who are already gravely ill and at risk of dying during the period of interest even without heat exposure.[42] This is followed by a decrease in overall mortality because the people most at risk have already died. This phenomenon is also known as a harvesting effect. Acclimatization and adaptation are also observed at the population level over longer periods. For example, in the United States, the increased risk of heatstroke hospitalizations on heat wave days compared to non–heat wave days decreased 20-fold from 1999 to 2010, even though 2010 had more heat wave days than other years.[41]

There are limits, however, to our ability to avoid extreme heat impacts through acclimatization and adaptation, and at a certain upper limit of heat exposure, human survivability is threatened regardless of health status and access to cooling mechanisms. For example, in the face of climate change, the southern Persian Gulf coast and parts of Pakistan and northern India already experience heat and humidity that approach the upper limit of survivability. The highest known heat index, 81°C (178°F), was recorded in Dhahran, Saudi Arabia, on July 8, 2003, and in 2018, the city of Quriyat, Oman, recorded the world's hottest *minimum* (low) temperature for a 24-hour period, 43°C (109°F).[43]

Mortality is perhaps the most commonly measured heat-related health outcome. Mortality data most often come from hospital records, death certificates, and vital statistics registries, although not all heat-attributable deaths happen in clinical settings or are recognized or recorded as caused directly or indirectly by heat. How heat-related deaths are reported depends on where the deaths occurred and who was involved in discovery and assessment—including family members, neighbors and friends, emergency response personnel, emergency room physicians and nurses, other medical staff, coroners, and public health officials. If data on heat-specific mortality are not attainable, researchers may measure total or excess all-cause mortality during a period of increased heat and compare it with a reference time period, allowing for estimation of the portion of excess deaths likely attributed to heat. Researchers may also collect data on hospital admissions or ED visits during or following a high-heat event. It is important to also monitor nonfatal conditions related to extreme heat exposure.

Heat Waves

Numerous particularly deadly heat waves have occurred over the past few decades, and heat wave mortality studies are illustrative of the devastating health impacts of heat exposure. For example, in 1995, Chicago experienced an extreme summer heat wave that contributed to a maximum recorded high temperature of 106°F and a heat index

of 126°F. More than 700 excess deaths occurred as compared with an "average" year, along with more than 3,300 hospital ED visits (**Figure 4.8**).[44] The average age of the victims was 76 years old, most lived in the city's poorest neighborhoods, and the death rate was 50% higher in Black people than white people. The odds of dying increased approximately 2.5 times for those living alone or living in an apartment building compared with a single-family home or duplex, and living on a top floor raised the odds 4.7-fold compared with living below the top floor. Other risk factors were being confined to bed or unable to care for oneself, and having a mental health problem or heart condition. Importantly, the *odds of dying were reduced 80% with a working air conditioner in the home*, as well as with visiting a cooling shelter or other air-conditioned place.[44]

In August 2003, Europe experienced a historic heat wave, with France suffering through its worst summertime heat since the year 1540 and the most summertime deaths since World War II[45] (**Figure 4.9a**). For a stretch of nine days, the average maximum temperature in France exceeded the seasonal average by 11–12°C (20–22°F). Cumulative excess mortality in France was nearly 15,000 deaths—55% higher than expected without a heat wave (**Figure 4.9b**)— and in total, 70,000 deaths across Europe were attributed to the heat wave. Two heat waves that struck Europe in the summer of 2019 killed more than 1,400 people in France and resulted in the highest temperature ever recorded in France up to that time, 46°C (114.8°F).[46]

A heat wave in Ahmedabad, India, in May 2010 resulted in hundreds of excess deaths (**Figure 4.10**)[47] and 3–6 times more admissions of newborns into hospital neonatal intensive care units than in the year before or the year after.[48] Neighboring

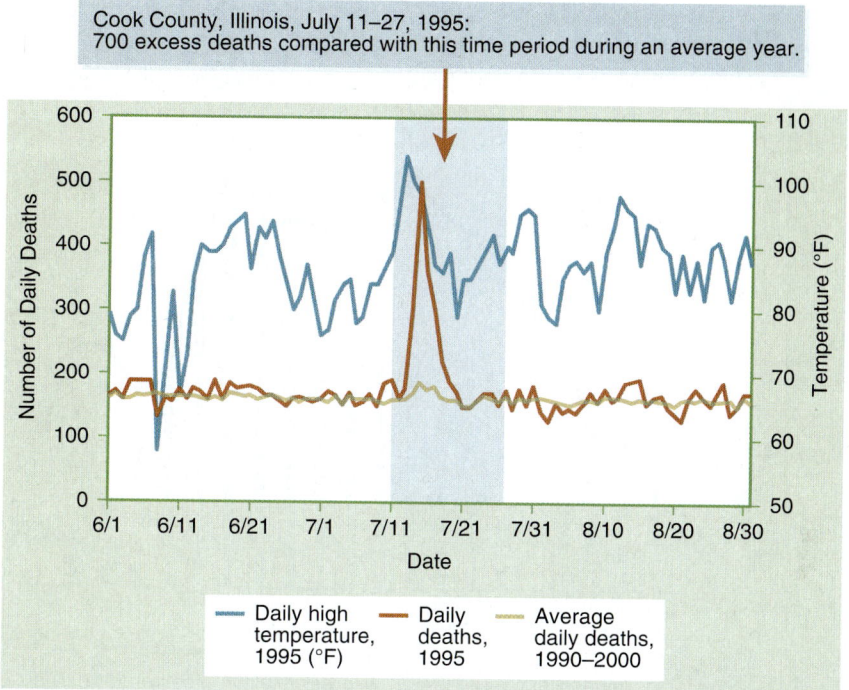

Figure 4.8 Time Series Data of High Temperature Readings and Mortality during the Chicago Heat Wave in July 1995. Excess all-cause mortality lags 2–3 days after the onset of extreme heat.

Reproduced from U.S. Global Change Research Program. *The Impacts of Climate Change on Human Health in the United States: A Scientific Assessment.* U.S. Global Change Research Program; 2016. https://health2016.globalchange.gov

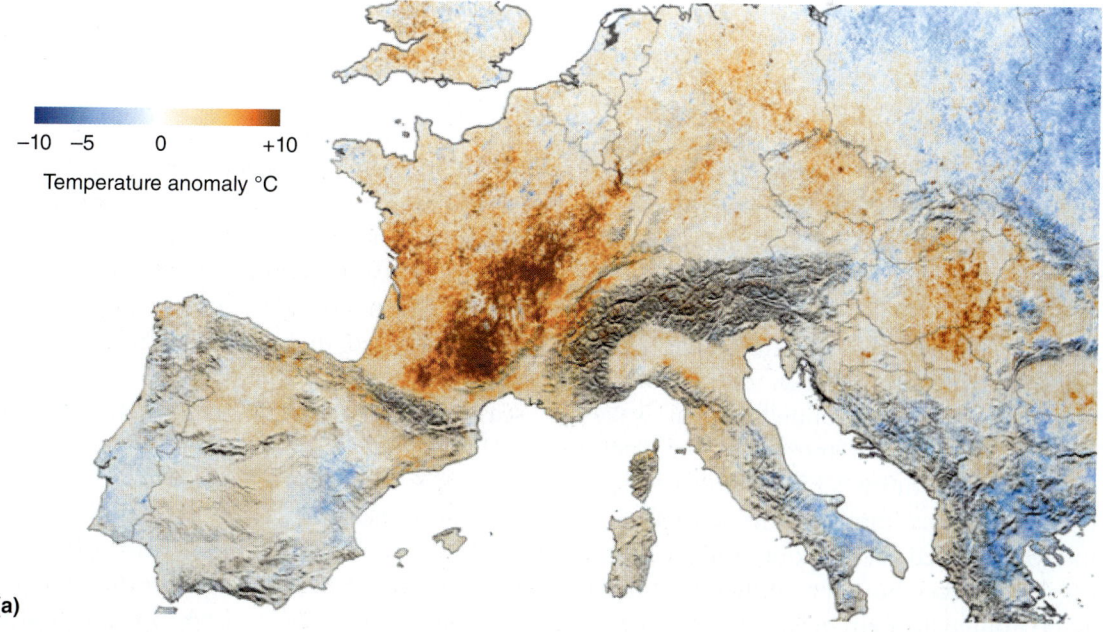

(a)

Figure 4.9 (Continued)

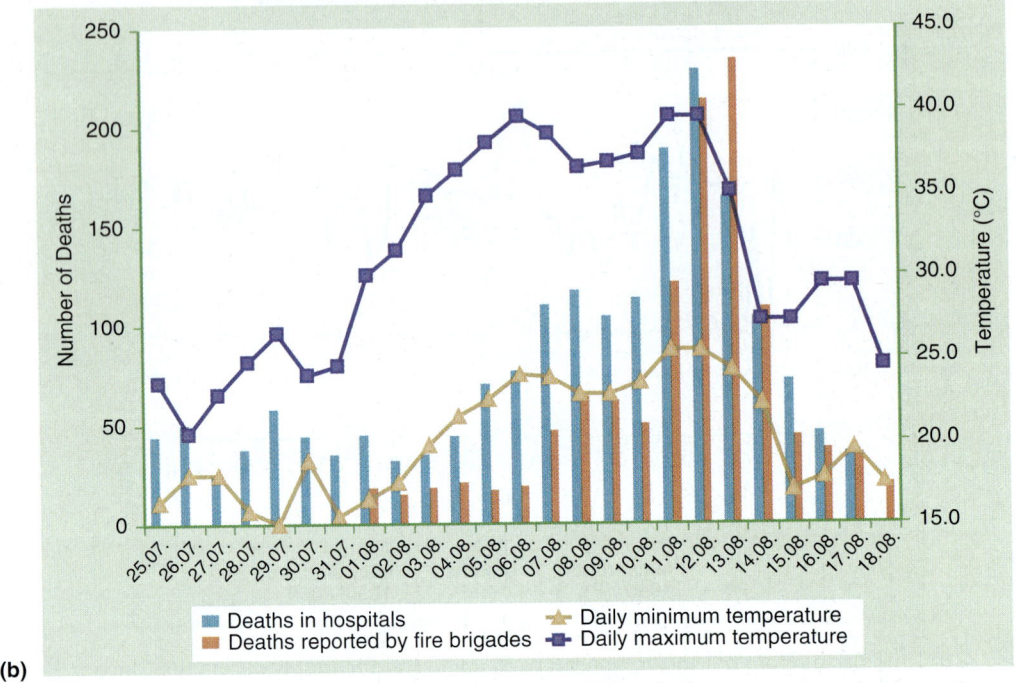

(b)

Figure 4.9 **(a)** Temperature Anomalies in Europe on August 16, 2003, Compared with 2001. Dark red indicates areas up to 10°C (18°F) hotter than average; dark blue indicates temperatures up to 10°C (18°F) lower than average. France in particular experienced heat extremes. **(b)** Temperatures and Mortality in Paris, France during the August 2003 European Heat Wave.

(a) Reproduced from National Aeronautics and Space Administration. European heat wave. August 16, 2003. Accessed August 10, 2020. https://earthobservatory.nasa.gov/images/3714/european-heat-wave; (b) Reproduced from European Environment Agency. cc-number-reported-deaths.eps. Created November 12, 2009. Updated September 5, 2011. https://www.eea.europa.eu/data-and-maps/figures/number-of-reported-deaths-and-minimum-and -maximum-temperature-in-paris-during-the-heatwave-in-summer-2003/cc-number-reported-deaths.eps

Pakistan, which currently experiences seven heat waves annually, is projected to experience 12 heat waves annually by 2030 and 26 by 2090.[4]

These heat waves illustrate the deadly health consequences of extreme heat exposure. The maximum effect of mortality is seen within 3 days following heat waves, suggesting that heat mortality is an acute event that requires rapid interventions. However, many more deaths and illnesses likely occurred at various times following heat waves but were difficult to detect as attributed to extreme heat exposure. In addition, most heat studies are done in urban areas, where populations and disease cases are higher than in nonurban areas; therefore, the impacts of extreme heat in nonurban areas are not well characterized.

In 50 major U.S. cities, heat waves have increased in frequency, duration, and intensity since the 1960s, and the heat wave season increased by 47 days in the decade 2010–2019 compared with the 1960s (**Figure 4.11**).[49] These shifts were associated with more heat-related illnesses and deaths, as well as increased preterm births. In releasing these heat wave and other climate indicators in 2021, EPA Administrator Michael Regan said, "There is no small town,

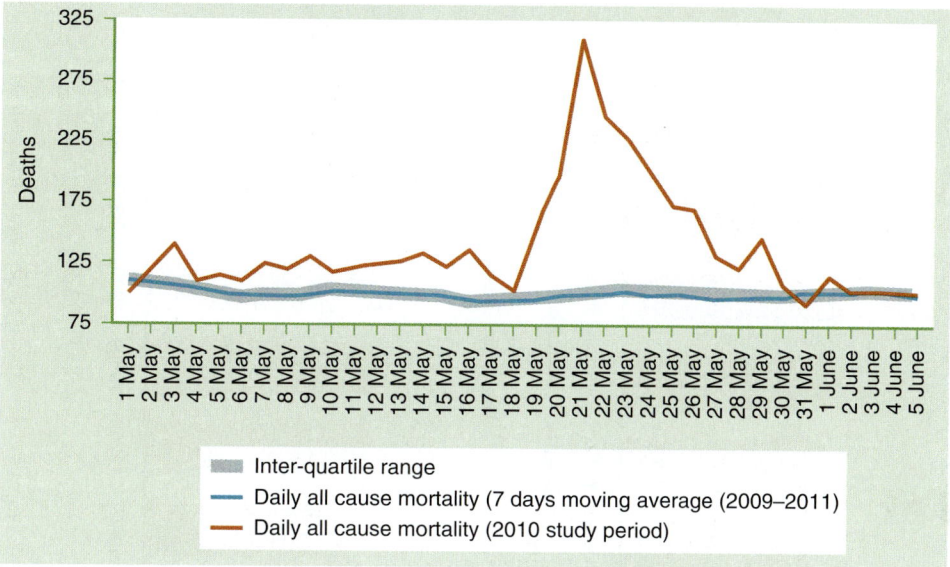

Ahmedabad, India, May–June 2010: 1,344 excess deaths.

Figure 4.10 Daily Mortality during a May 2010 Heat Wave in Ahmedabad, India. Red line, May 2010; blue line, average for 2009 and 2011.

Reproduced from Azhar GS, Mavalankar D, Nori-Sarma A, et al. Heat-related mortality in India: excess all-cause mortality associated with the 2010 Ahmedabad heat wave [published correction appears in *PLoS One*. 2014;9(9):e109457]. *PLoS One*. 2014;9(3):e91831. doi:10.1371/journal.pone.0091831

big city or rural community that is unaffected by the climate crisis."[50]

Adverse health impacts also occur upon exposure to increased temperatures not associated with heat waves, as indicated by the upward curve on the temperature mortality response curve at temperatures above the MMT (Figure 4.7). For example, in England and Wales, up to 1% of summertime deaths may be linked to heat. Hospital admissions for respiratory and kidney diseases increased with warming, and pediatric trauma increased 11% for every 5°C rise in maximum temperature.[51] Patients with neurological and psychiatric diagnoses, such as dementia and substance abuse, and those taking antipsychotic, antidepressant, and hypnotic medications were at highest risk of heat-related death.

Projecting Future Health Impacts of Heat

In the 2014 *Fifth Assessment Report* (AR5) of the Intergovernmental Panel on Climate Change, scientists concluded with *very high confidence* that the future will see "greater risk of injury, disease, and death due to more intense heat waves" and with *high confidence* that "the combination of high temperatures and high humidity will compromise normal human activities, including growing food or working outdoors in some areas for parts of the year." In addition, they acknowledged that "the capacity of the human body to thermoregulate may be exceeded on a regular basis, particularly during manual

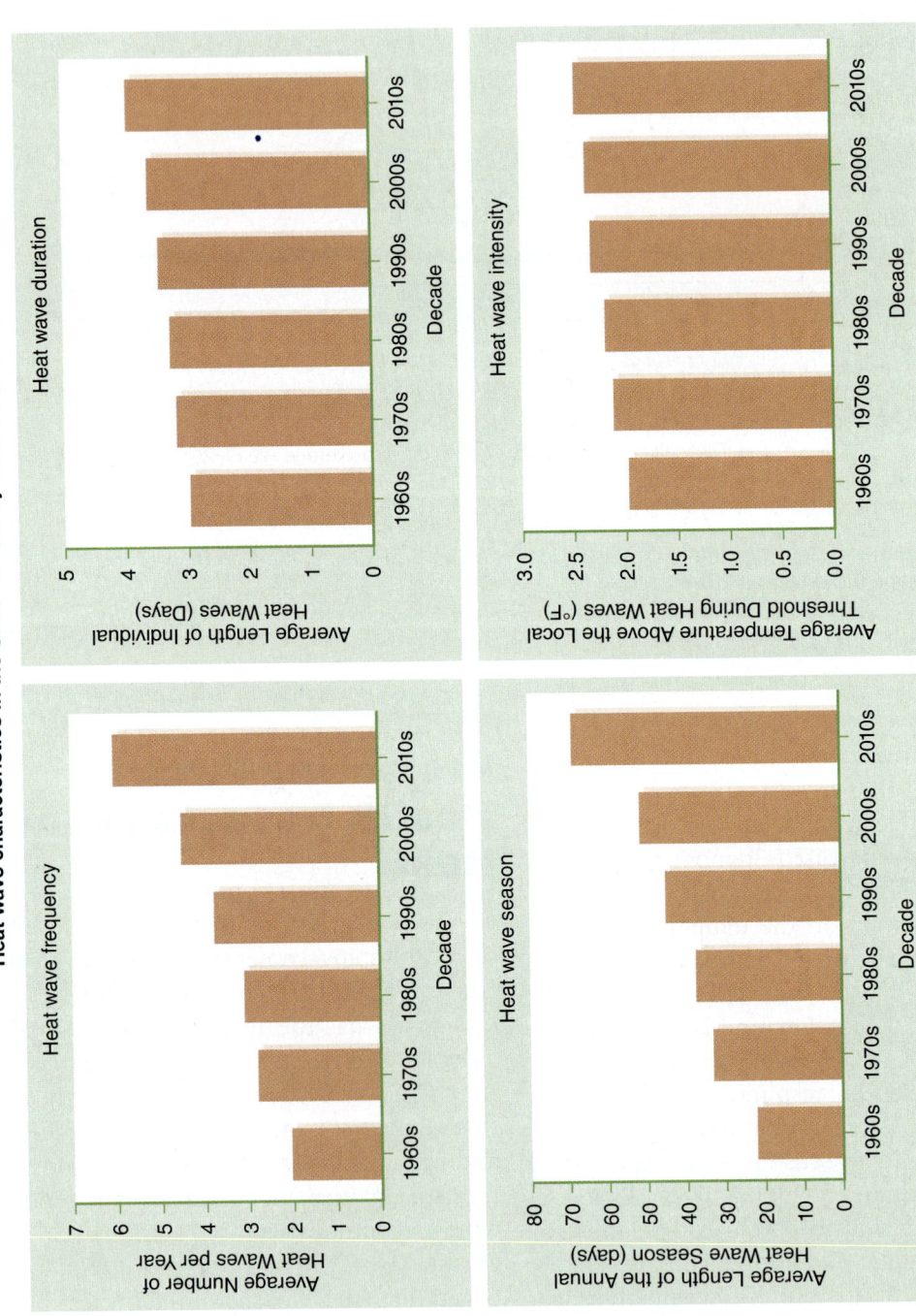

Figure 4.11 Heat Wave Indicators by Decade for 50 U.S. Cities, 1961–2019. Upper Left: Number of heat waves per year (*frequency*). Upper Right: Average length of heat waves in days (*duration*). Lower Left: Number of days between the first and last heat wave of the year (*season length*). Lower Right: Average temperature during the heat wave compared with the local temperature threshold for defining a heat wave (*intensity*).

labor, in parts of the world during this century."[52]

By 2100, even with drastic GHG emission reductions, nearly half of the world's population will be at risk of exposure to conditions that exceed a maximum heat threshold for at least 20 days per year, and with unmitigated GHG emissions, nearly three-quarters will be at risk.[53] Recent studies have shown that the highest future heat stress risk is projected for tropical regions of Africa and South Asia. By 2070, 1.5 billion people in these regions will be vulnerable to deadly heat waves, which will occur every year rather than every 25 years, as happened in the past.[1] Human migration is projected to increase as some of these regions become unlivable.

In India, for example, annual warming is projected to be between 2.2 and 5.5°C (4 and 10°F) by 2100.[54] Approximately one quarter of India's population may be at particular risk because they live in extreme poverty and have no electricity access, which constrains access to household cooling and increases reliance on outdoor occupations. In India, the probability of **mass heat-related mortality** events (*heat-related deaths of more than 100 people*) increases from 13 to 32% when mean summer temperature increases just one-half degree (from 27 to 27.5°C), and from 46 to 82% when the average number of heat wave days across India increases from 6 days to 8 days.

Even if warming is limited to the targets of the Paris Agreement, hundreds of thousands of additional heat deaths are projected to occur in just the United States and European countries by this century's end.[1] In the United States, EPA projects that mortality on extremely hot days may increase 10-fold or more in many cities, leading to as many as 11,000 additional premature heat deaths by 2030.[13] As with current heat vulnerability, northern U.S. cities are likely to be more vulnerable to increases in heat mortality in the future, although some will also see reductions in cold deaths. By midcentury, Milwaukee and New York City may experience triple the number of days of extreme heat (defined as daily maximum temperature above 90°F), whereas Atlanta may see only a doubling of extreme heat days.[55] During the deadly 1995 Chicago heat wave, the WBGT was 31°C, a level of heat that occurs in Chicago today on only one day every 35 years—but this level will occur in Chicago on *one day every 5 years* with 2°C warming and on *four days every year* with 4.5°C warming.[5] In Spain, heat-related deaths are projected to rise by nearly 8% by 2050 if the current pace of warming continues, and even if warming is limited to 1.5°C, deaths will still rise by nearly 4%.[1] Generally, future heat mortality will be strongly affected by urbanization, aging of the population, and the quality of healthcare services and climate adaptations.

The most extreme warming on Earth right now is in Arctic regions. In June 2020, parts of Arctic Siberia hit 38°C (100°F)—more than 30 degrees above the monthly average temperature.[56] Much of Alaska is experiencing average temperatures more than 4°F warmer than a century ago.[49] Data are lacking on how this recent extreme heat is affecting health burdens among Arctic populations, but impacts on Indigenous peoples in the Arctic are becoming clear. Loss of traditional ice routes for hunting, fishing, and travel affect health and livelihoods, and cultural and ecosystem losses cause physical and emotional stress, anxiety, grief, and despair. In addition, migration, which carries significant health risks, may be inevitable for communities facing sea level rise, land erosion, permafrost melt, shifting resource access, and destruction of infrastructure.[57]

Adaptations

All human health impacts of climate change, including heat-related illnesses and deaths, will be lessened by mitigating GHG emissions to limit warming and by effective context-specific climate adaptations, including access to AC and other cooling measures.[58] Other effective adaptations, best implemented at a regional or local level, are increased health monitoring, emergency preparedness, responsive healthcare systems and social services, more abundant trees, parks, and other green spaces, and sustainable urban planning to minimize heat island effects and maximize cooling. In general, accurate climate models with predictive capacity for extreme heat events will allow for the creation of effective preparation and response planning to reduce future health impacts.

Early warning interventions include heat watches, warnings, and advisories—like those issued by the NWS—that are used to spread awareness of the risks of extreme heat and actions to take, as well as to alert emergency responders in affected locations (**Figure 4.12**). Criteria vary by region, but generally in the U.S., a **Heat Advisory** is issued within 12 hours of the onset of extremely dangerous heat conditions (in the Northeast, when the heat index is forecast to reach 95–99°F for at least 2 consecutive days or 100–104°F for any length of time).[59] An **Excessive Heat Warning** is issued within 12 hours of the onset of even more extreme heat conditions (in the Northeast, when the heat index is forecast to reach or exceed 105°F for at least 2 consecutive hours). An **Excessive Heat Watch** is issued when conditions are favorable for an excessive heat event in the next

Heat Advisory

Issued: 11:19 AM Jul. 20, 2020 – National Weather Service

HEAT ADVISORY IN EFFECT UNTIL 8:00 PM EDT.

- WHAT...Heat index values up to 98.
- WHERE...Kennebec County.
- WHEN...Until 8 PM EDT this evening.
- IMPACTS...Hot temperatures and high humidity may cause heat illnesses to occur.

PRECAUTIONARY/PREPAREDNESS ACTIONS...

Drink plenty of fluids, stay in an air-conditioned room, stay out of the sun, and check up on relatives and neighbors.

Young children and pets should never be left unattended in vehicles under any circumstances.

Take extra precautions if you work or spend time outside. When possible reschedule strenuous activities to early morning or evening. Know the signs and symptoms of heat exhaustion and heat stroke. Wear lightweight and loose-fitting clothing when possible. To reduce risk during outdoor work, the Occupational Safety and Health Administration recommends scheduling frequent rest breaks in shaded or air-conditioned environments. Anyone overcome by heat should be moved to a cool and shaded location. Heat stroke is an emergency! Call 911.

Figure 4.12 Example of a Heat Advisory Issued by the U.S. National Weather Service for Kennebec County, Maine, on July 20, 2020.

Reproduced from U.S. National Weather Service for Kennebec County, Maine.

24–72 hours. An **Excessive Heat Outlook** is issued for a potential excessive heat event in the next 3–7 days.

Emergency measures put in place in Chicago after the 1995 heat wave resulted in lower mortality among older residents in a subsequent heat wave in late July 1999, in which daytime temperatures peaked at 105°F, with heat index values near 120°F (similar to the 1995 heat wave).[60] Chicago's heat wave response system included expanding Notify Chicago, the city's text and email emergency notification system, setting up cooling centers, and conducting disaster preparedness and response training.[61] The city also created a Call 311 program during heat waves for officials to conduct well-being checks for those who may need additional assistance, including the elderly and those who are unwell or socially isolated.

Chicago officials also performed a city-wide assessment of climate vulnerability to predict future extreme heat conditions in all its neighborhoods and adopted adaptation strategies that support vulnerable populations, such as partnering with the Chicago Field Museum of Natural History to develop outreach to at-risk communities.[61] The city identified areas where the urban heat island effect would be worsened by climate change and used this information to target green infrastructure and heat island mitigation efforts, such as green or cool roofs, cool pavements, and increased vegetation and trees. They also adopted a comprehensive Climate Change Action Plan whose first priority was to increase adaptations for extreme heat events.[62] Unfortunately, residents of many of the city's poorest neighborhoods, which suffered a disproportionate number of deaths during the 1995 heat wave and experience

the persistent ill effects of poverty, racism, and poor housing, have not benefited from Chicago's disaster preparedness planning and continue to be at particular risk.[63] Close attention to the ultimate causes of health inequities, which become fault lines during a heat wave or other disaster, needs to be incorporated into climate adaptation measures.

The U.S. government provides extreme heat preparedness guidelines to the public through the Ready Campaign, "a national public service campaign to educate and empower the American people to prepare for, respond to and mitigate emergencies, including natural and man-made disasters."[64] Its website offers recommendations for advance preparation and protections during an extreme heat event, including recognizing the signs of heat-related illnesses. Other countries have also implemented heat adaptation plans. For example, after experiencing recent heat wave mortality events, India has taken drastic measures to increase heat wave preparedness and adaptation. As a result, heat wave death tolls have dropped significantly.[65] These measures include adjusting worker hours to avoid extreme hot weather exposures, opening drinking water kiosks and shelters for those experiencing homelessness, making sure hospitals are adequately stocked with rehydration fluids and other supplies to treat heatstroke, and painting roofs white to reduce heat absorption.

The World Health Organization urges countries to strengthen their health sectors' heat wave preparedness and response plans and to build scientific capabilities to predict heat impacts on human health and well-being. Protecting people in the workplace, especially outdoor workers, requires that government agencies, such as the U.S. Occupational Safety and Health

Administration, review occupational heat exposure standards to ensure they are adequately protective now and in the future. Increased research and heat-related disease surveillance in countries around the world is critical. Because current and future extreme heat events are a major threat to human survival, well-being, and livelihoods, countries must commit to rapid climate mitigation and adaptation.

Discussion Questions

1. Discuss how the human body responds to high heat exposure.
2. Which illnesses are associated with heat stress?
3. What are the major causes of heat-related mortality?
4. Distinguish among temperature, heat index, and wet-bulb globe temperature.
5. Describe risk factors that positively or negatively modify the relationship between extreme heat exposure and health outcomes.
6. Which groups are at disproportionate risk of heat-related morbidity and mortality? Why?
7. What is the temperature mortality response curve, and how is it useful to study the health effects of heat? What is minimum mortality temperature, and how does it vary?
8. What heat adaptations should be incorporated into climate action plans to prevent heat-related illnesses and deaths?

References

1. Berwyn, B. Extreme heat risks may be widely underestimated and sometimes left out of major climate reports. *Inside Climate News.* Published May 16, 2021. Accessed June 18, 2021. https://insideclimatenews.org/news/16052021/extreme-heat-risks-climate-change/
2. The Weather Channel. Heat wave across Southwest turns deadly. June 23, 2017. Accessed August 10, 2020. https://weather.com/news/news/southwest-heatwave-june-2017
3. World Meteorological Organization. WMO verifies 3rd and 4th hottest temperature recorded on Earth. June 18, 2019. Accessed August 10, 2020. https://public.wmo.int/en/media/press-release/wmo-verifies-3rd-and-4th-hottest-temperature-recorded-earth
4. Nasim W, Amin A, Fahad S, et al. Future risk assessment by estimating historical heat wave trends with projected heat accumulation using SimCLIM climate model in Pakistan. *Atmospheric Res.* 2018;205:118–133. doi:10.1016/j.atmosres.2018.01.009
5. Li D, Yuan J, Kopp R. Escalating global exposure to compound heat-humidity extremes with warming. *Environ Res Lett.* 2020;15. doi:10.1088/1748-9326/ab7d04
6. National Park Service. Numerous heat records set in Death Valley this summer. September 22, 2020. Accessed January 13, 2021. https://www.nps.gov/deva/learn/news/summer-2020-heat-records.htm
7. Wong K, Associated Press. NWS: Phoenix hits 102°F, sets new record for number of days with triple-digit temperatures. *FOX10 Phoenix.* October 14, 2020. Accessed January 13, 2021. https://www.fox10phoenix.com/news/nws-phoenix-hits-102f-sets-new-record-for-number-of-days-with-triple-digit-temperatures
8. National Oceanic and Atmospheric Administration. It's official: July was Earth's hottest month on record. August 13, 2021. Accessed August 30,

2021. https://www.noaa.gov/news/its-official-july-2021-was-earths-hottest-month-on-record

9. National Oceanic and Atmospheric Administration. The new U.S. Climate Normals are here. What do they tell us about climate change? Published May 4, 2021. Accessed June 18, 2021. https://www.noaa.gov/news/new-us-climate-normals-are-here-what-do-they-tell-us-about-climate-change

10. United Nations Environment Programme. Emissions gap report 2020—executive summary. December 9, 2020. Accessed February 19, 2021. https://www.unep.org/emissions-gap-report-2020

11. Xu C, Kohler TA, Lenton TM, Svenning J-C, Scheffer M. Future of the human climate niche. *Proc Natl Acad Sci U S A*. 2020;117:11350–11355. doi:10.1073/pnas.1910114117

12. Watts N, Amann M, Arnell N, et al. The 2020 report of The Lancet Countdown on health and climate change: responding to converging crises. *The Lancet*. 2021;397(10269):129–170. doi:10.1016/S0140-6736(20)32290-X

13. U.S. Global Change Research Program. *The Impacts of Climate Change on Human Health in the United States: A Scientific Assessment*. U.S. Global Change Research Program; 2016. https://health2016.globalchange.gov

14. National Weather Service. Weather related fatality and injury statistics. 2021. Accessed June 18, 2021. https://www.weather.gov/hazstat/

15. Centers for Disease Control and Prevention. Heat stress. August 31, 2020. Accessed January 13, 2021. https://www.cdc.gov/niosh/topics/heatstress/default.html

16. World Health Organization. *ICD-10 Version:2010*. Accessed January 13, 2021. https://icd.who.int/browse10/2010/en

17. Anderson BG, Bell ML. Weather-related mortality: how heat, cold, and heat waves affect mortality in the United States. *Epidemiology*. 2009;20(2):205–213. doi:10.1097/EDE.0b013e318190ee08

18. Smith CJ. Pediatric thermoregulation: considerations in the face of global climate change. *Nutrients*. 2019;11(9):2010. doi:10.3390/nu11092010

19. KidsAndCars.org. Heatstroke. Accessed January 13, 2021. https://www.kidsandcars.org/how-kids-get-hurt/heat-stroke/

20. National Institute for Occupational Safety and Health. Reproductive health and the workplace. April 20, 2017. Accessed January 13, 2021. https://www.cdc.gov/niosh/topics/repro/heat.html

21. Borg MA, Xiang J, Anikeeva O, et al. Occupational heat stress and economic burden: a review of global evidence. *Environ Res*. 2021;195:110781. doi:10.1016/j.envres.2021.110781

22. Baptiste N. Farmworkers are dying from extreme heat. *Mother Jones*. August 24, 2018. Accessed January 13, 2021. https://www.motherjones.com/food/2018/08/farmworkers-are-dying-from-extreme-heat/

23. Taylor EV, Vaidyanathan A, Flanders WD, et al. Differences in heat-related mortality by citizenship status: United States, 2005–2014. *Am J Public Health*. 2018;108:S131–S136. doi:10.2105/AJPH.2017.304006

24. Hoffman J, Shandas V, Pendleton N. The effects of historical housing policies on resident exposure to intra-urban heat: a study of 108 U.S. urban areas. *Climate*. 2020;8:12. doi:10.3390/cli8010012

25. Namin S, Xu W, Zhou Y, et al. The legacy of the Home Owners' Loan Corporation and the political ecology of urban trees and air pollution in the United States. *Soc Sci Med*. 2020;246:112758. doi:10.1016/j.socscimed.2019.112758

26. Centers for Disease Control and Prevention. Health insurance coverage: early release of estimates from the National Health Interview Survey, January–June 2020. February 2021. Accessed June 19, 2021. https://www.cdc.gov/nchs/data/nhis/earlyrelease/insur202102-508.pdf

27. New York City Environmental Justice Alliance. Midway to 2030: building resiliency and equity for a just transition. April 2018. Accessed January 13, 2021. https://www.nyc-eja.org/wp-content/uploads/2018/04/NYC-Climate-Justice-Agenda-Final-042018-1.pdf

28. New York City Department of Health and Mental Hygiene. Heat-related deaths in New York City, 2013. *Epi Data Brief*. 2014;47. Accessed January 13, 2021. https://www1.nyc.gov/assets/doh/downloads/pdf/epi/databrief47.pdf

29. Weinberger KR, Haykin L, Eliot MN, et al. Projected temperature-related deaths in ten large U.S. metropolitan areas under different climate change scenarios. *Environ Int*. 2017;107:196–204. doi:10.1016/j.envint.2017.07.006

30. Olumhense E, Michel C. Looking for relief as summer heat wave hits Black and Brown neighborhoods hardest. *The City*. July 29, 2020. Accessed January 13, 2021. https://www.thecity.nyc/health/2020/7/29/21347387/new-york-city-summer-heat-wave-black-neighborhoods-pandemic

31. Madrigano J, Lane K, Petrovic N, et al. Awareness, risk perception, and protective behaviors for

extreme heat and climate change in New York City. *Int J Environ Res Public Health*. 2018;15:1433. doi:10.3390/ijerph15071433

32. National Weather Service National Oceanic and Atmospheric Administration. Heat forecast tools. Accessed August 10, 2020. https://www.weather.gov/safety/heat-index

33. National Weather Service National Oceanic and Atmospheric Administration. Excessive heat: Wet Bulb Globe Temperature & Heat Index. Accessed August 10, 2020. https://www.weather.gov/arx/wbgt4

34. Centers for Disease Control and Prevention. National Environmental Public Health Tracking. December 16, 2020. Accessed June 19, 2021. https://www.cdc.gov/nceh/tracking/index.html

35. Weinberger KR, Harris D, Spangler KR, et al. Estimating the number of excess deaths attributable to heat in 297 United States counties. *Environ Epidemiol*. 2020;4. doi:10.1097/EE9.0000000000000096

36. California Office of Environmental Health Hazard Assessment. Heat-related mortality and morbidity. February 11, 2019. Accessed August 10, 2020. https://oehha.ca.gov/epic/impacts-biological-systems/heat-related-mortality-and-morbidity

37. Li T, Horton RM, Kinney P. Future projections of seasonal patterns in temperature-related deaths for Manhattan. *Nat Clim Change*. 2013;3:717–721. doi:10.1038/nclimate1902

38. Curriero FC, Heiner KS, Samet JM, Zeger SL, Strug L, Patz JA. Temperature and mortality in 11 cities of the eastern United States. *Am J Epidemiol*. 2002; 155(1):80–87. doi:10.1093/aje/155.1.80

39. Yin Q, Wang J, Ren Z, et al. Mapping the increased minimum mortality temperatures in the context of global climate change. *Nat Commun*. 2019;10:4640. doi:10.1038/s41467-019-12663-y

40. Périard JD, Racinais S, Sawka MN. Adaptations and mechanisms of human heat acclimation: applications for competitive athletes and sports. *Scand J Med Sci Sports*. 2015;25:20–38. doi:10.1111/sms.12408

41. Wang Y, Bobb JF, Papi B, et al. Heat stroke admissions during heat waves in 1,916 U.S. counties for the period from 1999 to 2010 and their effect modifiers. *Environ Health*. 2016;15(1):83. doi:10.1186/s12940-016-0167-3

42. Cheng J, Xu Z, Bambrick H, et al. Heatwave and elderly mortality: an evaluation of death burden and health costs considering short-term mortality displacement. *Environ Int*. 2018;115:334–342. doi:10.1016/j.envint.2018.03.041

43. Samenow J. A city in Oman just posted the world's hottest low temperature ever recorded: 109 degrees. *The Washington Post*. June 27, 2018. Accessed August 10, 2020. https://www.washingtonpost.com/news/capital-weather-gang/wp/2018/06/27/a-city-in-oman-just-set-the-worlds-hottest-low-temperature-ever-recorded-109-degrees/

44. Semenza JC, Rubin CH, Falter KH, et al. Heat-related deaths during the July 1995 heat wave in Chicago. *N Engl J Med*. 1996;335:84–90. doi:10.1056/NEJM199607113350203

45. Fouillet A, Rey G, Laurent F, et al. Excess mortality related to the August 2003 heat wave in France. *Int Arch Occup Environ Health*. 2006;80:16–24. doi:10.1007/s00420-006-0089-4

46. Agence France-Presse. Summer heatwaves in France killed 1,500, says health minister. *The Guardian*. September 8, 2019. Accessed August 10, 2020. https://www.theguardian.com/world/2019/sep/09/summer-heatwaves-in-france-killed-1500-says-health-minister

47. Azhar GS, Mavalankar D, Nori-Sarma A, et al. Heat-related mortality in India: excess all-cause mortality associated with the 2010 Ahmedabad heat wave [published correction appears in *PLoS One*. 2014;9(9):e109457]. *PLoS One*. 2014;9:e91831. doi:10.1371/journal.pone.0091831

48. Kakkad K, Barzaga ML, Wallenstein S, et al. Neonates in Ahmedabad, India, during the 2010 heat wave: a climate change adaptation study. *J Environ Pub Health*. 2014;2014:946875. doi:10.1155/2014/946875

49. Environmental Protection Agency. Climate change indicators: heat waves. April 14, 2021. Accessed June 18, 2021. https://www.epa.gov/climate-indicatorsz/climate-change-indicators-heat-waves

50. Flavelle C. Climate change is making big problems bigger. *The New York Times*. May 12, 2021. Accessed June 18, 2021. https://www.nytimes.com/2021/05/12/climate/climate-change-epa.html

51. Arbuthnott KG, Hajat S. The health effects of hotter summers and heat waves in the population of the United Kingdom: a review of the evidence. *Environ Health*. 2017;16:119. doi:10.1186/s12940-017-0322-5

52. Smith KR, Woodward A, Campbell-Lendrum D, et al. Human health: impacts, adaptation, and co-benefits. In: Field CB, Barros VR, Dokken DJ, et al., eds. *Climate Change 2014: Impacts, Adaptation, and*

Vulnerability. Part A: Global and Sectoral Aspects. Cambridge University Press; 2014:709–754. https://www.ipcc.ch/site/assets/uploads/2018/02/WGIIAR5-PartA_FINAL.pdf

53. Mora C, Dousset B, Caldwell IR, et al. Global risk of deadly heat. *Nat Clim Change.* 2017;7:501–506. doi:10.1038/nclimate3322

54. Mazdiyasni O, AghaKouchak A, Davis S, et al. Increasing probability of mortality during Indian heat waves. *Sci Adv.* 2017;3:e1700066. doi:10.1126/sciadv.1700066

55. Patz JA, Frumkin H, Holloway T, et al. Climate change: challenges and opportunities for global health. *JAMA.* 2014;312:1565–1580. doi:10.1001/jama.2014.13186

56. Simon M. Why the Arctic is warming so fast, and why that's so alarming. *Wired.* June 23, 2020. Accessed August 10, 2020. https://www.wired.com/story/why-the-arctic-is-warming-so-fast/

57. Vincent WF. Arctic climate change: local impacts, global consequences, and policy implications. In: Coates K, Holroyd C, eds. *The Palgrave Handbook of Arctic Policy and Politics*: Palgrave Macmillan; 2020:507–526.

58. Gasparrini A, Guo Y, Sera F, et al. Projections of temperature-related excess mortality under climate change scenarios. *Lancet Planet Health.* 2017;1:e360–e367. doi:10.1016/S2542-5196(17)30156-0

59. National Weather Service. National Weather Service New York, NY excessive heat page. Accessed June 19, 2021. https://www.weather.gov/okx/excessiveheat

60. Centers for Disease Control and Prevention. Heat-related deaths—Chicago, Illinois, 1996-2001, and United States, 1979-1999. *MMWR Morb Mortal Wkly Rep.* 2003;52(26):610–613. July 4, 2003. Accessed June 19, 2021. https://www.cdc.gov/mmwr/preview/mmwrhtml/mm5226a2.htm

61. Environmental Protection Agency. Chicago, IL adapts to improve extreme heat preparedness. January 13, 2021. Accessed June 19, 2021. https://www.epa.gov/arc-x/chicago-il-adapts-improve-extreme-heat-preparedness

62. City of Chicago. Climate Action Plan. Accessed August 10, 2020. https://www.chicago.gov/city/en/progs/env/climateaction.html

63. Helfand J. *Cooked: Survival by Zip Code.* Judith Helfand Productions and Kartemquin Films. 2019.

64. Ready. Extreme heat. May 26, 2021. Accessed June 19, 2021. https://www.ready.gov/heat

65. Hess JJ, Sathish LM, Knowlton K, et al. Building resilience to climate change: pilot evaluation of the impact of India's first heat action plan on all-cause mortality. *J Env Public Health.* 2018:7973519. doi:10.1155/2018/7973519

CHAPTER 5

Extreme Weather

"Extreme weather is becoming terrifyingly ordinary as the climate changes."[1]

KEY TERMS

Extreme weather
Flooding
Tropical cyclones

Drought
Wildfires
Sea level rise

Planned relocation
Sendai Framework

LEARNING OBJECTIVES

- Describe various forms of extreme weather and how each is fueled by climate change.
- Understand the human health impacts of floods, storms, drought, and wildfires.
- Describe how and why certain populations are disproportionately impacted by extreme weather.
- Describe how sea level rise is forcing coastal communities to relocate.
- Describe examples of extreme weather adaptations that will reduce impacts on human health.

The year 2020 saw unprecedented extreme weather events that cost billions of dollars and caused untold human suffering. The 2020 Atlantic hurricane season set a record, with 30 named tropical storms—13 of which were hurricanes, the second highest number on record. All but one of these hurricanes made landfall in the United States, and nearly the entire U.S. Atlantic coast was under a tropical watch or warning at least once that year for only the second time ever.[2] The final hurricanes of the season, Iota and Eta, killed hundreds of people and displaced at least half a million in Central America. Overall, the 2020 Atlantic storm season saw at least 400 fatalities and $43 billion in damages.[3]

Also in 2020, western U.S. states experienced record-setting wildfires, including five of the six largest wildfires in California's history. Over 9,600 California wildfires burned more than 6,500 square miles, killed dozens, and destroyed or damaged more than 10,000 structures.[4] Colorado saw three of the largest wildfires in its history, which forced tens of thousands of people to evacuate. Drought conditions affected half

Figure 5.1 Floodwaters of the Yangtze River Pass through Chongqing, China, August 19, 2020.
© dyl0807/Shutterstock.

the United States in 2020, which contributed to extreme wildfire risks.[5] Devastating wildfires in Australia, which raged from September 2019 to February 2020, burned more than 93,000 square miles, killed at least 33 people, and sickened millions as a result of blanketing toxic smoke.[6]

Deadly and destructive flooding from extreme monsoon rainfall in China (**Figure 5.1**) and Cyclone Amphan in India and Bangladesh displaced millions of people and were two of the costliest disasters in 2020.[3] In southern Japan, record precipitation caused floods and landslides in 2020 that killed more than 80 people, destroyed dykes and bridges, and displaced over a million people.[3] Worldwide, natural disasters in 2020 claimed 8,200 lives and cost $210 billion, 61% of which were uninsured losses.[3]

Extreme weather has always occurred periodically, but climate change is making these events more frequent, more intense, and more destructive. Extreme weather incidence is on the rise, up to 6 times more common than in 1980.[3] Hundreds of extreme weather events occur each year, mostly hydrological (floods) and meteorological (precipitation, tropical cyclones, other storms, and drought), but also climatological (extreme temperatures and wildfires).[3] Over the past century, the continent of Asia experienced the greatest number of extreme weather events (nearly 4,000) and recorded by far the most deaths attributed to these disasters (17.8 million).[7] According to the United Nations (UN), in the two decades between 2000 and 2019, 1.65 billion people were affected by floods, 1.43 billion by droughts, and 727 million by storms (**Figure 5.2**).[8]

Though often referred to as "natural disasters," extreme weather events are anything but "natural" and only become "disasters"

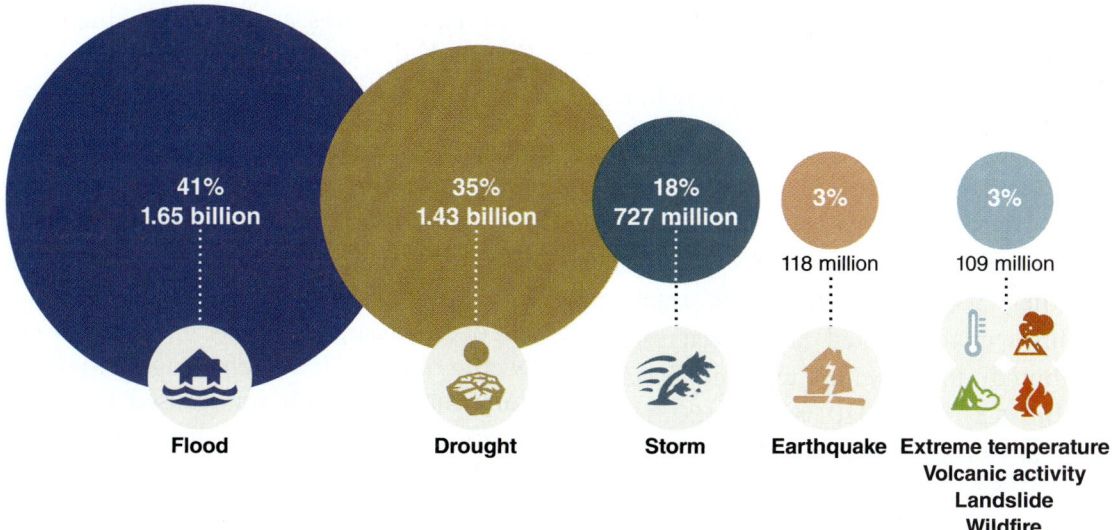

Figure 5.2 Number of People Affected Globally by Various Types of Disasters, 1998–2019. Floods, droughts, and storms affect the most people.

when they intersect with conditions that leave communities vulnerable and unprepared, leading to human suffering and widespread property damage. Certain populations are more susceptible to extreme weather hazards because of coastal location, local weather patterns, degraded land cover or ecosystems, fragile infrastructure, inadequate health systems, preexisting health conditions, and lack of rescue and emergency management operations. Individuals at particular risk include those in outdoor or rescue occupations, children, the elderly, women, people of color, people with disabilities, and those with low incomes and/or low social connectivity. In addition, people with chronic diseases are at particular risk when exposure to an extreme weather event disrupts essential medical treatments, including chemotherapy for cancer, dialysis for kidney disease, oxygen or medications for respiratory diseases, and medications for high blood pressure.[9] Disasters result from *cascading failures*—when many interconnected systems fail to protect affected populations. Inequalities that existed before extreme weather events are often magnified afterward.[10]

All forms of extreme weather are capable of causing significant human illness, injury, suffering, and death, along with economic hardships. Health impacts of extreme weather may be **direct**, such as drowning in floodwaters, blunt trauma from falling trees or floating debris, and motor vehicle accidents caused by storms, or **indirect**, such as infectious diseases, adverse pregnancy outcomes, and mental trauma. In 2016, the U.S. Global Change Research Program projected three key future health impacts of climate change–related extreme weather exposures[9]:

- Increased illness, injury, and death, exacerbation of preexisting health conditions, and adverse mental health impacts
- Disrupted infrastructure, including water systems, power, transport, and communications necessary to protect health and

maintain access to emergency services and health care

- Increased exposure to drought, wildfires, and flooding, particularly among vulnerable people

On average, more than 100 disasters hit the United States each year.[10] The National Weather Service (NWS) keeps track of fatalities in the United States that can be attributed to specific weather hazards (**Figure 5.3**).[11] In 2020, nearly 400 deaths were attributed to extreme weather, along with many injuries and billions of dollars in property damage.[11] Heat and floods had the highest 10- and 30-year average deaths. The year 2005 saw the most weather fatalities in the United States, including more than 1,000 deaths attributed to Hurricane Katrina.[11]

These data likely significantly underestimate actual mortality burdens or trends, however, because each death must be directly attributed to the specific weather event. For example, a death could be recorded using the *International Classification of Diseases* codes for "exposure to excessive natural heat," "victim of cataclysmic storm" (e.g., hurricane) or "victim of flood."[12] The challenge is to determine that the death was caused by direct forces of the disaster (e.g., heat, rain, flood, wind, structural collapse, or flying debris). In addition, most health effects are nonfatal and tend to go unreported as having resulted from extreme weather exposures.

This chapter summarizes the ways that human health is affected by four major

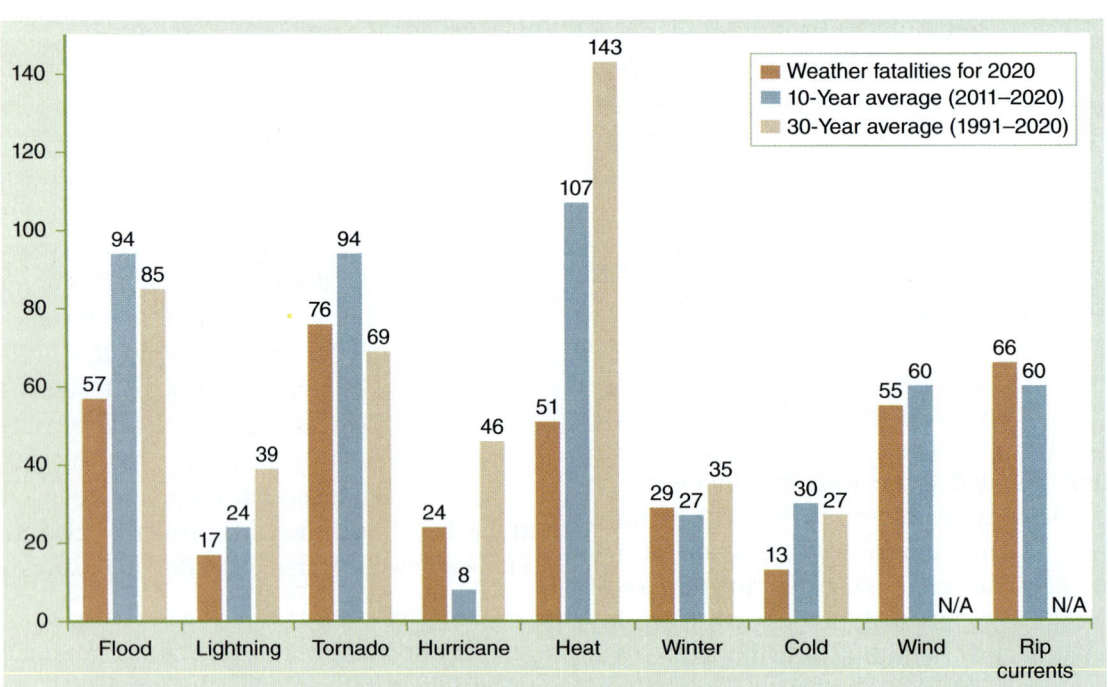

Figure 5.3 Fatalities in the U.S. from Extreme Weather in 2020.

types of extreme weather events: flooding, tropical cyclones, drought, and wildfires. Additional health impacts of extreme weather related to heat, water, air quality, agriculture, vector-borne diseases, mental health, and human displacement are discussed elsewhere in this text.

Flooding

Flooding is the natural hazard with the greatest public health, social, and economic costs. Flooding results mostly from extreme precipitation, hurricanes or other storms, sea level rise, coastal storm surges, rivers overflowing their banks, or sewer or stormwater system backups. In the United States in recent years, increasing amounts of precipitation have come in the form of intense single-day rain events. In fact, 9 of the top 10 extreme single-day precipitation events have occurred in the years since 1996.[13] In many regions of the world, increased precipitation during the wettest months of the year and heavy precipitation events are rising, and they raise the risk of flooding damage.[14] Natural and built environments; patterns of land development; protective land cover such as estuaries, wetlands, and mangroves; and urban water management systems play a large role in whether or not people are directly exposed to floodwaters.

People at disproportionate risk include the elderly. For example, half of the deaths attributed to Hurricane Katrina and Hurricane Sandy were in people over 65.[9] Others at risk include the poor, people of color in urban areas, renters, those with mobility challenges, those with low levels of social and political empowerment, and those without health insurance, who tend to be at greater risk of being admitted to the hospital as a result of extreme weather exposure.[9]

More fatalities occur in males, who make up a large proportion of first responders, emergency workers, and those in occupations with high flood exposure risk, such as law enforcement, public utilities restoration, and construction.[9]

Flooding is particularly destructive in major urban areas, and a recent analysis of urban flooding events in the United States revealed that Black and low-income neighborhoods face disproportionate harm (**Figure 5.4**).[15] Despite the well-known connection between climate change and flood risk, urban development continues to create more impervious surfaces and neglect aging sewer and stormwater systems that may become overwhelmed by increasing amounts of water. Residents of the lowest lying flood-prone areas or neighborhoods without green space to absorb floodwaters tend to be low-income and people of color.[15]

Many deaths in the United States attributed to flooding are due to drowning associated with flash floods, when people become stranded or are swept away when driving or walking near or through floodwaters.[9] Other *direct* impacts include blunt trauma injuries from falling debris or objects moving quickly in floodwaters; motor vehicle accidents when roads are wet, obstructed, or damaged as a result of the storm; electrocution by fallen live electrical wires; cuts, puncture wounds, sprains, and other injuries; hypothermia; and bites from animals in floodwaters.

Indirect health impacts of floods include infectious diseases, such as diarrheal diseases, skin rashes, and infections of the eye, ear, nose, throat, lungs, and skin that result from exposure to floodwaters contaminated with biological pathogens.[9] Standing floodwaters create habitats for mosquitoes, which may result in increased transmission

Figure 5.4 Flooding in an Urban Area, Showing the High Degree of Paved and Impervious Surfaces That Prevent Floodwaters from Being Absorbed.
© Matthew Davidson/EyeEm/Getty Images.

of vector-borne diseases such as dengue and malaria. In flooded homes and buildings, toxic molds may grow, exposure to which may cause or exacerbate allergy symptoms, asthma, other respiratory illnesses, and lung infections. If floodwaters become contaminated with chemical pollutants—as often happens when hurricanes strike the U.S. Gulf Coast, where fossil fuel refineries and petrochemical plants are concentrated—many acute and chronic health impacts associated with pollutant exposures may occur. Floods and storms also create numerous occupational health risks. Most storm-related

fatalities are associated with cleanup, followed by construction, public utilities restoration, and security.[9]

Flood exposure is linked to a number of adverse birth outcomes. For example, a study in river-dependent communities in Amazonia found that exposure of pregnant women to extremely intense rainfall was associated with preterm birth and lower mean birth weight.[16] Floods may create conditions in which children become separated from family members or caregivers, which increases child vulnerability to injury, disease, psychological trauma, abuse, and even death, particularly for children with disabilities or special needs. Flood-related adverse mental health impacts occur with direct exposure to floodwaters, loss of loved ones or property, displacement, and economic hardship.[9] In many disaster situations, adequate mental health support is often lacking or disrupted.

Floods often damage critical infrastructure needed for human safety, including roads, bridges, water and sewer systems, electricity grids, and hospitals and clinics. Power failures impact health if critical medical equipment fails or if indoor heat or food refrigeration is disrupted. In addition, use of generators, portable heaters, or stoves increases the risk of carbon monoxide poisoning. Interruption of public water supplies decreases the availability of clean water, and disrupted waste management increases the threat of exposure to unsafe water, which may lead to increased waterborne illnesses.[9]

One of the most severe flooding events in recent history occurred in Pakistan in July 2010. Monumental rainfall flooded more than one-fifth of the country's land area, resulting in over 1,100 casualties and affecting an estimated 20 million people (**Figure 5.5**). This event led to widespread destruction of infrastructure and disrupted access to electricity, clean drinking water, and sanitation. Reported health impacts included diarrheal diseases, acute malnutrition in children, respiratory infections, skin diseases, and outbreaks of malaria, dengue, cholera, measles, diphtheria, and polio.[17] Disrupted care resulted in increased tetanus, measles, encephalitis, and heat-related deaths in newborns. After experiencing flooding, many people suffered significant adverse mental health impacts, including anxiety, depression, helplessness, aggression, and certain phobias.[18,19] Also observed in some people was **posttraumatic growth**, the strengthening of mental resilience as people processed their experiences of the disaster, recognized personal strengths, and improved emotional responses, relationships, and opportunities for change.[18,19]

One of the largest flooding events in the United States in recent years occurred in 2017 in the greater area of Houston, Texas, as a result of the record-setting 50+ inches of rainfall from Hurricane Harvey (**Figure 5.6**). At least 70 people died as a direct result of the storm, 81% from drowning (mostly in or near vehicles) and the rest from electrocution, disrupted medical treatment for preexisting conditions, or physical trauma.[20] Physical health problems, including headaches, allergies, and skin, lung, eye, nose, and throat irritations, were more commonly reported among people who did not evacuate ahead of the storm.

Overall, the extent of flooding from Harvey and exposure to floodwaters were greater in neighborhoods with higher proportions of Black, Latino, and low-income residents.[21] Three weeks after the storm, nearly half of Greater Houston residents reported probable posttraumatic stress disorder, which decreased to 16% three months later.[22] Black residents had higher odds of posttraumatic symptoms than white residents, as did people living in households where someone lost a job after Harvey.[23] As with the 2010 flooding in Pakistan, posttraumatic growth was also observed among some people who experienced Harvey.[22] Reduced healthcare access was reported after the storm for people experiencing job loss and people with disabilities.[23] One year after the storm, food insecurity was higher among Black and Latino people compared with white people.[24]

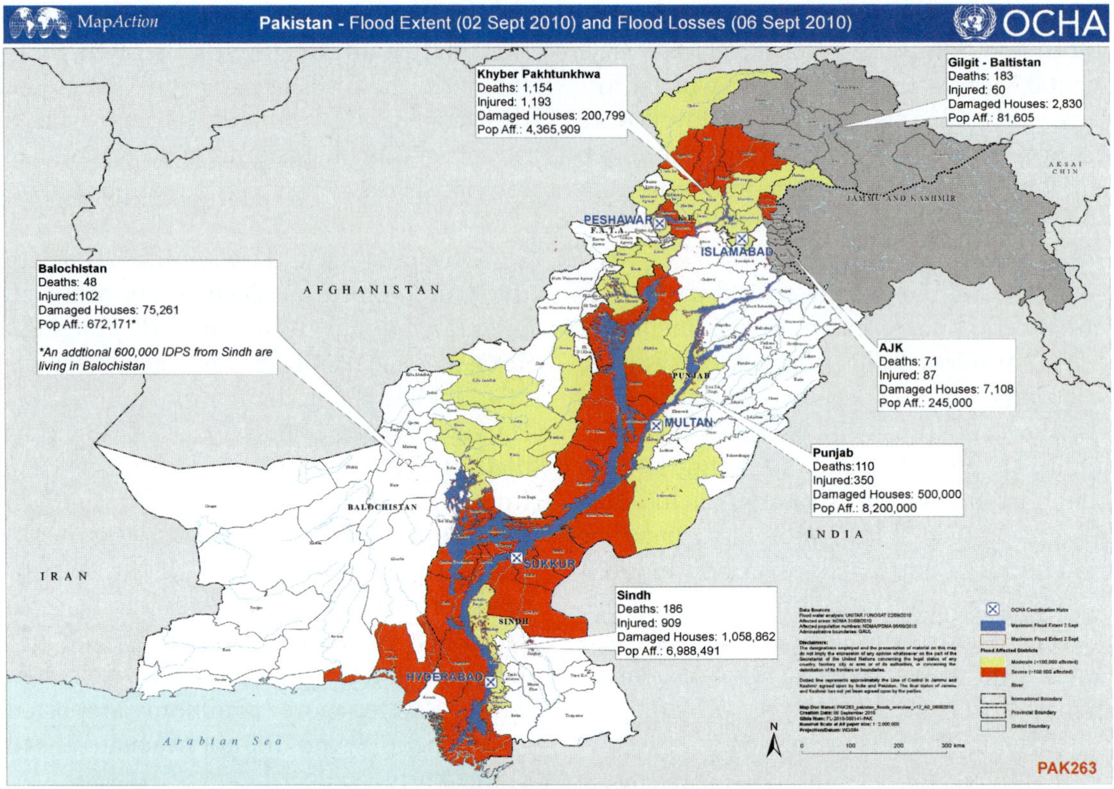

Figure 5.5 Top: Map of Pakistan Showing Extent of 2010 Floods and Human Impacts. Bottom: Aerial View of a Village Inundated by Floodwaters Near Multan, Pakistan, August 24, 2010.

Top: Courtesy of United Nations Office for the Coordination of Humanitarian Affairs (OCHA); Bottom: © Visual News Pakistan/Moment/Getty Images.

Figure 5.6 Left: People Navigate Flooded Streets of Houston, Texas, after Hurricane Harvey, August 27, 2017. Right: Debris Lines the Streets from Homes Damaged or Destroyed in Hurricane Harvey.

Left: © michelmond/Shutterstock; Right: © MDay Photography/Shutterstock.

Tropical Cyclones

Tropical cyclones, organized rotating systems with winds of at least 74 mph that form over warm ocean waters, are some of the most extreme storms that occur on Earth. These storms are called **hurricanes** in the North Atlantic, central North Pacific, and eastern North Pacific; **typhoons** in the Northwest Pacific; and **cyclones** in the South Pacific and Indian Ocean. Tropical cyclones are powered by warm, humid air over the ocean, and for every 1°C (1.8°F) the atmosphere warms, the air holds about 7% more water.[1] Rain falls within the cyclone when water vapor in the air condenses.

The strength of a storm is categorized from 1 to 5 on the Saffir-Simpson Hurricane Wind Scale, with Category 5 being the strongest. Tropical cyclone activity can be measured by frequency of storms in a single season or by storm intensity, including how many *major* storms occur (Category 3 or higher), the time duration of storms, the rate of intensification of storm strength, the rainfall rate in storms, or the **Accumulated Cyclone Energy** (**ACE**) index, which describes the maximum sustained surface wind speed over the duration of the storm when it is at tropical storm intensity or higher.[25]

Many ocean basins around the world are susceptible to tropical cyclone activity. According to the U.S. National Oceanic and Atmospheric Administration (NOAA) and the Japan Meteorological Agency, during the years 2011–2019, 40–50 tropical cyclones per year consistently formed in the busiest ocean basins around the world.[25,26]

One of the main drivers of the unusually active 2020 Atlantic hurricane season was very warm sea surface temperatures. In addition, 2020 saw La Niña conditions, the cold extreme of the El Niño–Southern Oscillation (ENSO) that creates atmospheric conditions conducive to North Atlantic cyclone formation.[3] Unique features of the hurricanes in 2020 were rapid intensification before landfall and formation of storms late in the season. Seven cyclones formed in October and November, including the most intense storm, Category 5 Hurricane Iota, which was the final storm of the season.[3]

Certain characteristics of tropical cyclones have shifted in recent decades, possibly as a result of climate change. **Figure 5.7** shows the Power Dissipation Index (PDI),

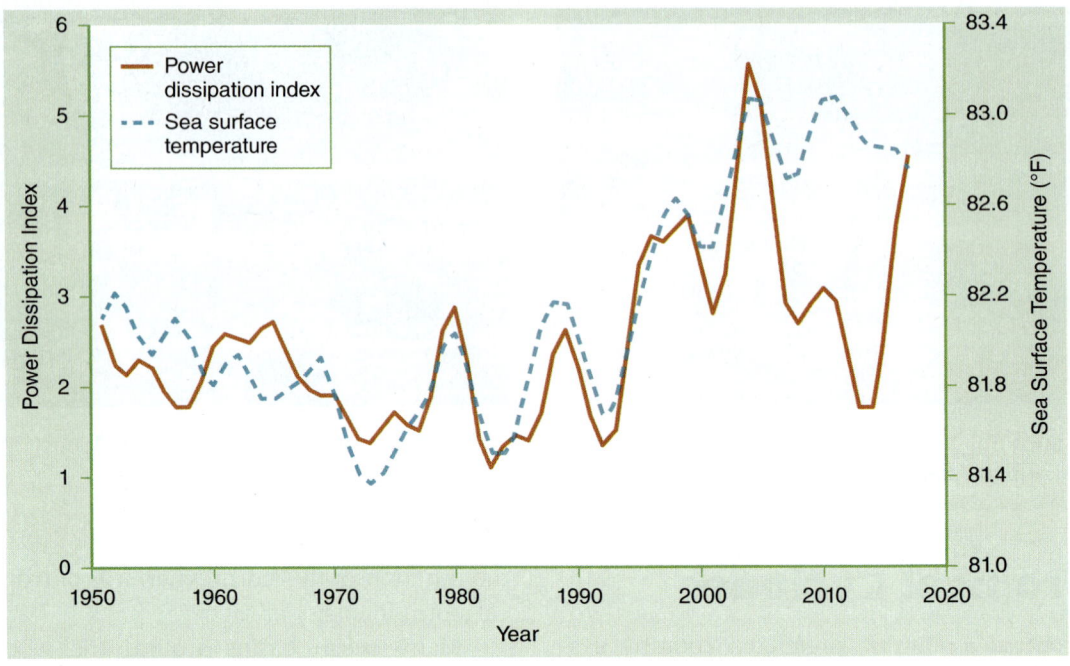

Figure 5.7 Solid Line: Power Dissipation Index Indicating Tropical Cyclone Activity in the North Atlantic, 1950–2020. Dashed Line: Sea Surface Temperatures in the Tropical North Atlantic over the Same Period. Lines have been smoothed using a five-year weighted average, plotted at the middle year.

Reproduced from Environmental Protection Agency. Climate change indicators: tropical cyclone activity. Updated April 14, 2021. Accessed June 19, 2021. https://www.epa.gov/climate-indicators/climate-change-indicators-tropical-cyclone-activity

which measures cyclone activity by accounting for cyclone strength, duration, and frequency, for tropical storms in the North Atlantic since 1950.[27] PDI varies from year to year but has increased since 1995, along with sea surface temperature and ACE, suggesting a trend toward higher cyclone *intensity*. In addition, the proportion of storms that reach at least Category 3 strength has increased over the past four decades.[28] The year 2020 was record-setting for the number of tropical cyclones in the North Atlantic, but it is not clear whether the *frequency* of storms is increasing. Many storms do not make landfall and are not as well characterized, and less is known about storm frequency in the years before satellite data were available.[28]

Globally, future hurricane *intensity* and *rainfall rates* are predicted to increase. According to NOAA, climate change will impact tropical cyclone dynamics in several ways[28]:

- The amount of rainfall in a tropical cyclone will increase, with models suggesting a 10–15% increase in rainfall rates within about 100 kilometers of the storm with 2°C warming.
- Global cyclone intensity will increase, perhaps by 1–10% with 2°C warming. There is some evidence, but much uncertainty, that there could be more Category 4 and 5 storms in the future.
- Sea level rise will increase coastal inundation during storms.

The two most destructive features of tropical storms are high rainfall levels and storm surges. Both contribute to flooding, which is responsible for most health impacts associated with these storms. Coastal flooding is particularly dangerous, but inland impacts also occur when storms drop an excessive amount of rainfall and when rainwater flows back toward the ocean, overwhelming riverbanks and drainage channels.[1] In addition to floodwater exposure, older Americans who experience hurricanes are at increased risk of emergency hospitalizations for respiratory diseases, infectious diseases, and injuries.[29]

Among the countries most vulnerable to tropical cyclones are small island developing states (SIDS), which are formally recognized by the UN as highly disaster-prone. The 29 Caribbean SIDS face extraordinarily high risks of harm from tropical storms in the Atlantic. The warm ocean waters and low wind shear where they are located create favorable conditions for tropical storms. In addition, as mostly small islands, their entire coastlines are vulnerable to storm surges and destructive waves regardless of the direction storms approach from. SIDS are very prone to the flooding effects of sea level rise, which causes prolonged submergence and lowered ability of storm floodwaters to drain back into the ocean. Many SIDS are geographically remote and disconnected, which makes evacuation, rescue, and recovery difficult, particularly in countries made up of many inhabited islands, each in need of relief. SIDS tend to have vulnerable infrastructure and limited economic resources, so they often cannot commit large amounts of money to disaster preparedness and instead rely significantly on external aid.[30]

In 2017, Caribbean SIDS were hit hard by a highly active Atlantic hurricane season that produced 17 named storms, 10 hurricanes, and 6 major hurricanes. Hurricanes Maria and Irma each reached Category 5 status and impacted 16 and 14 SIDS, respectively.[30] In particular, Puerto Rico was significantly damaged by Maria, and Barbuda was rendered uninhabitable by Irma. Residents of these islands experienced many health problems, including drownings, heat-related illnesses and deaths during prolonged power outages that limited artificial cooling measures, exacerbation of chronic diseases requiring ongoing medications and treatments, increased waterborne and mosquito-borne diseases, significant mental health impacts, and population displacement and outmigration to other places (**Box 5.1**).[30] In Puerto Rico, many deaths occurred weeks or months after Hurricane Maria and went unrecorded or unattributed to the storm, so the actual death toll may be much higher than the official tally of more than 2,000.[28] Tens of thousands of Puerto Ricans left the island after Hurricane Maria and came to the mainland United States, many permanently.

One of the deadliest storms of the 2020 Atlantic season was Hurricane Laura, which made landfall in southwestern Louisiana as a Category 4 storm (**Figure 5.8**). Laura tied for the strongest storm ever to hit Louisiana and was the most costly hurricane to hit the U.S. mainland in 2020.[3] About 20 million people were in the path of the storm, and 500,000 evacuated. Two weeks after the storm, the death toll in Louisiana was 26. Most deaths were not related to floodwater exposure, but to carbon monoxide poisoning and excessive heat exposure that resulted from prolonged lack of power.[35] Additional causes of death were falling trees, cleanup injuries, falls, drowning, electrocution, and a house fire. Dozens of people also perished when this storm hit Haiti, the Dominican Republic, and other U.S. states.

Measuring health outcomes in populations exposed to extreme storms is more common in high-income countries and those well prepared for disasters, and resulting data may not always accurately capture the full

---

Box 5.1 Stories from Puerto Rico after Hurricane Maria

Hurricane Maria caused untold suffering among Puerto Ricans, and health impacts were often unrecorded, unseen, or ignored. Many people also created opportunities to support one another, build relationships, and enhance community health resilience. The stories here were told in media reports one or two years after the storm hit.

Marta Rivera lived at the bottom of a hill in Arecibo, and her home was swamped by a big wave during the hurricane. She was rescued and taken to the hospital, which collapsed while she was there. Rivera suffered from diabetes and cancer, and she had medical setbacks when she couldn't get the medication she needed. She reported being hungry and having nightmares about the storm.[31]

Ramona Gonzalez was 59 years old and disabled. She died not as a direct result of the hurricane but from sepsis due to an infected bedsore that was not treated. Her siblings took her to hospitals and a medical ship, but these facilities were overwhelmed. She was bedridden, and her sores worsened at home without power or air conditioning. In general, hospitals ran out of oxygen and lacked fresh water and fuel for generators to run during the prolonged power outage. They often discharged ill patients whom they simply could not treat.[32]

Marilian Vázquez lived in Toaville, and her house and her husband's ice cream truck were heavily damaged. As a result, her husband became depressed and didn't work for two years after the storm. She still felt unsafe two years later because authorities had not made repairs or improvements to the water drainage system in her town.[33]

Madeline Gotay lived in the suburbs of the capital city, San Juan. Her house was destroyed by the hurricane, and her appliances were washed away. She had two sons serving prison sentences, but she had no contact with them for a month, causing her to suffer from grief and anxiety.[31]

Camille Mercado Rivera was a student at the University of Puerto Rico. Her studies were disrupted for more than two months, and she returned to classes amid persistent power outages. She spent much of her time volunteering at a nonprofit kitchen that provided meals to those in need.[31]

In the town of **Mameyes**, residents responded to the devastation of local infrastructure and healthcare services by opening a solar-powered health clinic with support from grants and charitable donations. Previously, residents, many of whom are elderly, had to travel an hour to get care, but this clinic now serves seven nearby towns.[33]

Jackeline Soto Perez relocated to Florida with her husband and two children, one of whom has an immune disorder and needs medicine that must be refrigerated. The extended lack of electricity and threat to her son's health led to the family's decision to leave Puerto Rico. Perez found a job as a home health aide, and her husband was hired as a paramedic. They planned to stay in Florida.[34]

---

range of impacts and experiences in developing countries. In storm-prone regions, tropical cyclone preparedness measures, the most important of which is evacuating people out of harm's way before the storm strikes, are key to reducing morbidity and mortality. In addition, addressing the factors that prevent people from evacuating will result in better protection of at-risk populations. In all situations, effective emergency response and healthcare services are needed for all exposed to extreme storms.

Drought

People have experienced droughts throughout history, but they are increasing in frequency and severity with climate change-driven

Figure 5.8 Top: Satellite Image of Hurricane Laura as It Made Landfall in Louisiana, August 27, 2020. Bottom: Damage Caused by Hurricane Laura in Holly Beach, Louisiana.

Top: Courtesy of National Ocean Service. https://oceanservice.noaa.gov/news/aug20/ngs-storm-imagery -laura.html; Bottom: © Eric Gay/AP/Shutterstock.

warming, extreme heat, altered precipitation, and land desertification. In recent years, locations in East Africa, Australia, and the western United States, to name a few, have experienced unprecedented droughts, with devastating consequences for food production, wildfire risk, and human safety and survival.[36]

Drought can be defined in several ways.[36] A meteorological drought results from a lack of rainfall. A hydrological drought occurs when aquifers, rivers, and other surface water reservoirs become depleted. An agricultural drought occurs when a lack of precipitation reduces soil moisture and crop yields. In the United States, drought conditions are continuously monitored by the National Drought Mitigation Center at the University of Nebraska–Lincoln and made publicly available. The map in **Figure 5.9** shows that in January 2021, much of the western U.S. was under "extreme" or "exceptional" drought conditions.[37] From 2002 to 2005 and from 2012 to 2020, nearly the entire Southwestern United States was under drought conditions.[38]

Persistent drought causes crop failures, wildfires, dust storms, and degraded air and water quality; each of these has human health consequences, such as increased food insecurity, malnutrition, water-related diseases, respiratory illnesses, and adverse mental health conditions.[9] Certain regions, notably East Africa, have repeatedly faced crop failures, food insecurity, and famine due to prolonged drought. In addition, drought may increase certain vector-borne diseases, such as West Nile Virus, and rodent-borne hantavirus infections, as vector and host populations, habitats, and behaviors shift under drying conditions.[9] The health effects of drought tend to be slow-onset, accumulating over time as drought conditions persist, in contrast with rapid-onset impacts that often occur with extreme storms.[9]

Three years of severe drought in Madagascar led to a humanitarian crisis in 2020–2021 in which 135,000 children became acutely malnourished, and over 1 million people were food insecure and in need of emergency food aid.[39] Madagascar is one of the most vulnerable countries to climate stressors, and in the south, seasonal rains have failed for three years.[39] Adding to the problem, the UN World Food Programme (WFP) warned that COVID-19 caused

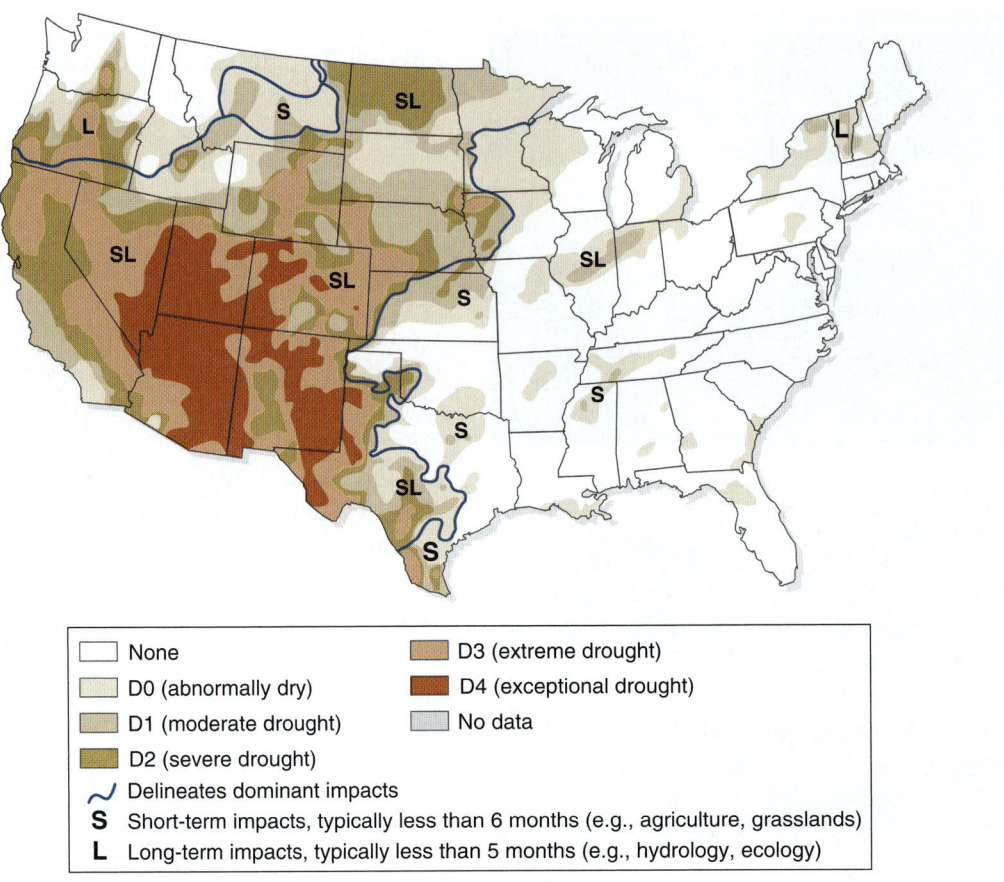

Figure 5.9 Map from U.S. Drought Monitor on January 12, 2021. Red and brown colors indicate extreme or exceptional drought.

Reproduced from National Drought Mitigation Center at University of Nebraska-Lincoln. Map released: January 12, 2021. Updated January 14, 2021. Accessed January 15, 2021. https://droughtmonitor.unl.edu

significant disruptions in food access because lockdown restrictions and the pandemic's economic downturn reduced employment, so people sold their possessions to survive and left their homes to look for food and work. Three-quarters of children in this region dropped out of school to forage for food or beg in the streets. People were forced to eat "cactus or tamarind mixed with mud, roots, whatever they can find, leaves, seeds, whatever is available," said regional WFP director Lola Castro.[39] In early 2021, WFP called for raising $35 million to pay for food aid, cash distribution schemes, and school feeding programs so that children could stay in school (**Figure 5.10**).

Wildfires

Warming temperatures, prolonged drought, drying vegetation, and increased human settlement at the wildland–urban interface have increased the risk of **wildfires** and their deadly consequences. More people are moving into fire-prone wild areas, with more cars, power lines, and other ignition sources, and clearing land has eliminated forest undergrowth that would naturally burn in

Figure 5.10 The UN World Food Programme Aimed to Set Up a School Feeding Program in 2020 Similar to One Created in 2018 in the Amboasary-South District of Southern Madagascar.
© RIJASOLO/AFP/Getty Images.

smaller fires. Over the past four decades in the United States, the amount of land burned by wildfires has roughly quadrupled,[40] and in some parts in the West, wildfires are the most visible manifestation of climate change and a significant health risk.

Wildfires in California now burn 5 times as much area as fires in the 1980s.[41] More than 4% of California's land burned in wildfires in 2020, an area of about 4 million acres.[42] Poor wildland fire management has been a factor, but warmer temperatures have also been significant drivers of wildfires. Warm air holds more moisture and draws water out of vegetation and soil, leading to conditions that increase combustibility and fire risk.

The most destructive and deadly California wildfire was the Camp Fire in 2018, which killed 85 people and destroyed nearly 19,000 structures, including most of the buildings in the towns of Paradise and Concow (**Figure 5.11**).[42] The fire scorched Paradise in just a few hours, killing dozens of people who were caught in their burning vehicles as they tried to escape. Few people

reached the local hospital, and those who did had to evacuate as the flames approached rapidly, and the hospital was threatened. The Camp Fire was ignited by a 100-year-old faulty electric transmission line and spread by high winds through an area suffering from severe drought.[42]

Devastating wildfires struck other western U.S. states in 2020, including Oregon, where about 4,000 homes were damaged or destroyed, and the highest air pollution levels ever recorded affected the majority of the state's population.[3] At least 38 million people in California, Oregon, Washington, Idaho, and Montana were exposed to dangerous levels of wildfire smoke for at least five days in 2020, and they suffered difficulty breathing, asthma attacks, suppressed immune function, and even premature death.[43]

Wildfire smoke is a complex mixture of particulate matter, carbon monoxide, and volatile chemicals, many of which are carcinogens and respiratory irritants. The precise composition depends on which vegetative materials are burning and at what temperatures, and smoke may also contain toxic chemicals from building materials.

Figure 5.11 Fire Rages through Neighborhoods of Paradise, California, as the Camp Fire Burns Out of Control, November 8, 2018.
© PETER DASILVA/EPA-EFE/Shutterstock.

Particulate matter increases respiratory diseases and raises the risk of respiratory and cardiovascular mortality. Carbon monoxide exposure causes headaches, dizziness, visual and cognitive impairment, reduced dexterity, chest pain, cardiac arrhythmias, and, at lethal concentrations, asphyxiation and death.[44] During wildfires in San Diego in 2007, excess emergency department visits were reported for asthma, chronic obstructive pulmonary disease, and chest pain.[9]

Wildfire smoke may also impair driving visibility, which raises the risk of motor vehicle accidents. Following wildfires, drinking water supplies may become unsafe if runoff from charred land brings contaminants into water reservoirs.[9] Wildfires near populated areas often require large-scale evacuations, which lead to increased need for shelter, medical treatment, and public health services. People may also be affected by wildfire smoke even if they live far from the burning fires. For example, forest fires in Quebec in 2002 resulted in up to a 30-fold increase in airborne fine particulate concentrations in Baltimore, Maryland—a city nearly 1,000 miles downwind.[9] The 2020 western U.S. wildfires blew smoke plumes across the entire country.

People who work outdoors and around wildfire smoke are also at elevated risk. Firefighters working on wildlands now spend on average 100 days each summer fighting wildfires, and retired wildland firefighters are at higher risk of lung cancer and heart disease.[45] Overall, reducing wildfire smoke exposure is estimated to avert thousands of premature deaths each year in the United States.[40] The U.S. Clean Air Act has been very successful over the past 50 years in significantly reducing outdoor air pollutant levels, but increasing wildfire frequency and intensity threaten this progress.

As with other types of disasters, wildfires can cause significant adverse mental health conditions. The 2016 Fort McMurray Wildfire in Alberta, one of the most destructive Canadian wildfires in history, forced the evacuation of 90,000 residents while burning more than 1 million acres and 2,500 homes. After the wildfire, an eightfold spike in prevalence of generalized anxiety disorder and a fourfold spike in major depressive disorder were observed among fire victims compared with the general population.[46]

The 2017 Tubbs Fire, one of the deadliest California wildfires in history, destroyed more than 2,500 structures in the city of Santa Rosa and was followed a year later by new wildfires.[42] This repeated fire exposure caused significant stress in residents. Said one resident, Danielle Bryant, "The orange-tinged sky is just enough to set off my anxiety and feelings of fear. Just seeing the smoke . . . I get this sense of dread. You can just feel it. There's a sense of tension here in Santa Rosa."[47] Francis Fuchs, a psychologist and counselor in Santa Rosa, reported that fire victims were "having more difficulty with sleeping, . . . a heightened sense of anxiety and unease, . . . [and] some flashbacks of their fire experience from last October. Also mood changes—more anxious or tearful."[47] Local therapist Mary Wood commented, "It was an absolute trauma for everybody involved. The fire is over, but the grief may last a long time."[48]

In Santa Rosa, thousands of residents took advantage of a website and phone app to self-assess potential trauma symptoms and connect to mental health resources. In addition, volunteers and mental health professionals in the area responded quickly by setting up pop-up clinics and offering mental health education and free counseling services. Christian Burgess, director of the National Disaster Distress Helpline (a confidential 24-7 call and text service), reported that during wildfires, the hotline heard "deeper mental health concerns from callers and texters, such as persistent anxiety, depression, and substance abuse, which can be related to traumatic exposure during the event, loss of loved ones, including pets, and financial strain."[48]

Sea Level Rise

Global mean sea level has risen 8–9 inches (21–24 cm) since 1880,[49] and in 2020, sea level was 3.8 inches (97 cm) higher than the average in 1993, when the satellite record began.[50] Seas are rising mostly because of meltwater from glaciers and ice sheets, and thermal expansion of warming sea water.[46] **Sea level rise** is causing increased storm surges, coastal flooding, saltwater intrusion, and salinization of drinking water and agricultural soils—all of which have significant human health impacts, including mental illnesses, adverse pregnancy outcomes, malnutrition, and high blood pressure.

When Hurricane Sandy hit New York City in 2012, the city experienced its highest water level in at least 300 years.[51] Climate change may not have made this storm more intense, but sea level rise associated with anthropogenic climate change expanded the flood area to affect an additional 71,000 people and caused $8 billion in damage that likely would not have occurred in a world without climate change.[51] In general, the flooding impacts of tropical cyclones are being amplified by sea level rise.

Human retreat from the lowest lying coastal areas around the world is all but inevitable in the not-too-distant future, and it is estimated that 13 million people in the United States alone will have to move away from the coast by 2100.[52] Migration inland is happening today in many parts of the world, largely unmanaged and unassisted, especially in poor communities. This movement often comes with increased physical and mental health burdens and disruptions to health care.

A safer approach is **planned relocation**, a coordinated process in which people or communities are moved preemptively with the assistance of governments to protect them from the risks and impacts of environmental disasters. For example, planned relocation (also known as managed retreat) has occurred in Oakwood Beach along the east shore of Staten Island in New York City, which was destroyed by Hurricane Sandy. Nearly half of the deaths due to the hurricane occurred on Staten Island, and many homes in this neighborhood were destroyed. Instead of rebuilding, residents decided to demolish homes and return the area to a natural storm buffer with state and federal funds. In 2017, the city rezoned the former housing area as a Special Coastal Risk District, which limits development.[53] Maui County, Hawaii, which consists of the islands of Maui, Lanai, most of Molokai, and two uninhabited islands, has begun working on a plan for managed retreat and new infrastructure to protect communities from the impacts of rising sea levels.[54] In late 2020, Maui County initiated a lawsuit against major oil companies, *County of Maui v. Chevron*, in an attempt to force them to pay for this costly retreat program. In Florida, six million residents will need to move inland by century's end to avoid inundation by rising seas. Florida has 40% of the riskiest coastal land in the United States, but lags behind other states in relocating people away from the coast.[52]

Adaptations

As with other climate change impacts, communities at risk of exposure to extreme weather will benefit from aggressive greenhouse gas emissions mitigation to minimize the greenhouse effect and resulting warming that fuels extreme weather. Building individual- and community-level adaptive capacity to prepare for extreme weather can moderate health effects, particularly for those most at risk. The U.S. government defines **adaptive capacity** as "the ability of communities, institutions, or people to adjust to potential hazards, to take advantage of opportunities, or to respond to consequences."[9] Strong adaptive capacity, notably as a result of strong psychosocial support and financial resources for affected communities, contributes to **resilience**, "the ability to prepare and plan for, absorb, recover from, and more successfully adapt to adverse events."[9] This is particularly important in developing countries, which might need assistance for disaster risk reduction measures. Key adaptive measures that may reduce exposure to flood hazards include emergency preparedness systems, effective public communications, and flood risk management.[55]

In 2020, the U.S. Federal Emergency Management Agency collaborated with other government agencies and the private sector to release the **National Risk Index**, an online tool to identify communities in the United States at highest risk of natural hazards (**Figure 5.12**).[56] This index is designed to help planners, emergency managers, other decision-makers, and the public take action to reduce disaster risk at local, regional, state, and federal levels. This visual tool shows that Southern California and much of Florida are at the highest risk.

At the international level, the **Sendai Framework** "aims to achieve the substantial reduction of disaster risk and losses in lives, livelihoods and health and in the economic, physical, social, cultural and environmental assets of persons, businesses, communities and countries over the next 15 years."[57] The Framework was adopted at the Third UN World Conference on Disaster Risk Reduction in Sendai, Japan, on March 18, 2015. It is designed to work together with the Paris Agreement and the Sustainable Development Goals to provide a roadmap for countries to reduce global disaster risk and resulting impacts on people's health, well-being, and livelihoods.

The Sendai Framework sets seven targets to achieve by 2030 (**Figure 5.13**), which will require significant international cooperation, climate change mitigation, and adaptation.[57] These targets include lowering mortality and the total number of people affected by disasters, reducing economic losses and infrastructural damage, improving country-level disaster risk reduction and early-warning systems, and catalyzing international cooperation. A region-specific collaboration created to guide local- and national-level implementation of the Sendai Framework, along with achieving the Sustainable Development Goals and Paris Agreement commitments, is the 2016 **Framework for Resilient Development in the Pacific.** This collaboration among Pacific island states, headquartered in Fiji, is working toward disaster risk reduction, low carbon development, and improved response and reconstruction.[58]

Many cities and regions around the world that are vulnerable to extreme weather are initiating disaster risk reduction planning to improve preparedness and resilience. For example, Miami, Florida, one of the most hurricane-prone cities in the world and also highly threatened by sea level rise, has developed a 44-page hurricane readiness guide

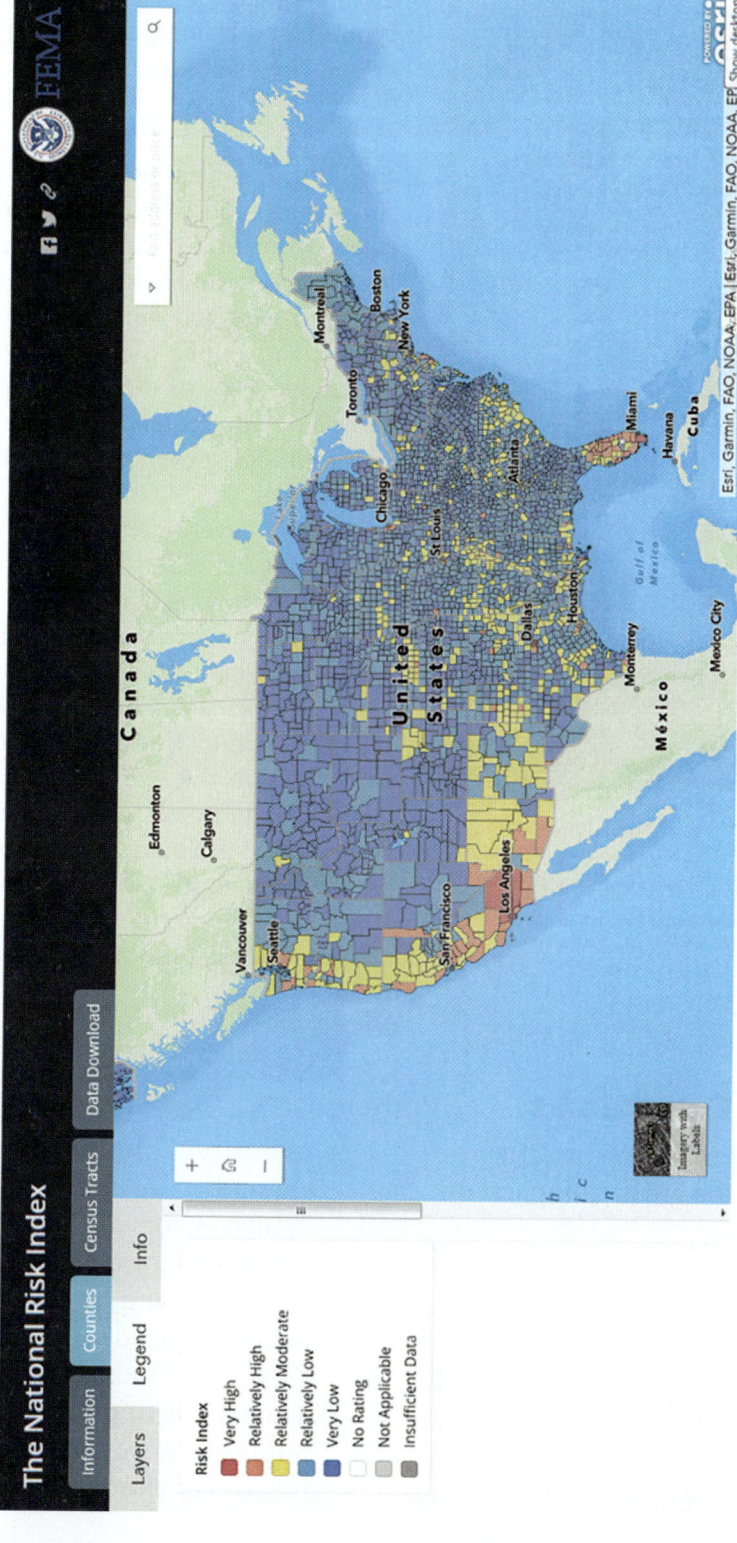

Figure 5.12 Map of the U.S. National Risk Index for Disasters. Counties in orange and red are at high risk.

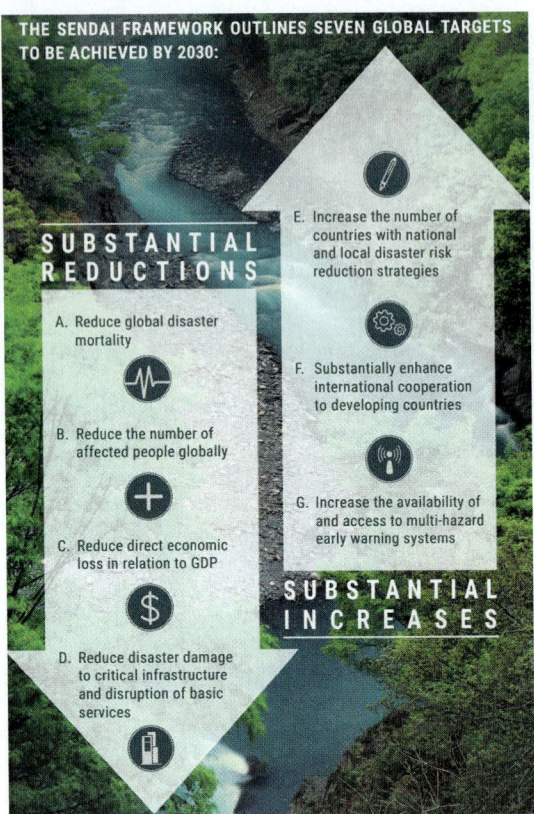

THE SENDAI FRAMEWORK OUTLINES SEVEN GLOBAL TARGETS TO BE ACHIEVED BY 2030:

SUBSTANTIAL REDUCTIONS

A. Reduce global disaster mortality

B. Reduce the number of affected people globally

C. Reduce direct economic loss in relation to GDP

D. Reduce disaster damage to critical infrastructure and disruption of basic services

E. Increase the number of countries with national and local disaster risk reduction strategies

F. Substantially enhance international cooperation to developing countries

G. Increase the availability of and access to multi-hazard early warning systems

SUBSTANTIAL INCREASES

Figure 5.13 Seven Targets of the UN Sendai Framework to Reduce Disaster Impacts on Vulnerable Populations around the World.

Reproduced from United Nations Office for Disaster Risk Reduction. What is the Sendai Framework for Disaster Risk Reduction? Accessed January 15, 2021. https://www.undrr.org/implementing-sendai-framework/what-sendai-framework

for the public in three languages (English, Spanish, and Haitian Creole) that is available on the city's website.[59] Miami is currently in the planning stages of a $4.6 billion project to protect residents from coastal flooding and storm surges by building floodwalls, surge gates, and water pumps, and planting extensive natural protective barriers, including mangroves and oyster reefs.[60] The city also hopes to raise and flood-proof homes, businesses, and public buildings in particularly vulnerable areas of the city. When finalized, the plan will require approval and funding from local governments and the federal government.

Many governmental and nongovernmental stakeholders play critical roles in disaster preparedness and emergency response, communications, recovery, and delivery of medical and public health services that increase adaptive capacity at local, regional, national, tribal, and international levels.[9] Given the enormous human health and economic costs of extreme weather, a wide range of climate adaptations must be rapidly developed and deployed in ways that effectively and equitably protect communities all over the world.

Discussion Questions

1. Which types of extreme weather events have the greatest impacts on health? Why?

2. How is mental health affected by experiences of extreme weather?

3. Discuss why certain populations are disproportionately impacted by extreme weather.

4. What do we currently know about how hurricane intensity is shifting with climate change?

5. Which factors are creating optimal conditions for increased wildfire formation, and how are people being impacted?

6. Discuss the most effective adaptations to protect at-risk populations from the health impacts of extreme weather.

References

1. Naranjan A. Extreme weather explained: How climate change makes storms stronger. November 6, 2020. Accessed January 15, 2021. https://www.dw.com/en/climate-storms-cyclones-hurricanes-typhoons-explained/a-55521226

2. Masters J. A look back at the horrific 2020 Atlantic hurricane season. *Yale Climate Connections*. December 1, 2020. Accessed January 15, 2021. https://yaleclimateconnections.org/2020/12/a-look-back-at-the-horrific-2020-atlantic-hurricane-center/

3. Munich Re. Record hurricane season and major wildfires—The natural disaster figures for 2020. January 7, 2021. Accessed January 15, 2021. https://www.munichre.com/en/company/media-relations/media-information-and-corporate-news/media-information/2021/2020-natural-disasters-balance.html

4. CalFire. 2020 Incident Archive. 2021. Accessed January 5, 2021. https://www.fire.ca.gov/incidents/2020/

5. Fountain H, Migliozzi B, Popovich N. Where 2020's record heat was felt the most. *The New York Times*. January 14, 2021. Accessed January 15, 2021. https://www.nytimes.com/interactive/2021/01/14/climate/hottest-year-2020-global-map.html

6. Brackett R. 2020's worst environmental disasters, and how climate change played a role. *The Weather Channel*. December 20, 2020. Accessed January 15, 2021. https://weather.com/news/news/2020-11-30-2020-year-in-review-worst-environmental-disasters-climate-change

7. Rataj E, Kunzweiler K, Garthus-Niegel S. Extreme weather events in developing countries and related injuries and mental health disorders—a systematic review. *BMC Public Health*. 2016;16(1):1020. doi:10.1186/s12889-016-3692-7

8. Center for Research on the Epidemiology of Disasters and United Nations Office for Disaster Risk Reduction. Human cost of disasters: An overview of the last 20 years (2000-2019). October 13, 2020. Accessed January 15, 2021. https://www.undrr.org/publication/human-cost-disasters-2000-2019

9. United States Global Change Research Program. *The Impacts of Climate Change on Human Health in the United States: A Scientific Assessment*. 2016. Accessed March 8, 2021. https://health2016.globalchange.gov

10. Boustan LP, Kahn ME, Rhode PW, et al. The effect of natural disasters on economic activity in U.S. counties: a century of data. National Bureau of Economic Research working paper 23410. May 2017. doi:10.3386/w23410

11. National Weather Service. Weather related fatality and injury statistics. Accessed June 18, 2021. https://www.weather.gov/hazstat

12. World Health Organization. *ICD-10 Version:2010*. 2010. Accessed January 13, 2021. https://icd.who.int/browse10/2010/en

13. Environmental Protection Agency. Climate Change Indicators: Heavy Precipitation. April 14, 2021. Accessed June 20, 2021. https://www.epa.gov/climate-indicators/climate-change-indicators-heavy-precipitation

14. Davenport FV, Burke M, Diffenbaugh NS. Contribution of historical precipitation change to U.S. flood damages. *Proc Nat Acad Sci*. 2021;118(4):e2017524118. doi:10.1073/pnas.2017524118

15. Frank T. Flooding disproportionately harms Black neighborhoods. *E&E News*. June 2, 2020. Accessed January 15, 2021. https://www.eenews.net/stories/1063295449

16. Chacón-Montalván EA, Taylor BM, Cunha MG, et al. Rainfall variability and adverse birth outcomes in Amazonia. *Nat Sustain*. 2021. https://doi.org/10.1038/s41893-021-00684-9

17. Warraich H, Zaidi AK, Patel K. Floods in Pakistan: a public health crisis. *Bull World Health Organ*. 2011;89(3):236–267. doi:10.2471/blt.10.083386

18. Aslam N, Kamal A. Light at the end of the tunnel: post-traumatic growth among individuals exposed to Flood 2010 in Pakistan. *J Pak Psychiatr Soc*. 2013;10(1):34–37.

19. Fatima N, Rana S. Repercussion of Flood of 2010 on the mental health of Pakistani victims. *Pak J Soc Clin Psychol*. 2017;15(1):42–52.

20. Jonkman SN, Godfroy M, Sebastian A, et al. Brief communication: Loss of life due to Hurricane Harvey. *Nat Hazards Earth Syst Sci*. 2018;18(4):1073–1078. doi:10.5194/nhess-18-1073-2018

21. Chakraborty J, Collins TW, Grineski SE. Exploring the environmental justice implications of Hurricane Harvey flooding in Greater Houston, Texas. *Am J Public Health*. 2018;109(2):244–250. doi:10.2105/AJPH.2018.304846

22. Long LJ, Bistricky SL, Phillips CA, et al. The potential unique impacts of hope and resilience on mental health and well-being in the wake of Hurricane Harvey. *J Trauma Stress*. 2020;33(6):962–972. doi:10.1002/jts.22555

23. Flores AB, Collins TW, Grineski SE, et al. Disparities in health effects and access to health care among Houston area residents after Hurricane Harvey. *Public Health Rep.* 2020;135(4):511–523. doi:10.1177/0033354920930133

24. Clay LA, Ross AD. Factors associated with food insecurity following Hurricane Harvey in Texas. *Int J Environ Res Public Health.* 2020;17(3):762. doi:10.3390/ijerph17030762

25. National Oceanic and Atmospheric Administration. Tropical cyclones—annual 2020. January 2021. Accessed January 15, 2021. https://www.ncdc.noaa.gov/sotc/tropical-cyclones/202013

26. Japan Meteorological Agency. Climatology of tropical cyclones. Accessed January 15, 2021. https://www.jma.go.jp/jma/jma-eng/jma-center/rsmc-hp-pub-eg/climatology.html

27. Environmental Protection Agency. Climate change indicators: tropical cyclone activity. Updated April 14, 2021. Accessed June 19, 2021. https://www.epa.gov/climate-indicators/climate-change-indicators-tropical-cyclone-activity

28. National Oceanic and Atmospheric Administration. Global warming and hurricanes. Updated September 23, 2020. Accessed January 14, 2021. https://www.gfdl.noaa.gov/global-warming-and-hurricanes/

29. Parks RM, Anderson GB, Nethery RC, et al. Tropical cyclone exposure is associated with increased hospitalization rates in older adults. *Nat Commun.* 2021;12(1):1545. doi:10.1038/s41467-021-21777-1

30. Shultz JM, Kossin JP, Shepherd JM, et al. Risks, health consequences, and response challenges for small-island-based populations: observations from the 2017 Atlantic hurricane season. *Disaster Med Public Health Prep.* 2019;13(1):5–17. doi:10.1017/dmp.2018.28

31. Holpuch A, Laughland O, Milman O. My Maria story: six Puerto Ricans on surviving after the hurricane. *The Guardian.* August 9, 2018. Accessed March 14, 2021. https://www.theguardian.com/world/2018/aug/09/hurricane-maria-puerto-rico-six-stories

32. Weissenstein M. Telling the stories of Puerto Rican victims of Maria. *AP News.* September 14, 2018. Accessed March 14, 2021. https://apnews.com/article/8b61b8f167204d2cb41cf997db959ebb

33. Allen G, Penaloza M. "I Don't Feel Safe": Puerto Rico preps for next storm without enough government help. *NPR.* July 3, 2019. Accessed March 14, 2021. https://www.npr.org/2019/07/03/737001701/i-don-t-feel-safe-puerto-rico-preps-for-another-maria-without-enough-government

34. Sesin C. "I'm staying": Months after Maria, Puerto Ricans settle in Florida. *NBC News.* March 14, 2018. Accessed March 14, 2021. https://www.nbcnews.com/news/latino/i-m-staying-months-after-maria-puerto-ricans-settle-florida-n851826

35. Umholtz K. Hurricane Laura-related death toll now at 26 after man dies in Calcasieu Parish house fire. *The Times-Picayune/The New Orleans Advocate.* September 8, 2020. Accessed January 14, 2021. https://www.nola.com/news/hurricane/article_206f4b1c-f234-11ea-b23f-c33e5287def4.html

36. Cook B. Guest post: climate change is already making droughts worse. *CarbonBrief.* May 14, 2018. Accessed January 15, 2021. https://www.carbonbrief.org/guest-post-climate-change-is-already-making-droughts-worse

37. National Drought Mitigation Center at University of Nebraska-Lincoln. Map released: January 12, 2021. Updated January 14, 2021. Accessed January 15, 2021. https://droughtmonitor.unl.edu/

38. Environmental Protection Agency. Climate change indicators: drought. April 14, 2021. Accessed June 20, 2021. https://www.epa.gov/climate-indicators/climate-change-indicators-drought

39. United Nations. Humanitarian crisis looms in Madagascar amid drought and pandemic. *UN News.* January 12, 2021. Accessed January 15, 2021. https://news.un.org/en/story/2021/01/1081892

40. Burke M, Driscoll A, Heft-Neal S, et al. The changing risk and burden of wildfire in the United States. *Proc Nat Acad Sci.* 2021;118(2):e2011048118. doi:10.1073/pnas.2011048118

41. Williams AP, Abatzoglou JT, Gershunov A, et al. Observed impacts of anthropogenic climate change on wildfire in California. *Earth Future.* 2019;7(8):892–910. doi:10.1029/2019EF001210

42. California Department of Forestry and Fire Protection. Stats and events. 2021. Accessed January 15, 2021. https://www.fire.ca.gov/stats-events/

43. Brown M. Study: Wildfires produced up to half of pollution in U.S. West. *AP News.* January 11, 2021. Accessed January 15, 2021. https://apnews.com/article/climate-climate-change-san-diego-health-wildfires-387f17ec8658335a7da5b9fe81c65abc

44. Environmental Protection Agency. Wildfire smoke and your patients' health. September 30, 2019. Accessed January 15, 2021. https://www.epa.gov/wildfire-smoke-course/why-wildfire-smoke-health-concern

45. Frederick E. What we don't know about wildfire smoke is likely hurting us. *Science*. February 14, 2020. doi:10.1126/science.abb3141

46. Hrabok M, Delorme A, Agyapong VIO. Threats to mental health and well-being associated with climate change. *J Anxiety Disord*. 2020;76:102295. doi:10.1016/j.janxdis.2020.102295

47. McClurg L, Snow K. California wildfires reignite old trauma for survivors of last year's blazes. *NPR*. August 5, 2018. Accessed January 15, 2021. https://www.npr.org/sections/health-shots/2018/08/05/635475707/wildfire-trauma-revisited

48. Burlison D. What wildfires do to our minds. *Yes! Magazine*. August 7, 2018. Accessed January 15, 2021. https://www.yesmagazine.org/issue/mental-health/2018/08/07/what-wildfires-do-to-our-minds

49. Lindsey R. Climate change: global sea level. Climate.gov. January 18, 2020. Accessed February 19, 2021. https://www.climate.gov/news-features/understanding-climate/climate-change-global-sea-level

50. National Aeronautical and Space Administration. Sea level. Updated June 10, 2021. Accessed February 19, 2021. https://climate.nasa.gov/vital-signs/sea-level/

51. Strauss BH, Orton PM, Bitterman K, et al. Economic damages from Hurricane Sandy attributable to sea level rise caused by anthropogenic climate change. *Nat Commun*. 2021;12:2720. doi:10.1038/s41467-021-22838-1

52. Gopal P. America's great climate exodus is starting in the Florida Keys. *Bloomberg*. September 20, 2019. Accessed February 19, 2021. https://www.bloomberg.com/news/features/2019-09-20/america-s-great-climate-exodus-is-starting-in-the-florida-keys

53. Rush E. Buy high, sell low. *Anthropocene*. August 2020. Accessed February 19, 2021. https://www.anthropocenemagazine.org/2020/08/buy-high-sell-low/

54. Hasemyer D. Maui has begun the process of managed retreat. It wants big oil to pay the cost of sea level rise. *Inside Climate News*. October 14, 2020. Accessed February 19, 2021. https://insideclimatenews.org/news/14102020/maui-big-oil-lawsuit/

55. National Academies of Sciences, Engineering, and Medicine. *Framing the Challenge of Urban Flooding in the United States*. 2019. https://doi.org/10.17226/25381.

56. Federal Emergency Management Agency. National Risk Index for Natural Hazards. January 8, 2021. Accessed January 18, 2021. https://www.fema.gov/flood-maps/products-tools/national-risk-index

57. *Sendai Framework for Disaster Risk Reduction 2015-2030*. United Nations Office for Disaster Risk Reduction; 2015:abstract. https://www.undrr.org/publication/sendai-framework-disaster-risk-reduction-2015-2030

58. Pacific Islands Forum. Framework for Resilient Development in the Pacific. 2016. Accessed January 15, 2021. https://www.forumsec.org/the-framework-for-resilient-development-in-the-pacific/

59. Miami-Dade. Are You Ready? 2020. Accessed January 15, 2021. https://www.miamidade.gov/hurricane/library/guide-to-hurricane-readiness.pdf

60. Allen G. A $4.6 Billion Plan to Storm-Proof Miami. *NPR*. June 13, 2020. Accessed January 15, 2021. https://www.npr.org/2020/06/13/875725714/a-4-6-billion-plan-to-storm-proof-miami

CHAPTER 6

Air Pollution

KEY TERMS

Ground-level ozone
Particulate matter (PM)
$PM_{2.5}$
Air quality index (AQI)
Aeroallergens

Respiratory system
Alveoli
Chronic obstructive pulmonary
   disease (COPD)
Asthma

Emphysema
Chronic bronchitis
Wood smoke
Allergies
Clean Air Act

LEARNING OBJECTIVES

- Understand the structure and function of the lungs and how they are affected by exposure to air pollutants.
- Become familiar with asthma and chronic obstructive pulmonary disease, recent trends in disease burdens, and how they are caused or exacerbated by air pollutants.
- Define $PM_{2.5}$ and ground-level ozone, describe how they are impacted by climate change, and discuss the disease burdens associated with exposure to each.
- Describe how allergies are worsened by climate change.
- Describe how and why certain populations are disproportionately impacted by air pollution.

In 2020, the hazards of air pollution came into focus for unexpected reasons as the COVID-19 pandemic unfolded worldwide. Lockdowns early on restricted human activities and resulted in temporarily reduced air pollutant levels. These reductions varied regionally and with seasonal and weather-related conditions, but many people around the world reported seeing clearer air, due primarily to severely curtailed road traffic. For example, in India, a country that struggles with extreme air pollution, two highly toxic pollutants, particulate matter and nitrogen dioxide (NO_2), declined in the air 50% and nearly 70%, respectively.[1] From March to July in New York City, NO_2 levels decreased by more than 20% compared with the previous year.[2] We do not yet know the full extent of the human health benefits of these reductions, and pandemics are no solution to air pollution or climate change, but lives were likely saved due to temporarily lower air pollutant exposures.

During the pandemic, researchers also discovered a link between air pollution exposure and susceptibility to the virus that causes COVID-19. People who lived in places with high levels of airborne fine particulate matter faced higher risks of COVID-19 mortality.[3] Inhaled fine particles travel deep into the lungs, weaken immune function, and cause widespread inflammation, including in the heart and lungs—two organs targeted by the virus.

In general, air pollution causes a great deal of human suffering and mortality. Two recent and unprecedented legal cases revealed just how hazardous air pollution can be. In one case, the death of a London girl who lived near a busy highway and suffered from severe asthma was determined to be caused by air pollution exposure.[4] Lawyers for the girl's mother argued that air pollution needs to be recognized as not only a public health emergency but also an actual cause of death in order to draw attention to the need for action to reduce people's exposures. In the other case, a man from Bangladesh facing deportation from France successfully argued that if he was forced to return home, he would risk premature death from air pollution because of his severe asthma.[5]

The World Health Organization (WHO) estimates that each year, 7 million people die because of air pollution, and 9 out of 10 people in the world breathe polluted air.[6] Climate change makes air pollution worse, mainly because the human activities that produce greenhouse gases (GHGs) also produce air pollutants, which only increases this health burden. Besides asthma, some of the major impacts of air pollutant exposures are other respiratory illnesses, heart disease, adverse pregnancy outcomes, and mental health problems.

What Is Air Pollution?

Air pollution is a complex mixture of hazardous airborne substances that arise from natural and anthropogenic sources. The combustion of fossil fuels and biomass, largely for power generation, vehicles, heating and cooling, agriculture, and manufacturing, constitute the major anthropogenic source of air pollutants, and many of the adverse health impacts of climate change are mediated through exposure to air pollutants.

Air pollution contains gases, aerosols (solid or liquid particles suspended in a gas), and solid matter, and its chemical composition varies by source and location. Components of air pollution include several forms of carbon, oxides of nitrogen and sulfur, **volatile organic compounds** (**VOCs**), **ground-level ozone**, and **particulate matter (PM)**. PM is classified by size as coarse (PM_{10}), fine ($PM_{2.5}$), and ultrafine ($PM_{0.1}$), with the number indicating maximum particle diameter (**Figure 6.1**). $PM_{2.5}$ are particularly hazardous and are produced from incomplete combustion of fossil fuels, wood, and vegetation, which makes them a major component of diesel exhaust and wildfire smoke. $PM_{2.5}$ are about 30 times thinner than a human hair and are inhaled deeply into the lungs, causing serious health problems. In addition, heavy metals and other toxic chemicals can adhere to the surface of PM and be delivered into the lungs.

Another hazardous air pollutant, ground-level ozone (O_3), is also associated with climate change and is a major component of urban "smog" (**Figure 6.2**). Ground-level ozone forms in a chemical reaction between VOCs and nitrogen oxides

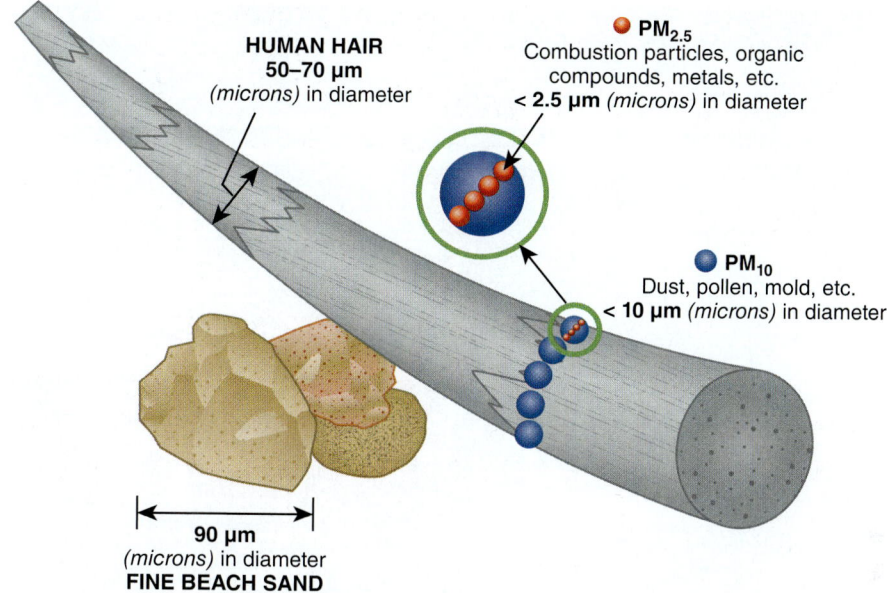

HUMAN HAIR
50–70 µm
(microns) in diameter

● **PM$_{2.5}$**
Combustion particles, organic
compounds, metals, etc.
< 2.5 µm *(microns)* in diameter

● **PM$_{10}$**
Dust, pollen, mold, etc.
< 10 µm *(microns)* in diameter

90 µm
(microns) in diameter
FINE BEACH SAND

Figure 6.1 Comparative Sizes of Coarse (PM$_{10}$) and Fine (PM$_{2.5}$) Particulate Matter.

Reproduced from United States Environmental Protection Agency. Particulate matter (PM) pollution: particulate matter (PM) basics. https://www.epa.gov/pm-pollution/particulate-matter-pm-basics

(NOx) from vehicle tailpipe emissions and industrial sources in the presence of heat and sunlight (**Figure 6.3**). Ozone formation is sensitive to meteorological and climatological conditions, including temperature, humidity, precipitation, wind, cloud cover, and vertical mixing in the atmosphere. Certain regions are ozone hot spots because of warm, sunny weather, high traffic density and/or industrial activities, and being surrounded by mountains or prone to hot air inversion layers, both of which trap and concentrate air pollutants close to the ground, where people are exposed. Climate change creates more favorable conditions for ozone formation, including higher temperatures, and many sources of GHG emissions also produce ozone's chemical precursors.

Air pollutant levels vary widely around the world and depends on many factors, including pollution sources, geography and weather, energy consumption, industrial

activity, and clean air regulations. Air monitoring stations record pollutant levels that are used to calculate a localized composite **air quality index (AQI)** made available to the public in real time (**Figure 6.4**).[7] Many of the world's most polluted cities are in India and China, indicated on the AQI map in red and purple, signifying "unhealthy" or "hazardous" conditions that are likely to have significant adverse health impacts. For example, at one site in Delhi, India, on January 26, 2021, the AQI was 628—an extremely hazardous level. At other locations on that day, AQI was low, indicated in green. In the United States, government agencies maintain AirNow, a public database of real-time air quality information across the country.

AQI is useful to identify air pollution problems in specific areas and inform strategies to improve air quality and protect people's health. AQI will be important to monitor as climate change worsens air pollution and erodes progress on clean air

Figure 6.2 Top: Soot Containing Particulate Matter from Diesel Vehicle Exhaust. Bottom: Ground-Level Ozone Blankets the City of Denver, Colorado.

Top: © Photodisc; Bottom: Courtesy of National Ocean Service. https://www.noaa.gov/news/study-people-in-parts-of-us-and-beyond-breathe-unhealthy-ozone-pollution-more-than-2-weeks-year

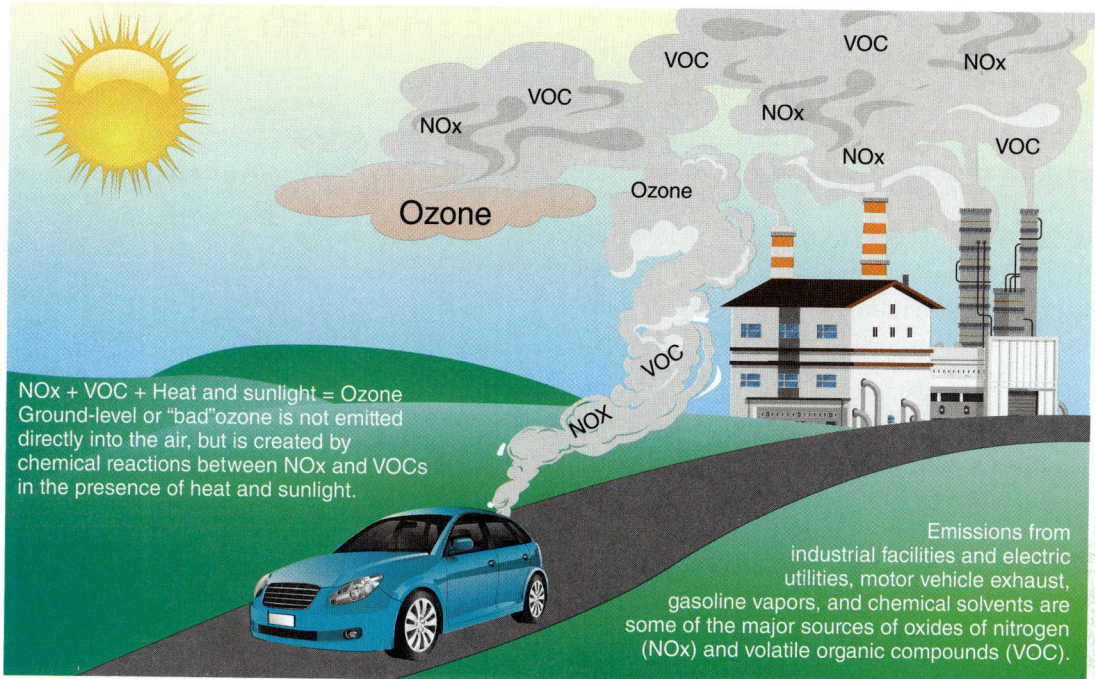

Figure 6.3 Diagram Depicting the Process of Ground-Level Ozone Formation.

Reproduced from United States Environmental Protection Agency. Ground-level ozone pollution: ground-level ozone basics. https://www.epa.gov/ground-level-ozone-pollution/ground-level-ozone-basics

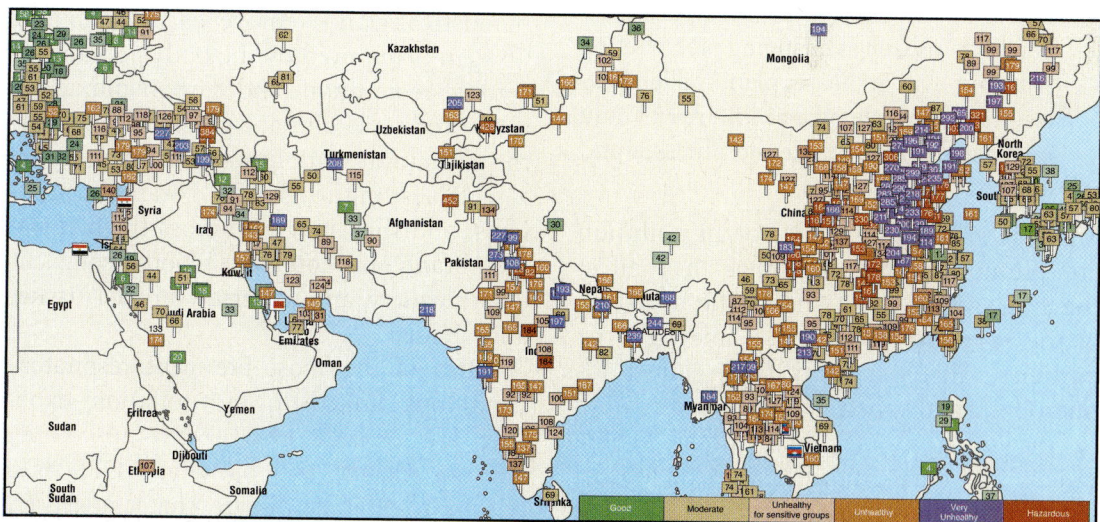

Figure 6.4 Global Real-Time Air Quality Index Values for Much of Asia, the Middle East, and the Eastern Mediterranean, January 26, 2021. All data are collected from each country's official government environment agency.

Reproduced from World Air Quality Index. World's air pollution: real-time air quality index. Accessed January 26, 2021. https://waqi.info

initiatives. Even at low levels, air pollution is dangerous for human health and can be a "silent killer" without being visibly present.

Health Impacts

In 2016, the U.S. Global Change Research Program reported three key findings about future climate change impacts on air pollution and human health[8]:

- Meteorological conditions will become more suitable for ground-level ozone production and make it more difficult to meet existing clean air standards for ozone. Health impacts include more premature deaths, hospitalizations, lost school days, and adverse respiratory disease outcomes.

- Climate change is expected to increase the number and severity of wildfires, which emit fine particulates and ozone precursors and will increase premature deaths and adverse cardiovascular and respiratory disease outcomes.

- Rising atmospheric CO_2 levels and temperatures, as well as changing precipitation patterns, will increase plant pollen levels that may act as **aeroallergens** and increase the risk of allergic illnesses and asthma.

Many of the health impacts of air pollution are pulmonary, meaning that they affect the lungs. **Figure 6.5** shows the structure of the **respiratory system**. Large airborne particles, allergens, and some infectious microbes are trapped in the upper airways of the nose, mouth, and throat, where mucus and cilia capture and clear foreign substances and prevent them from entering the lungs. Small particles, gases, aerosols, and many infectious agents enter the lungs when inhaled and may penetrate deep into the ever-narrowing airways. At the ends of

RESPIRATORY SYSTEM

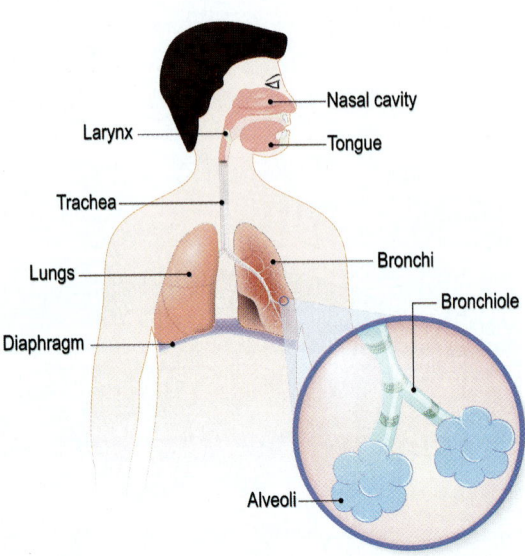

Figure 6.5 Structure of the Respiratory System.
© Designua/Shutterstock.

the airways are **alveoli**, tiny balloonlike sacs with an elastic membrane across which gas exchange occurs to bring oxygen into the bloodstream and remove carbon dioxide. Oxygen is required by the body's organs for aerobic respiration, and carbon dioxide is a by-product. Pollutant gases may damage the structural and functional integrity of alveoli, leading to chronic breathing problems, and they may also cross the alveolar membrane and enter the bloodstream, circulate widely, and adversely impact tissues and organs.

Two of the most prevalent respiratory conditions linked to air pollution exposure are **chronic obstructive pulmonary disease (COPD)** and **asthma**. COPD is a serious respiratory disease characterized by difficulty breathing that becomes progressively worse over time and can be fatal. Most COPD patients suffer from emphysema, chronic bronchitis, or both (**Figure 6.6**).

Chronic Obstructive Pulmonary Disease (COPD)

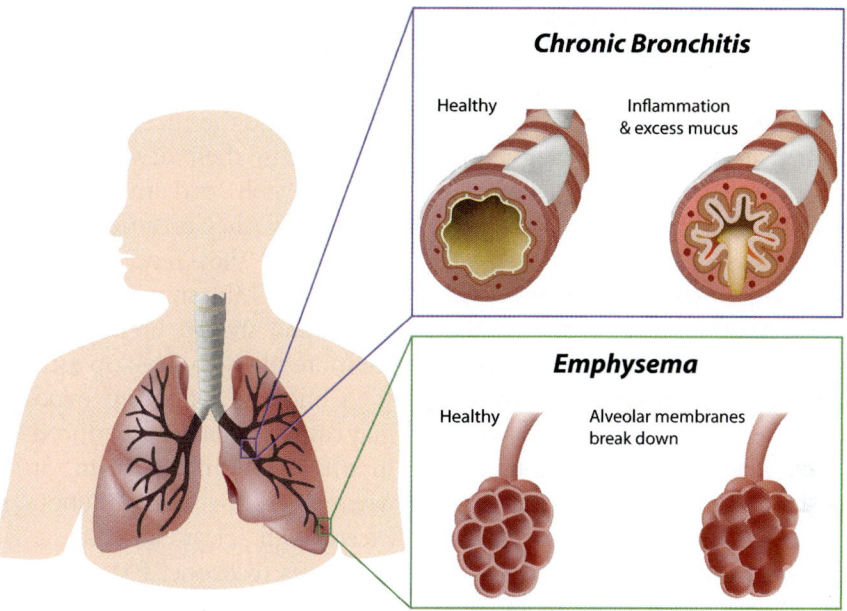

Figure 6.6 Structure of Airways and Alveoli in a Healthy Lung (Left) and in Someone with COPD (Right).
© Alila Sao Mai/Shutterstock.

In **emphysema**, lung alveoli become dysfunctional, mostly because of long-term exposure to the hazardous components of tobacco smoke or air pollution. This exposure damages alveolar membrane elasticity and breaks down alveolar walls, greatly impeding gas exchange. **Chronic bronchitis** results from persistently inflamed, swollen, and irritated airways that have high levels of mucus buildup. COPD affects mainly older adults after years of toxic exposures. The disease is not diagnosed in childhood. Treatment for COPD includes medications (steroids and bronchodilators), oxygen therapy, counseling for disease management, education, and exercise. Prevention measures include smoking cessation and avoiding exposure to air pollution.

Asthma is a chronic respiratory condition that affects people of all ages. Airways of asthmatics are prone to inflammation and narrowing, making it hard to breathe. Asthma symptoms are often triggered by exposure to dust, allergens, cold air, tobacco smoke, and indoor and outdoor air pollutants. These exposures may cause the lungs' immune system to release inflammatory signals, the airways to swell and narrow, and muscles around the airways to tighten in bronchospasms. Over time, the walls of the airways thicken and cause persistent airway obstruction and dysfunction.[9] Asthma often arises in childhood and is a leading cause of morbidity and lost school and work productivity.

Symptoms of asthma include tightness in the chest, coughing, wheezing, and shortness of breath. These symptoms may also indicate other conditions, but they are likely to lead to an asthma diagnosis if they are triggered by exercise, cold air, or allergies;

are worse in the early morning or at night; are episodic; or worsen with a cold. Early-life exposures and a family history of asthma are risk factors, as are childhood viral respiratory infections, allergies, and obesity.[8] Numerous medications are used to treat asthma for both short-term relief of breathing problems and long-term control of the disease. Preventing exposure to asthma triggers is also a key part of disease treatment.

Asthmatics often seek emergency care for asthma exacerbations ("attacks"). In the United States, asthma exacerbations cause more than 1.8 million hospital emergency department (ED) visits each year, which account for the majority of asthma-related healthcare costs.[10] Asthma is the third leading cause of ED visits by children, and as many as 20% of child asthmatics in the United States visit the ED each year.[11] Most asthma exacerbations among young children visiting the ED are due to respiratory infections, and among older children, allergic triggers are also common.[11] ED visits are a particularly significant part of asthma care for people who lack health insurance or a primary healthcare provider. Approximately 14 million missed school days each year are attributed to asthma.[10]

Air pollutant exposure is a major risk factor for chronic respiratory diseases. $PM_{2.5}$ exposure increases hospital admissions and adult mortality because of respiratory and cardiovascular diseases. Fine particulates damage the structural integrity and function of lung alveoli, increasing COPD risk, and increase inflammation and oxidative stress in the lungs and other tissues. "Particulate matter in outdoor air pollution" is classified as a known human carcinogen, and $PM_{2.5}$ exposures induce cellular changes that lead to lung tumor formation.[12]

Ozone is a powerful oxidant that damages epithelial cells lining the airways, reduces lung function, and increases airway inflammation, exacerbating preexisting asthma and COPD and raising the risk of respiratory mortality. Exposure to ground-level ozone increases the rate of hospital ED admissions for asthma in children in a dose-dependent manner.[13] Young children may be particularly vulnerable to the effects of air pollution exposure because their airways are narrower, their lungs, alveoli, and immune system are still developing, they breathe at a higher rate than adults, and they tend to spend more time outdoors. Children who play several outdoor sports and live in high-ozone communities are more likely to develop asthma.[14]

Children's long-term exposure to ozone may cause permanently reduced lung function. In studies of children living in Mexico City, where outdoor air pollution levels often exceed U.S. clean air standards, long-term exposures to ozone, PM_{10}, and NO_2, primarily from traffic emissions, were associated with reduced lung growth.[15] Lung function deficits in childhood may lead to increased risks of developing COPD in adulthood and dying from cardiovascular or respiratory diseases.[15] Children exposed to high levels of air pollutants are also at higher risk of lower respiratory infections, the leading cause of death in children under 5 worldwide, as well as childhood cancers, impaired cognitive and motor development, and behavioral disorders.[14]

The Global Burden of Disease (GBD) report produced by the Institute for Health Metrics and Evaluation (IHME) summarizes evidence for the health impacts of three forms of air pollution: ambient (outdoor) particulate matter pollution, ambient ozone pollution, and household (indoor) air pollution from solid fuel use for cooking or heating. In total, these three risk factors accounted for nearly 7 million global deaths in 2019, with 4 million deaths attributed to PM pollution, 2.3 million to household air pollution, and 365,000 to ozone.[16] Total PM pollution deaths doubled between 1990 and 2019.

A recent study estimated a much higher global mortality burden attributed to $PM_{2.5}$ pollution from fossil fuel combustion alone: 8.7 million global premature deaths in 2018 (1 in 5 deaths).[17] Most of these premature deaths, as well as most $PM_{2.5}$ pollution from fossil fuels, occurred in China and India. Mortality by region in 2018 attributed to $PM_{2.5}$ is shown in **Table 6.1**. In addition, more than 2,000 deaths in children under 5 from lower respiratory infections in Europe and the Americas alone were linked to $PM_{2.5}$ exposure from fossil fuels.

Specific diseases attributed to PM air pollution in the GBD report include cardiovascular diseases; chronic respiratory diseases; maternal and neonatal disorders; cancers (neoplasms); diabetes; kidney diseases; and respiratory infections, including tuberculosis (TB) (**Figure 6.7**). Ozone is attributed only to chronic respiratory diseases in the GBD report.[16]

In 2019, a total of 117 million **disability-adjusted life years** (**DALYs**) were attributed to ambient PM pollution, with 61% due to cardiovascular diseases and 15% due to chronic respiratory diseases (primarily COPD and asthma). DALYs reflect disease burden in a population and account for years of life lost to premature mortality and years of healthy life lost to disability. In addition, nearly 12 million DALYs due to chronic respiratory diseases were attributed to ambient ozone pollution exposure. For household air pollution exposure, DALYs totaled 90 million, most due to cardiovascular diseases, respiratory infections, including TB, and maternal and neonatal disorders.[16]

The global disease burden attributed to ambient PM pollution has been level since 1990, but that masks very different trends among countries with high and low sociodemographic indices (SDI). SDI is a composite measure created by IHME that takes into account a country's income per capita, educational attainment, and fertility rate.[16] Among high-SDI countries (those with high income and socioeconomic development), DALYs rates due to PM pollution are decreasing, but in low-SDI countries, they are increasing (**Figure 6.8a**). DALYs rates due to ozone pollution have decreased globally and in high-SDI countries, but they have increased in low SDI countries (**Figure 6.8b**). Death rates for PM and ozone pollution show approximately the same trends as DALYs rates.

Understanding where and for whom respiratory diseases are a major burden allows for the identification of populations at high risk of exacerbated health problems attributed to air pollution. In addition, this knowledge highlights the need for targeted climate adaptations to protect people from increasing air pollution and to improve health care for the broad range of adverse health impacts attributed to air pollutants.

Table 6.1 Premature Mortality in 2018 from Exposure to $PM_{2.5}$ Attributed to Fossil Fuel Combustion.

Region	Premature Deaths
Asia	6.37 million
India	*2.46 million*
China	*2.36 million*
Europe	1.45 million
North America	483,000
Africa	194,000
South America	187,000
TOTAL	8.69 million

Data from Vohra K, Vodonos A, Schwartz J, Marais EA, Sulprizio MP, Mickley LJ. Global mortality from outdoor fine particle pollution generated by fossil fuel combustion: results from GEOS-Chem. *Environ Res.* 2021;195:110754. doi:10.1016/j.envres.2021.110754

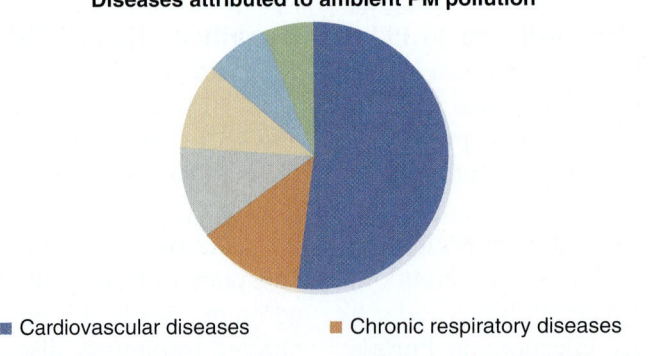

Diseases attributed to ambient PM pollution

- ■ Cardiovascular diseases
- ■ Chronic respiratory diseases
- ■ Respiratory infections and TB
- ■ Maternal and neonatal disorders
- ■ Diabetes and kidney diseases
- ■ Neoplasms

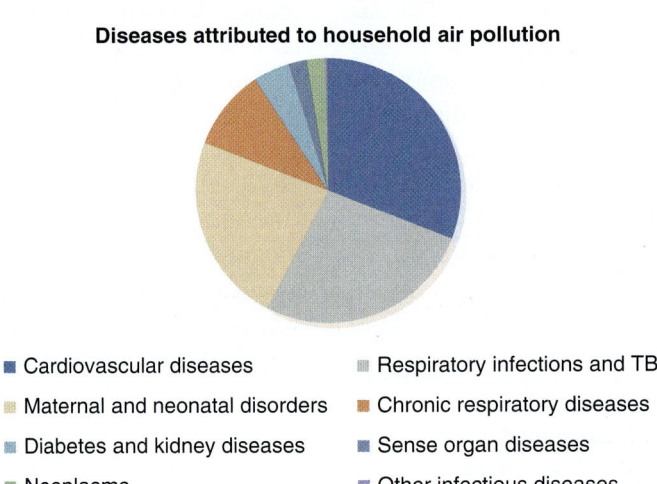

Diseases attributed to household air pollution

- ■ Cardiovascular diseases
- ■ Respiratory infections and TB
- ■ Maternal and neonatal disorders
- ■ Chronic respiratory diseases
- ■ Diabetes and kidney diseases
- ■ Sense organ diseases
- ■ Neoplasms
- ■ Other infectious diseases

Figure 6.7 DALYs Rates for Diseases Attributed to PM Pollution (Top) and Household Air Pollution (Bottom). Only chronic respiratory diseases are attributed to ozone pollution in the GBD report.

Data from Institute for Health Metrics and Evaluation. GBD Compare. Accessed October 15, 2020. https://vizhub.healthdata.org/gbd-compare

In 2019, Nepal had by far the highest disease burden due to ozone pollution, with an age-adjusted DALYs rate 7 times the global average.[16] Nepal has experienced high population growth in recent years, especially in the capital, Kathmandu, which is located in a valley surrounded by high mountains that restrict wind movement and trap air pollutants.[21] Private vehicle use has increased over 30-fold in the past 15 years, and these vehicles, mostly diesel-powered, are the main source of VOCs that serve as precursors for ozone formation.

Air in Nepal is also polluted with very high levels of carbon monoxide, NO_2, and $PM_{2.5}$.[21] The government of Nepal has established a national public health policy that prioritizes reducing air pollution, but to date, implementation of specific policies has been ineffective.[21] Other countries in South Asia—India (**Box 6.1**), Pakistan, Bhutan, and Bangladesh—also have very high ozone-related health burdens.[16] Pakistan also had one of the highest disease burdens due to PM pollution in 2019, along with Egypt, Uzbekistan, and Iraq.[16]

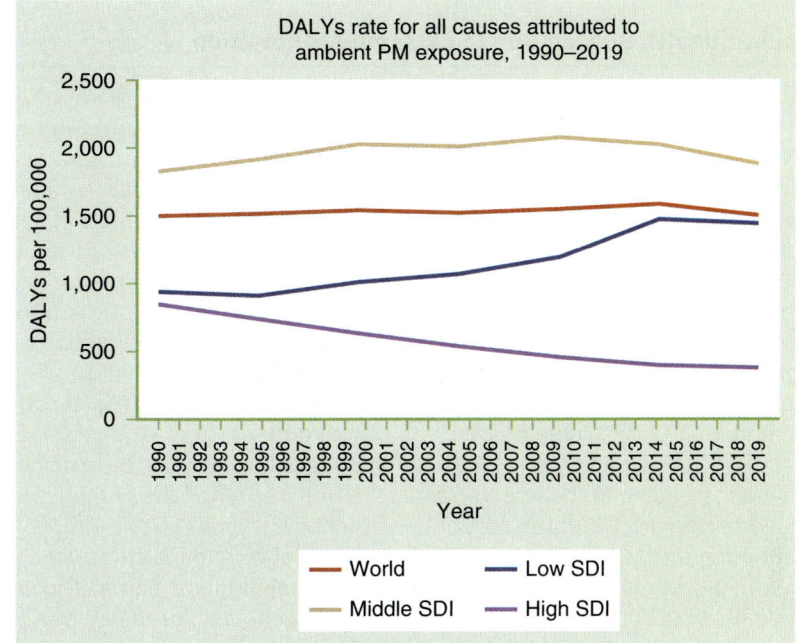

(a)

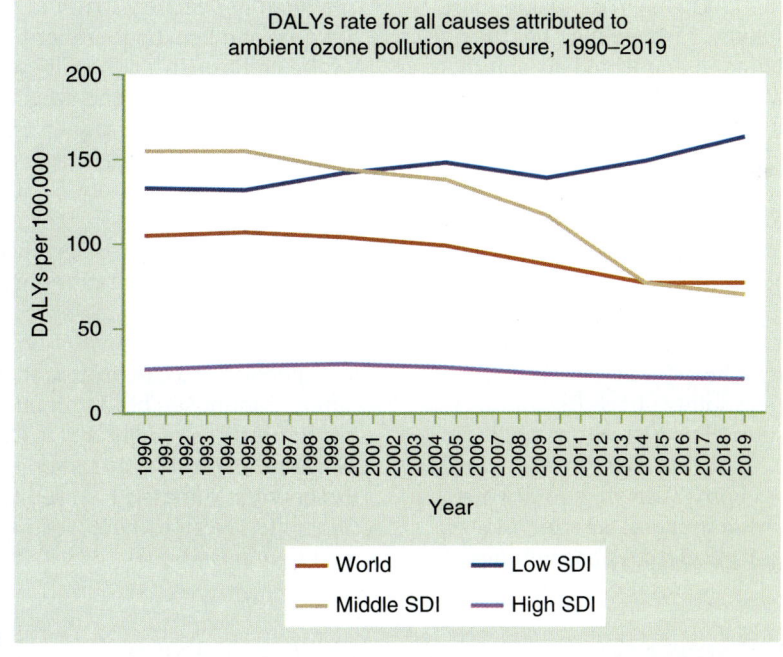

(b)

Figure 6.8 DALYs Rates (Age-Standardized) Attributed to **(a)** Ambient PM Pollution Exposure and **(b)** Ambient Ozone Pollution Exposure for the World and Countries of Different Sociodemographic Index (SDI) Values, 1990–2019.

Data from Institute for Health Metrics and Evaluation. GBD Compare. Accessed October 15, 2020. https://vizhub.healthdata.org/gbd-compare

Box 6.1 India's Health Burden Due to Extreme Air Pollution

India bears a significant portion of the global disease burden attributed to air pollution exposure. Nearly half of all global deaths and DALYs due to ozone pollution and one-quarter of all global deaths and DALYs due to PM pollution and household air pollution in 2019 occurred in India alone. These three air pollution risk factors together accounted for nearly 20% of all of India's deaths.[16] Significant morbidity, premature mortality, and lost economic productivity result from air pollution and in 2019 cost India nearly $37 billion (1.4% of its GDP).[18]

Air pollution is trending in the wrong direction in India compared with the rest of the world, although this is not surprising given India's rapidly developing economy. Since 1990, death and DALYs rates attributed to PM pollution increased in India, while remaining level globally. Deaths and DALYs attributed to ozone pollution increased about 20% in India, while decreasing more than 20% globally.[16] Fortunately, deaths and DALYs attributed to household air pollution decreased about 70% in India and the world, due mostly to a shift to cleaner fuels and stove technologies.

Delhi, India, is one of the most polluted cities in the world, and its residents suffer such extraordinarily high exposures to air pollutants that health warnings are frequently issued. In the past 5 years, air pollution has become a priority for the Indian government, but so far, clean air measures have fallen short. Residents with financial means are able to shift behaviors to minimize exposures, including working from home, traveling in private vehicles, staying indoors, wearing masks, and buying air purifiers. Those without

means and those who live or work in poor neighborhoods or outdoors without access to public health protections remain vulnerable to the health impacts of air pollution. People may be exposed to vehicle exhaust, factory emissions, and smoke from burning agricultural fields, and without a regular fuel source, some may be forced to burn toxic materials such as plastic or rubber for cooking and heating.[19]

A study of thousands of schoolchildren found that 44% of students in Delhi schools suffered from "poor or restrictive lungs," compared with 22% of those in rural schools (though 22% is still a very high prevalence and indicates widespread air pollution impacts across the country). Delhi schoolchildren had much higher levels of lung immune cells called alveolar macrophages, which spike on exposure to particulates and microbes, as well as more frequent headaches, irritated eyes, nausea, fatigue, and heart palpitations. Overall, half of the children living in Delhi were presumed to have irreversible lung damage.[20]

According to Dr. Pradeep Bijalwan, a physician who treats the poor and homeless in Delhi for free, "Pollution colludes with issues like malnutrition, lack of access to clean water, and unhygienic living conditions to compound [disease burdens]." Few measures have been taken to protect those most at risk from air pollution exposure, and people who can protect themselves often "cannot see beyond their own discomfort," according to Dr. Bijalwan.[19] Sarath Guttinkinda, a leading Indian pollution researcher, remarked, "If you have the option to live elsewhere, you should not raise children in Delhi."[20]

Asthma Burden

Asthma occurs in people of all ages but is most prevalent among older adults and young children (**Figure 6.9**).[16] In the United States,

children 5–9 years of age have the highest prevalence of any age group, and *asthma is the leading chronic disease in children in the United States and the world.*

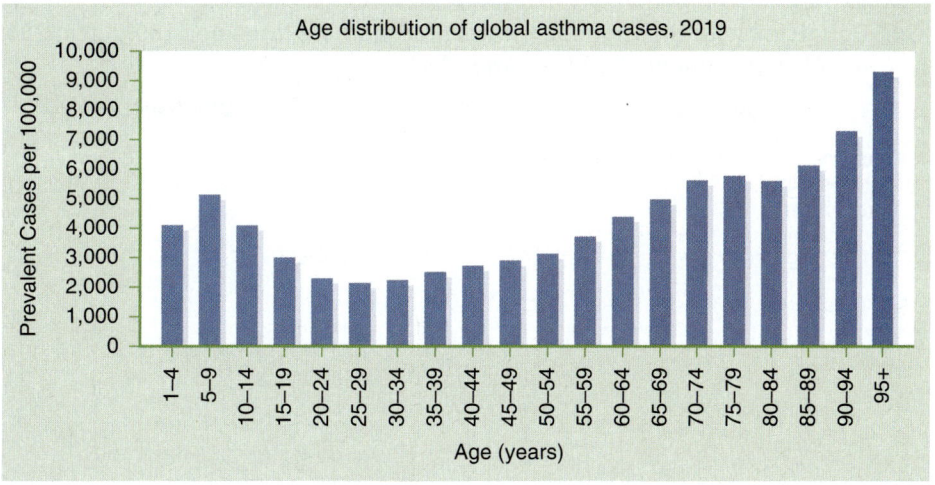

Figure 6.9 Global Asthma Prevalence by Age in 2019.

Data from Institute for Health Metrics and Evaluation. GBD Compare. Accessed October 15, 2020. https://vizhub.healthdata.org/gbd-compare

Global asthma *prevalence* rates have decreased since 1990 but has leveled off or increased in recent years (**Figure 6.10**). In children 5–9, prevalence decreased 22% between 1990 and 2005 but has *increased* 18% since 2005. In the United States, prevalence has risen 50% since 2000 in children 5–9 and people of all ages.[16]

Global asthma *death* rates have decreased steadily since 1990—down 33% in people of all ages and 61% in children 5–9.[16] In the United States, overall asthma mortality has declined 48%, but child asthma mortality has remained unchanged, indicating that more needs to be done to reduce child asthma deaths.

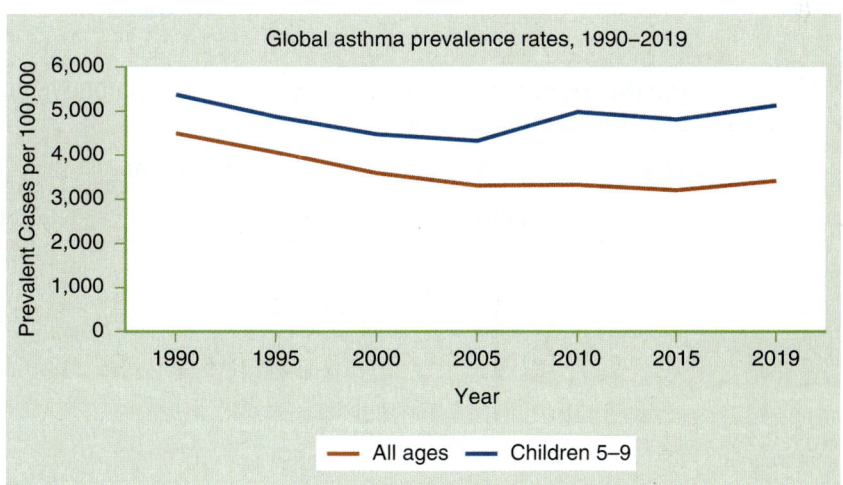

Figure 6.10 Global Asthma Prevalence Rates in People of All Ages (Age-Standardized) and in Children Ages 5–9 from 1990 to 2019.

Data from Institute for Health Metrics and Evaluation. GBD Compare. Accessed October 15, 2020. https://vizhub.healthdata.org/gbd-compare

Table 6.2 Countries with the Highest Asthma Prevalence Rates per 100,000 in 2019 for All Ages and for Children Ages 5–9.

Rank	All ages (age-standardized)		Children 5–9	
	Country	Prevalence rate	Country	Prevalence rate
1	United States	10,399	United States	17,545
2	United Kingdom	9,167	Puerto Rico	15,804
3	Portugal	9,106	Cuba	13,233
4	Australia	8,768	Grenada	13,101
5	Puerto Rico	8,571	Haiti	12,744
	World	3,416	World	5,132

Data from Institute for Health Metrics and Evaluation. GBD Compare. Accessed October 15, 2020. https://vizhub.healthdata.org/gbd-compare

Asthma burdens vary widely among countries. Reasons include disparate smoking rates, air pollution levels, other environmental exposures, disease surveillance capacity, and public health measures. Based on available data, the United States has the highest asthma *prevalence* in adults and children 5–9 (**Table 6.2**). In many countries, including the United States, asthma prevalence rates are *higher in children 5–9* than in people of all ages. Different countries, mostly Pacific Island nations and least developed countries, experience the highest asthma *mortality* (**Table 6.3**). Kiribati and Haiti have the highest asthma mortality rates for adults and children, respectively, which are 12–14 times higher than the global rate. In contrast to prevalence rates, mortality rates for asthma are *much lower in children 5–9* than the overall population.

High adult mortality in many of these countries may be explained at least in part by

Table 6.3 Countries with the Highest Asthma Mortality Rates per 100,000 in 2019 for All Ages and for Children Ages 5–9.

Rank	All ages (age-standardized)		Children 5–9	
	Country	Mortality rate	Country	Mortality rate
1	Kiribati	80.5	Haiti	3.21
2	Papua New Guinea	55.8	Philippines	1.60
3	Fiji	43.3	Papua New Guinea	1.58
4	Sri Lanka	40.3	Madagascar	1.50
5	Solomon Islands	36.5	Lao PDR	1.47
	World	5.8	World	0.26

Data from Institute for Health Metrics and Evaluation. GBD Compare. Accessed October 15, 2020. https://vizhub.healthdata.org/gbd-compare

high smoking rates. In 2019, the countries with the highest death and DALYs rates for asthma *attributable to smoking* were Kiribati, Solomon Islands, Papua New Guinea, and Fiji—four of the five top countries for overall asthma mortality.[16] In Papua New Guinea, remote villages and lack of infrastructure constrain access to public health services such as chronic disease management, and asthma may be of lower priority than communicable diseases (dengue, malaria, HIV, and tuberculosis) that cause enormous suffering and death.[22]

In the United States, significant disparities exist in asthma burdens among different populations.[23] Overall, asthma is more prevalent in females (8.9%) than males (6.6%), although the opposite is true for children. Americans with household incomes below the poverty threshold have twice the burden as those with incomes 4.5 times the poverty threshold. A close look at differences by race and ethnicity shows that Black people and American Indians/Alaska Natives have higher burdens than white people (**Table 6.4**). Those

who identified as "multiple" or "other" race(s) had the highest prevalence overall. These data count persons who answered yes to both of the following questions on the 2016–2018 U.S. National Health Interview Survey:

- Have you *ever been* told by a doctor or other health professional that you had asthma?
- Do you still have asthma?

The accuracy of these data may be affected by differences in asthma awareness, stigma, or medical care among racial and ethnic groups.

Black children have particularly high asthma prevalence, although disparities are likely attributed to living conditions rather than race or ethnicity.[24] Risk factors for child asthma prevalence include poverty, living in an urban area, lack of preventive health care, and having had a low birth weight (LBW). Black LBW children have higher asthma prevalence than white LBW children overall, but racial/ethnic differences disappear when comparing rates among LBW children of all

Table 6.4 Differences in Asthma Prevalence among Racial and Ethnic Groups in the U.S., 2017–2019.

Race/Ethnicity	Percent Total Population with Asthma	Percent Adult Population with Asthma	Percent Child Population with Asthma
Multiple NH	12.6	13.7	11.2
American Indian/Alaska Native NH	10.7	11.6	8.2
Black NH	10.6	9.7	13.5
White NH	7.7	8.1	6.4
Hispanic	6.6	6.1	7.5
Asian NH	3.8	3.8	3.7

Note. NH = non-Hispanic.
Data from Centers for Disease Control and Prevention. Most recent asthma data. Accessed June 22, 2021. https://www.cdc.gov/asthma/most_recent_data.htm

races and ethnicities living in predominantly Black neighborhoods.[24]

The United States is highly racially segregated, and predominantly Black neighborhoods tend to be more densely populated and have more young children, more renters, and more residents living in poverty or with lower educational attainment.[24] These areas tend to be located closer to sources of outdoor air pollution, including major highways, factories, and refineries (see **Box 6.2**). $PM_{2.5}$ levels from vehicle traffic are much lower in predominantly white neighborhoods than those with a majority of people of color.[25] In addition, older houses and vacant properties, both of which are associated with increased indoor air pollution from mold and fecal matter from insect and rodent infestations (all asthma triggers), are more common in poor and Black neighborhoods.[24] All these characteristics mean that residents of Black neighborhoods suffer multiple disadvantages that likely increase the risk of indoor and outdoor air pollution exposure and asthma risk.

Among U.S. states, asthma prevalence varies widely (**Table 6.5**). Maine and Vermont

---

Box 6.2 Air Pollution and Environmental Justice in Oakland, California

Air pollution concerns are a major impetus for community organizing, especially in areas with disproportionately high exposures and fewer benefits from clean air regulations. Favianna Rodriguez, the daughter of immigrants from Peru, is an artist and social justice activist who grew up in East Oakland, California—"a neighborhood of dead cement and abandoned industrial buildings."[26] Interstate 880 ran right by her house, carrying 10,000 trucks per day, the most truck traffic of any highway in the area, and high air pollution exposure contributes to a large asthma burden in her community. Rodriguez leads art projects around the country to change the stories being told about climate change and the "reasons why [her] community was exposed to pollution while White communities just a few miles away had better air quality."[26]

Another local environmental justice activist is Margaret Gordon, a lifelong community organizer and cofounder of the West Oakland Environmental Indicators Project (WOEIP). She noticed that her family's asthma problems flared after they moved to a house in West Oakland near the highway and close to the busy Port of Oakland.[27] Poor communities of color here face diesel exhaust pollution 3 times the regional average, and West Oakland has a higher number of hospital ED visits for asthma compared with the rest of the county where it is located.

Community volunteers began monitoring the air and collaborating with an environmental organization and university researchers to systematically document air pollution at the neighborhood level. As a result of their work, California's legislature passed AB 617 in 2017, which requires the California Air Resources Board to develop an air quality monitoring plan and, among other things, to work to reduce emissions of toxic air pollutants in communities affected by high exposures. In West Oakland, the state's first-of-its-kind community action plan was developed in 2019 with WOEIP's participation, and most of the recommended strategies came from residents who saw firsthand the factors impacting emissions in their neighborhoods.[27] WOEIP is also collaborating on the city of Oakland's Energy and Climate Action Plan. "We have to figure out how we could coexist to make this better," said Gordon. "This is all about collaboration and problem solving and having real, authentic equity in doing what needs to be done."[27]

Table 6.5 Ten U.S. States with the Highest Adult Asthma Burden in 2019, According to the CDC.

Rank	U.S. State	% Adults with Current Asthma, 2019
1	Maine	11.8
2	Vermont	11.6
3	West Virginia	11.5
4	District of Columbia	11.4
5	Rhode Island	11.2
6	Michigan/Ohio/Oregon	11.1
7	Pennsylvania	10.9
8	New Hampshire	10.8
	U.S.	*7.8*

Data from Centers for Disease Control and Prevention. Most recent asthma data. Accessed June 22, 2021. https://www.cdc.gov/asthma/most_recent_data.htm

have the highest percentage of adults with diagnosed asthma.[23] Prevalence in the top states is 50% higher than the U.S. average, although in Maine, even though adult asthma prevalence is high, child asthma prevalence is similar to the U.S. average.[28]

It is unclear what accounts for these disparities in asthma burden. The top two states, Maine and Vermont, are not among the states with the worst outdoor air pollution or highest smoking rates.[29] In Vermont, asthmatics are much more likely than nonasthmatics to be smokers, to live in a household that burns wood, and to report suffering from arthritis, obesity, and diabetes.[30] Many of the top states have high rates of wood burning, including Vermont (#1 for wood burning), Maine (#4), New Hampshire (#5), and Oregon (#7).[31] In states with cold winter climates, wood

burning is a major source of heat and may be perceived as a more climate-friendly energy source than fossil fuels, although awareness of health risks is often low.

Wood burning contributes significantly to localized and community-wide air pollutants, such as fine particulates, aerosols, and carcinogens.[32] **Wood smoke** is very hazardous to human health. Exposure damages lung development in children, with lifetime consequences for lung function, and children are at higher risk for asthma and serious lower respiratory infections.[32] Inhaling wood smoke leads to inflammation, suppressed immune function, damaged airways, and cardiovascular disease. The U.S. Environmental Protection Agency estimates that wood stove smoke exposure causes a lifetime cancer risk 12 times higher than exposure to an equal volume of secondhand tobacco smoke.[32]

COPD Burden

Unlike asthma, COPD is prevalent primarily in older people and is not a health concern in children, although exposures in childhood can lead to increased risk of COPD in adulthood. Globally, COPD is the third leading cause of death and the fifth leading cause of DALYs.[16] Trends in COPD differ in the United States compared to the rest of the world (**Figure 6.11**).[16] The global COPD *prevalence* rate decreased 8% from 1990 to 2019, but U.S. prevalence has not changed significantly since 1990, and in 2019 was higher than the global rate. Global COPD *death* rates declined 42% from 1990 to 2019, but *increased* 22% in the U.S. Over this period, the U.S. death rate decreased 3.4% among men but *increased* 49% among women, even though globally, no difference between men and women was seen. This sex difference in the United States is most likely explained by

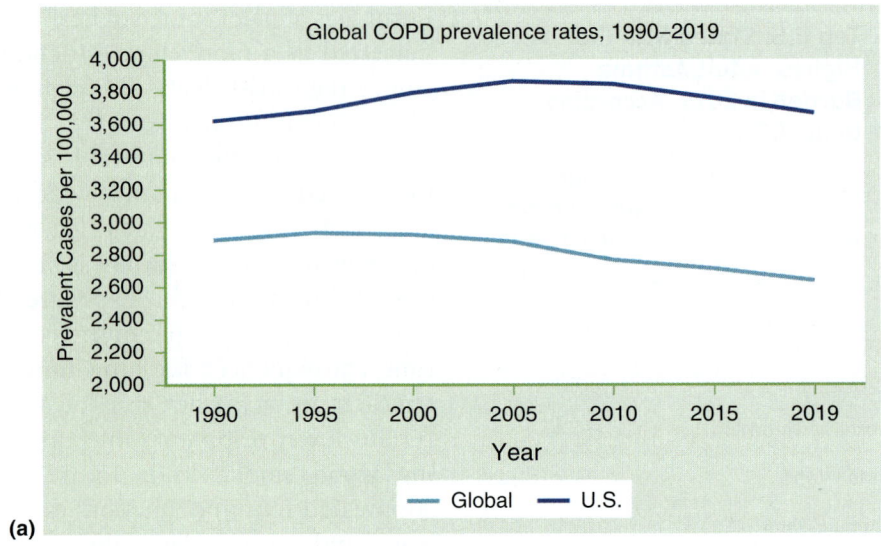

(a)

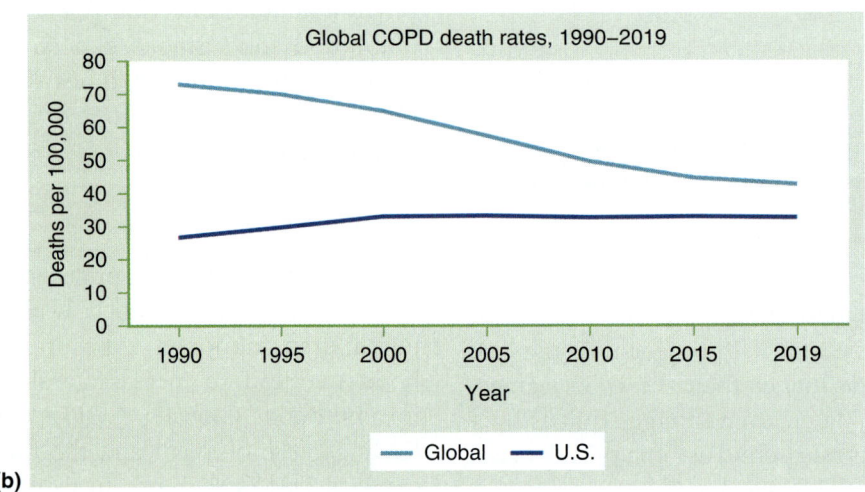

(b)

Figure 6.11 **(a)** COPD Prevalence Rates (Age-Standardized) for the World and the U.S., 2019. **(b)** COPD Death Rates for the World and the U.S., 2019.

Data from Institute for Health Metrics and Evaluation. GBD Compare. Accessed October 15, 2020. https://vizhub.healthdata.org/gbd-compare

a time lag in COPD onset in women because they took up smoking later than men.

India's COPD burden is more than double the global burden, and DALYs rates for COPD attributed to air pollution are much higher than both the global and U.S. rates (**Figure 6.12**).[16] A higher burden of COPD is attributed to PM exposure than ozone exposure in India and globally. In the United

States, with much stronger clean air laws than India, the two air pollutants are approximately equally harmful.

Denmark had the highest reported COPD *prevalence* rate in 2019, and 4 of the top 5 rates were countries in Europe (**Table 6.6**).[16] COPD *death* rates were highest in developing countries, mostly least developed countries and Pacific Island nations.

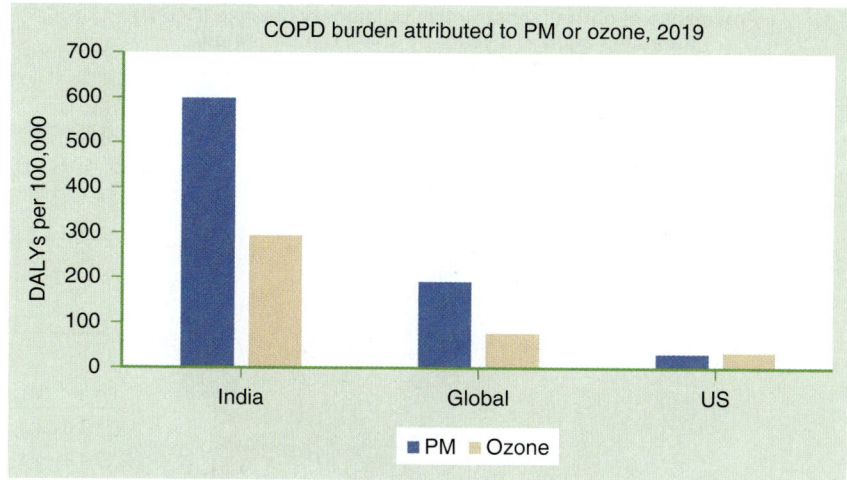

Figure 6.12 DALYs Rate (Age-Standardized) for COPD Cases Attributed to Ambient PM or Ozone Pollution in 2019 in India, the U.S., and the World.

Data from Institute for Health Metrics and Evaluation. GBD Compare. Accessed October 15, 2020. https://vizhub.healthdata.org/gbd-compare

Nepal had the highest death rate, more than 4 times higher than the world average, and by far the highest COPD burden attributable to smoking.[16]

COPD burden also varies among U.S. states (**Figure 6.13**). States with the highest COPD prevalence are clustered along the Ohio River and the lower part of the Mississippi River, and they are also among the top states for disease burdens attributed to smoking and PM pollution.[33]

Other Health Impacts

In addition to respiratory diseases, other health problems result from air pollution exposure. This includes allergies, adverse

Table 6.6 Countries with the Highest COPD Prevalence Rates and Death Rates (Age-Standardized) in 2019.

Rank	Country	Prevalent Cases per 100,000	Country	Deaths per 100,000
1	Denmark	4,299	Nepal	182
2	Myanmar	3,964	Papua New Guinea	145
3	Belgium	3,928	North Korea	105
4	United Kingdom	3,902	Solomon Islands	103
5	Netherlands	3,864	Bhutan	103
	World	*2,638*	*World*	*43*

Data from Institute for Health Metrics and Evaluation. GBD Compare. Accessed October 15, 2020. https://vizhub.healthdata.org/gbd-compare

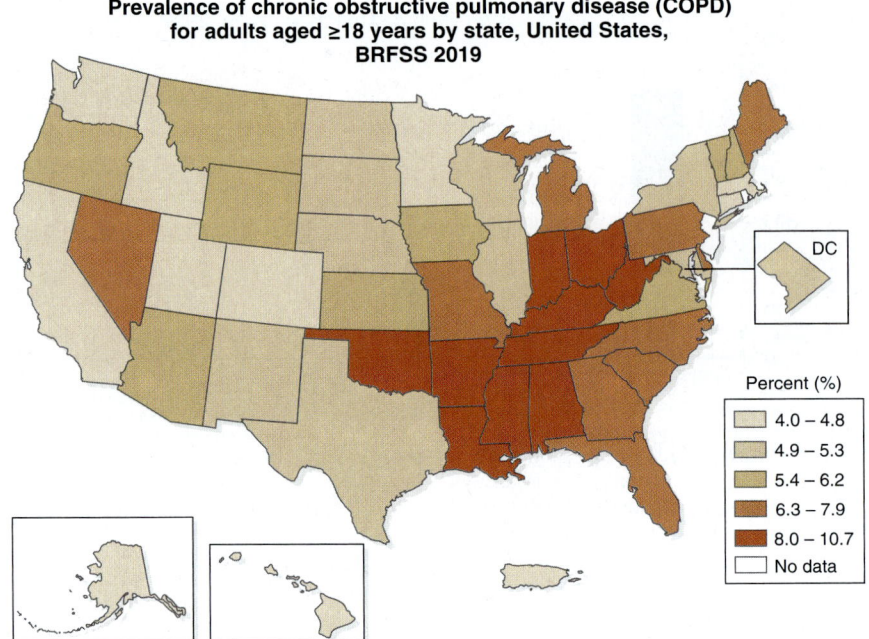

Figure 6.13 COPD Prevalence (Percent Adult Population) among U.S. States in 2019.

Data from Centers for Disease Control and Prevention. Chronic obstructive pulmonary disease (COPD): data and statistics. Accessed June 21, 2021. https://www.cdc.gov/copd/data.html

pregnancy outcomes, neurodegenerative diseases, and mental health conditions. **Allergies** are a common health condition characterized by the immune system over-reacting to exposure to a foreign substance. Many people are allergic to pollen from various plant species, levels of which are increasing with climate change (**Figure 6.14**). At higher temperatures and atmospheric CO_2 levels, the growing season for plants is lengthening and plants are growing faster, producing more airborne pollen, which can trigger allergies and allergic asthma. In North America over the past three decades, the pollen season has grown by 20 days, and pollen concentrations have increased 21%—both coupled strongly with observed temperature increases.[34] The amount and types of pollen in the air depend on the region and specific species of plants, trees, grasses, and weeds growing there. Pollen counts are

Figure 6.14 A Young Man Suffers a Seasonal Allergy to Ragweed, Which Produces More Allergenic Pollen With Climate Change.

© Elizaveta Galitckaia/Shutterstock.

higher in warmer months, although some plants pollinate year-round.[35]

Exposure to pollen may trigger allergic rhinitis ("hay fever"), with symptoms of congestion, runny nose, sneezing, and allergic conjunctivitis, inflammation of the eye lining that causes red, watery, itchy eyes. People with asthma may be more sensitive to allergens, which can trigger asthma exacerbations and hospital admissions, and allergies lead to diminished attendance and productivity at work and school.[35] In the United States, more than 19 million adults and 5 million children (7% of the population) were diagnosed with hay fever in 2018, leading to 12 million visits to the doctor.[36]

Adverse pregnancy outcomes are linked to air pollutant exposures (**Figure 6.15**). For example, high exposure to $PM_{2.5}$ is associated with increased risk of pregnancy loss, preterm birth, low birth weight, intrauterine growth restriction, and infant mortality.[37] Particulates have been detected on both the maternal and fetal sides of the placenta, the organ in the uterus that delivers oxygen and nutrients from the maternal to the fetal blood supply and removes fetal waste products. Proximity of a pregnant woman's residence to high-traffic roadways and overall neighborhood traffic density are linked to low birth weight in newborns, even after controlling for socioeconomic factors.[37] Long-term exposure to diesel soot causes blood vessel dysfunction in ways similar to heavy smoking, and in pregnant women, damaged blood vessels could lead to impaired circulation and inhibition of nutrient transfer to the fetus.[37] Exposure to wildfire smoke also causes pregnancy complications.

Air pollution exposure increases the risk of certain neurodegenerative conditions and mental illnesses. Fine particulates cross the

Figure 6.15 A Pregnant Woman Walks with Her Family along a Polluted Street in Bangkok, Thailand.
© Peter Charlesworth/LightRocket/Getty Images.

blood–brain barrier, deposit in brain cells, and induce inflammation and oxidative damage, and exposure, even at low doses, is linked to increased dementia incidence in large cohort studies.[38] A study of U.S. military veterans found increased risks of not only dementia, but also cardiovascular disease, strokes, chronic kidney disease, COPD, type 2 diabetes, hypertension, lung cancer, and pneumonia associated with $PM_{2.5}$ exposure.[39] Even if $PM_{2.5}$ do not reach brain tissue, they may still impact brain health by triggering inflammatory pathways that affect the brain and other organs.

Inflammatory and oxidative processes in brain cells play a role in the development of neuropsychiatric disorders, and this may be a mechanistic explanation for the link between air pollution and adverse mental health outcomes.[40] Among older adults, high $PM_{2.5}$ exposure raises the risk of moderate to severe depression and anxiety symptoms, and this association is the strongest in individuals with lower socioeconomic conditions and those with underlying stroke or respiratory illnesses.[40] A cohort study in Korea found that among adults, high stress, poor perceived quality of life, depression, and suicide ideation were positively associated with exposure to high concentrations of particulates, NO_2, and CO.[41] A study in London, England, found that exposure to $PM_{2.5}$ and NO_2 in children at age 12 was associated with the risk of major depressive disorder at age 18.[42]

Other health impacts related to climate change result from exposure to airborne molds. Extreme rainfall, flooding, and high humidity lead to indoor mold growth from higher levels of water or moisture inside homes and buildings. Certain mold exposures are hazardous and may cause or worsen respiratory conditions, particularly for asthmatics and those allergic to molds, and may lead to skin and respiratory infections.[35]

Dangerous air pollutant exposure is often a risk before and during tropical cyclones and other extreme storms that hit regions with facilities that emit chemical pollutants. This is the case along the U.S. Gulf Coast, where many petrochemical plants, oil refineries, and other polluting facilities are located and where most land-falling hurricanes hit the United States. For example, during Hurricane Harvey in 2017, more than 8 million pounds of unpermitted toxic air pollutants, including benzene and other carcinogenic gases, were released by industrial facilities in the Greater Houston area—emissions directly tied to the hurricane.[43]

Mitigation and Adaptations

Burning fossil fuels contributes significantly to GHG emissions and also produces much of the air pollution that harms human health. Reducing fossil fuel burning to mitigate GHG emissions and combat climate change has important **health co-benefits** by reducing air pollutant levels that cause respiratory and other diseases.

Countries with strong clean air regulations, such as the United States, with its **Clean Air Act**, have seen significant reductions in air pollutants in recent decades, including PM down approximately 40% (although PM_{10} increased in recent years) and ozone down 33% (**Figure 6.16**). Numerous studies show that an observed decline in cardiovascular disease deaths in the United States and an increase in life expectancy has resulted from $PM_{2.5}$ air standards under the Clean Air Act being set to more stringent levels.[44] Because it is difficult for most people to avoid exposure to air pollutants, reducing levels with strong laws is critical.

Clean air regulations target sources of air pollution for cleanup, including coal-burning power plants and diesel- or gasoline-powered trucks and other vehicles, which also produce GHGs. Use of cleaner fuels and electric vehicles and engaging in more active transport such as walking and biking improve both GHG emissions and air pollution. Active transport is particularly beneficial because it promotes physical activity, which has many important health benefits, including reducing risks of cardiovascular disease, stroke, cancers, obesity, and diabetes. Without aggressive GHG mitigation, changing climatological and meteorological conditions threaten to increase air pollution, eroding progress made from clean air laws. Also, the health effects

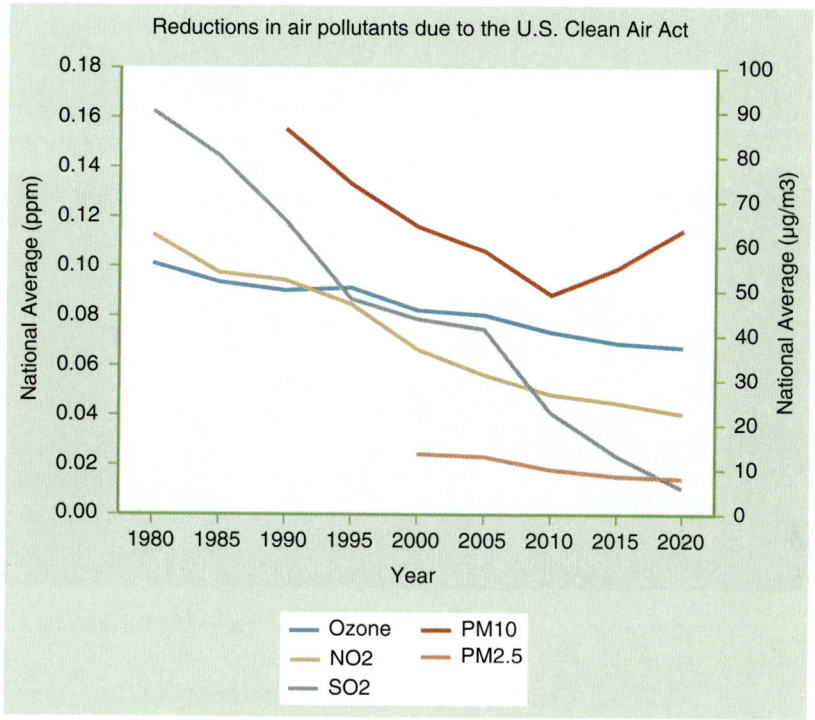

Figure 6.16 Reductions in Criteria Air Pollutants as a Result of Quantitative Standards Set by the U.S. Clean Air Act.

Data from Environmental Protection Agency. National air quality: status and trends of key air pollutants. Accessed June 21, 2021. https://www.epa.gov/air-trends

of air pollution occur even at low levels, so continued improvements in air quality are required to further protect health. Other beneficial air pollution regulations include a 2020 U.S. rule to reduce the smoke emission limit for wood stoves, which requires people who burn wood for heat to use safer and more efficient stove technologies.[45]

Until recently, China had the worst air pollution in the world. However, in 2013, the government enacted the Air Pollution Prevention and Control Action Plan, which has resulted in improved air quality, including reducing by one-third the annual $PM_{2.5}$ levels in 74 cities.[46] Other countries with high burdens, like India, are in the early stages of developing effective clean air policies. Enacting clean air laws is an important climate adaptation strategy.

Non-fossil fuel forms of energy are beneficial to reduce GHG emissions and air pollutants, but some choices may actually worsen air pollution and public health. For example, switching to biomass burning reduces GHGs but still produces high localized levels of hazardous air pollutants. Switching from coal to natural gas for power generation has resulted in not only higher methane GHG emissions but also toxic air pollutants near hydraulic fracturing operations where natural gas extraction occurs, sometimes at levels that exceed regulatory standards (**Figure 6.17**). Natural gas operations also require truck traffic to and from well sites, leading to increased diesel exhaust.[47] Major health problems associated with hydraulic fracturing include respiratory illnesses, cancers, adverse birth outcomes, mental illness, and substance abuse.[47]

Figure 6.17 Workers Tend to an Oil Pump behind a Natural Gas Flare in the Bakken Oil Field Near Watford City, North Dakota.
© Eric Gay/AP/Shutterstock.

To replace coal, countries in the European Union turned to wood pellets as one alternative fuel source to fulfill their commitments to mitigate GHG emissions. Unfortunately, this shift has had unintended health consequences. Many wood pellet manufacturing facilities supplying these countries are in the southern United States. One is Enviva Biomass, the world's largest producer of industrial wood pellets, which built a plant in Ahoskie, North Carolina (**Figure 6.18**). According to one community member, "They promised jobs, economic development, and minimal impacts. What Ahoskie got was . . . local tree loss, noise, heavy traffic, air pollution, and combustible dust from wood drying and processing."[48] These wood pellet mills emit thousands of tons of dust, PM, carbon monoxide, nitrogen oxides, and VOCs each year, as well as several million tons of GHGs. Many plants regularly violate their air permit limits, and fires and explosions are common

because of highly combustible wood dust. Fortunately, the European Union is backing away from wood pellet use and rapidly expanding solar and wind power.[48]

Toxic air pollutant emissions and their associated health burdens from indoor use of solid fuels for cooking and heating have declined as a result of people switching to cleaner fuels and using safer stoves. This has also resulted in lower GHG emissions from these sources. The Clean Cooking Alliance, a global network of partners from the public and private sectors, including environmental, humanitarian, academic, and philanthropic organizations, is working toward the goal of making clean cooking accessible to the 3 billion people worldwide who lack it.[49]

Air quality will improve as countries undergo decarbonizing energy transitions to meet commitments made under the **Paris Agreement**. Air pollution reduction is also

Figure 6.18 A Logging Truck Loaded with Freshly Cut Hardwoods Enters a Wood Pellet Plant at Ahoskie, North Carolina, May 26, 2015.
© Joby Warrick/The Washington Post/Getty Images.

included in the **Sustainable Development Goals (SDGs)**. Target 3.9 under SDG 3 (*Ensure healthy lives and promote well-being for all at all ages*) calls for substantially reducing mortality from ambient and household air pollution by 2030, and Target 11.6 under SDG 11 (*Make cities and human settlements inclusive, safe, resilient and sustainable*) calls for reducing particulate matter pollution in cities.[50]

In 2018, WHO convened an international conference on air pollution and health. As a result, the **Geneva Action Agenda to Combat Air Pollution** was agreed upon, with the goal to reduce global deaths from air pollution by two-thirds by 2030.[51] Specific recommendations included:

- Protecting the most vulnerable, especially children, and preventing childhood illnesses related to air pollution
- Increasing gender equity in access to clean fuel and technologies
- Improving disease surveillance
- Creating educational programs to raise awareness about the health impacts of air pollution
- Strengthening health systems and access to health care for all
- Training a health workforce capable of addressing health impacts related to air pollution related
- Supporting cities in developing policies to reduce urban air pollution and protect health
- Protecting outdoor workers
- Engaging a broad partnership of stakeholders from governments, organizations, businesses, and civil society

Air pollution creates an enormous global health burden, and climate change threatens to make it worse and erode the effectiveness of clean air initiatives. Mitigating GHG emissions will reduce air pollutants and the risks of climate-fueled drivers of air pollution like wildfires, extreme heat, and aeroallergen production, which cause significant health problems. Countries must commit to reducing GHG emissions while also strengthening air pollution control measures and public health responses to prevent, prepare for, and manage the health impacts of air pollution.

Discussion Questions

1. How do the lungs respond to exposure to $PM_{2.5}$ and ozone?
2. Discuss the burdens of asthma and COPD, trends, who is at highest risk, and how these burdens can be reduced.
3. What is AQI, and how is it useful to protect human health?
4. Discuss the health dangers of wood smoke.
5. How are allergies to plant pollen worsened by climate change?
6. Discuss how clean air regulations benefit human health.

References

1. Garg A, Kumar A, Gupta NC. Impact of lockdown on ambient air quality in COVID-19 affected hotspot cities of India: need to readdress air pollution mitigation policies. *Environ Claims J.* 2021;33:65–76. doi:10.1080/10406026.2020.1822615

2. Gramling C. What the pandemic can teach us about ways to reduce air pollution. *Science News.* January 4, 2021. Accessed January 27, 2021. https://www.sciencenews.org/article/covid19-coronavirus-pandemic-air-pollution-ozone-shutdown

3. Wu X, Nethery RC, Sabath MB, et al. Air pollution and COVID-19 mortality in the United States: strengths and limitations of an ecological regression analysis. *Sci Advance.* 2020;6:eabd4049. doi:10.1126/sciadv.abd4049

4. Laville S. Air pollution a cause in girl's death, coroner rules in landmark case. *The Guardian.* December 16, 2020. Accessed June 22, 2021. https://www.theguardian.com/environment/2020/dec/16/girls-death-contributed-to-by-air-pollution-coroner-rules-in-landmark-case

5. Taylor D. Air pollution will lead to mass migration, say experts after landmark ruling. *The Guardian.* January 15, 2021. Accessed January 24, 2021. https://www.theguardian.com/environment/2021/jan/15/air-pollution-will-lead-to-mass-migration-say-experts-after-landmark-ruling

6. World Health Organization. 9 out of 10 people worldwide breathe polluted air, but more countries are taking action. May 2, 2018. Accessed January 27, 2021. https://www.who.int/news/item/02-05-2018-9-out-of-10-people-worldwide-breathe-polluted-air-but-more-countries-are-taking-action

7. World Air Quality Index. World's air pollution: real-time air quality index. Accessed January 26, 2021. https://waqi.info

8. United States Global Change Research Program. *The Impacts of Climate Change on Human Health in the United States: A Scientific Assessment.* 2016. Accessed January 27, 2021. https://health2016.globalchange.gov

9. National Heart, Lung and Blood Institute. Asthma. December 3, 2020. Accessed January 27, 2021. https://www.nhlbi.nih.gov/health-topics/asthma

10. Johnson LH, Chambers P, Dexheimer JW. Asthma-related emergency department use: current perspectives. *Open Access Emerg Med.* 2016;8:47–55. doi:10.2147/OAEM.S69973

11. Dondi A, Calamelli E, Piccinno V, et al. Acute asthma in the pediatric emergency department: infections are the main triggers of exacerbations. *Biomed Res Int.* 2017:9687061. doi:10.1155/2017/9687061

12. International Agency for Research on Cancer. IARC: Outdoor air pollution a leading environmental cause of cancer deaths. Press release 221. October 17,

2013. Accessed March 13, 2021. https://www.iarc
.who.int/wp-content/uploads/2018/07/pr221_E
.pdf

13. Lin S, Liu X, Le Linh H, et al. Chronic exposure
to ambient ozone and asthma hospital admissions
among children. *Environ Health Perspect.*
2008;116:1725–1730. doi:10.1289/ehp.11184

14. National Institute of Environmental Health
Sciences. Air pollution and your health. Updated
May 20, 2021. Accessed June 21, 2021. https://
www.niehs.nih.gov/health/topics/agents/air
-pollution/index.cfm

15. Rojas-Martinez R, Perez-Padilla R, Olaiz-Fernandez
G, et al. Lung function growth in children with
long-term exposure to air pollutants in Mexico
City. *Am J Respir Crit Care Med.* 2007;176:377–384.
doi:10.1164/rccm.200510-1678OC

16. Institute for Health Metrics and Evaluation. GBD
Compare. Accessed October 15, 2020. https://vizhub
.healthdata.org/gbd-compare

17. Vohra K, Vodonos A, Schwartz J, Marais EA,
Sulprizio MP, Mickley LJ. Global mortality from
outdoor fine particle pollution generated by fossil
fuel combustion: Results from GEOS-Chem.
Environ Res. 2021;195:110754. doi:10.1016/j.envres
.2021.110754

18. Pandey A, Brauer M, Cropper ML, et al. Health and
economic impact of air pollution in the states of
India: the Global Burden of Disease Study 2019.
Lancet Planet Health. 2021;5:e25–e38. doi:10.1016
/S2542-5196(20)30298-9

19. Dasgupta P. Delhi's pollution is putting the poorest
at risk. *HuffPost.* January 23, 2018. Accessed
January 27, 2021. https://www.huffpost.com/entry
/india-poor-smog_n_5a60e659e4b05085b607905

20. Slutsky R. Delhi's air pollution and its effects on
children's health. *Yale Global Health Rev.* May
14, 2017. Accessed January 27, 2021. https://
yaleglobalhealthreview.com/2017/05/14/delhis
-air-pollution-and-its-effects-on-childrens-health/

21. Saud B, Paudel G. The threat of ambient air pollution
in Kathmandu, Nepal. *J Environ Public Health.*
2018:1504591. doi:10.1155/2018/1504591

22. Heath R, Tomaszewski P, Kuri M, et al. Message
in a bottle: how evidence-based medicine and
a programme change model improved asthma
management in a low-income emergency department in Papua New Guinea. *Emerg Med Australas.*
2019;31:97–104. doi:10.1111/1742-6723.13212

23. Centers for Disease Control and Prevention. Most
recent asthma data. Accessed June 22, 2021.
https://www.cdc.gov/asthma/most_recent_data
.htm

24. Alexander D, Currie J. Is it who you are or where
you live? Residential segregation and racial gaps
in childhood asthma. *J Health Econ.* 2017;55:
186–200. doi:10.1016/j.jhealeco.2017.07.003

25. Union of Concerned Scientists. Inequitable
exposure to air pollution from vehicles in the
Northeast and Mid-Atlantic. June 21, 2019.
Accessed January 27, 2021. https://www.ucsusa.
org/resources/inequitable-exposure-air-pollution
-vehicles

26. Rodriguez F. Harnessing cultural power. In:
Johnson AE, Wilkinson KK, eds. *All We Can Save:
Truth, Courage, and Solutions for the Climate Crisis.*
New York, NY: One World; 2020:121–127.

27. Uennatornwaranggoon F. How new data is helping
West Oakland clear the air. January 21, 2020.
Accessed January 27, 2021. http://blogs.edf.org
/health/2020/01/21/how-new-data-is-helping
-west-oakland-clear-the-air/

28. Valigra L. Why does Maine have one of nation's
highest asthma rates? The reasons vary. *Portland
Press Herald.* June 4, 2017. Accessed January 24,
2021. https://www.pressherald.com/2017/06/04
/why-does-maine-have-one-of-highest-asthma
-rates-in-u-s-the-reasons-vary/

29. Centers for Disease Control and Prevention. Current
cigarette smoking among adults in the United States.
December 10, 2020. Accessed January 24, 2021.
https://www.cdc.gov/tobacco/data_statistics/fact
_sheets/adult_data/cig_smoking/index.htm#states

30. King M. Vermont asthma rates surprisingly among
the highest in the U.S. *UVM OutReach.* January
8, 2015. Accessed January 24, 2021. https://
learn.uvm.edu/blog/blog-health/asthma-rates
-in-vermont

31. Associated Press. Top states for residential wood
burning, by emissions. March 9, 2015. Accessed
January 25, 2021. https://federalnewsnetwork
.com/health-news/2015/03/top-states-for
-residential-wood-burning-by-emissions/

32. Utah Physicians for a Healthy Environment. *Report
on the Health Consequences of Wood Smoke, 2015.*
2015. Accessed January 25, 2021. https://www
.uphe.org/report-on-the-health-consequences-of
-wood-smoke/

33. Centers for Disease Control and Prevention.
Chronic obstructive pulmonary disease (COPD):
data and statistics. Accessed June 21, 2021. https://
www.cdc.gov/copd/data.html

34. Anderegg WRL, Abatzoglou JT, Anderegg LDL, et al. Anthropogenic climate change is worsening North American pollen seasons. *Proc Nat Acad Sci.* 2021;118:e2013284118. doi:10.1073/pnas.2013284118

35. Centers for Disease Control and Prevention. Allergens and pollen. August 21, 2020. Accessed January 27, 2021. https://www.cdc.gov/climateandhealth/effects/allergen.htm

36. Centers for Disease Control and Prevention. Allergies and hay fever. March 1, 2021. Accessed March 10, 2021. https://www.cdc.gov/nchs/fastats/allergies.htm

37. Fleisch AF, Rifas-Shiman SL, Koutrakis P, et al. Prenatal exposure to traffic pollution: associations with reduced fetal growth and rapid infant weight gain. *Epidemiology.* 2015;26:43–50. doi:10.1097/EDE.0000000000000203

38. Chen H, Kwong JC, Copes R, et al. Exposure to ambient air pollution and the incidence of dementia: a population-based cohort study. *Environ Int.* 2017;108:271–277. doi:10.1016/j.envint.2017.08.020

39. Bowe B, Xie Y, Yan Y, et al. Burden of cause-specific mortality associated with $PM_{2.5}$ air pollution in the United States. *JAMA Netw Open.* 2019;2:e1915834. doi:10.1001/jamanetworkopen.2019.15834

40. Pun VC, Manjourides J, Suh H. Association of ambient air pollution with depressive and anxiety symptoms in older adults: results from the NSHAP Study. *Environ Health Perspect.* 2017;125:342–348. doi:10.1289/EHP494

41. Shin J, Park JY, Choi J. Long-term exposure to ambient air pollutants and mental health status: a nationwide population-based cross-sectional study. *PLoS One.* 2018;13:e0195607. doi:10.1371/journal.pone.0195607

42. Roberts S, Arseneault L, Barratt B, et al. Exploration of NO_2 and $PM_{2.5}$ air pollution and mental health problems using high-resolution data in London-based children from a UK longitudinal cohort study. *Psychiatr Res.* 2019;272:8–17. doi:10.1016/j.psychres.2018.12.050

43. Phillips A. *Preparing for the Next Storm: Environmental Integrity Project.* August 16, 2018. Accessed March 10, 2021. https://environmentalintegrity.org/reports/preparing-for-the-next-storm/

44. Corrigan AE, Becker MM, Neas LM, et al. Fine particulate matters: the impact of air quality standards on cardiovascular mortality. *Environ Res.* 2018;161:364–369. doi:10.1016/j.envres.2017.11.025

45. Environmental Protection Agency. Burn wise: choosing the right wood-burning stove. November 23, 2020. Accessed January 27, 2021. https://www.epa.gov/burnwise/choosing-right-wood-burning-stove#emission-limits

46. Huang J, Pan X, Guo X, et al. Health impact of China's Air Pollution Prevention and Control Action Plan: an analysis of national air quality monitoring and mortality data. *Lancet Planet Health.* 2018;2:e313–e323. doi:10.1016/S2542-5196(18)30141-4

47. Physicians for Social Responsibility. *Compendium of Scientific, Medical, and Media Findings Demonstrating Risks and Harms of Fracking (Unconventional Gas and Oil Extraction).* December 14, 2020. Accessed February 23, 2021. https://www.psr.org/blog/resource/fracking-compendium/

48. Purifoy D. How Europe's wood pellet appetite worsens environmental racism in the U.S. South. *Environ Health News.* October 5, 2020. Accessed January 27, 2021. https://www.ehn.org/wood-pellet-energy-environmental-racism-2647890088.html

49. Clean Cooking Alliance. What we do. Accessed June 21, 2021. https://www.cleancookingalliance.org/what-we-do/index.html

50. United Nations. The 17 Goals. Accessed March 10, 2021. https://sdgs.un.org/goals

51. World Health Organization. Clean air for health: Geneva Action Agenda. November 1, 2018. Accessed June 22, 2021. https://www.who.int/news/item/01-11-2018-clean-air-for-health-geneva-action-agenda

CHAPTER 7

Vector-Borne Diseases

KEY TERMS

Vector-borne diseases (VBDs)
Pathogens
Vectors
Hosts
Reservoir host
Parasites
Endemic
Vectorial capacity (VC)
Extrinsic incubation period (EIP)

Malaria
Plasmodium
Anopheles mosquito
Severe malaria
Cerebral malaria (CM)
Malarial anemia
The Global Fund to Fight AIDS,
    Tuberculosis and Malaria
Dengue

Arbovirus
Aedes mosquito
Severe dengue
Lyme disease
Borrelia burgdorferi
Ixodes tick
Leishmaniasis
Leishmania
Phlebotomine sand fly

LEARNING OBJECTIVES

- Understand what vector-borne diseases are.
- Be familiar with the pathogens, vectors, and hosts for malaria, dengue, Lyme disease, and leishmaniasis, and how each disease is transmitted.
- Describe the symptoms of each disease.
- Understand trends for each disease and how each is climate sensitive.
- Describe international efforts to reduce the burden of malaria.

In 2017, Sri Lanka, a tropical island nation in the Indian Ocean, experienced its most severe outbreak of dengue, a mosquito-borne viral disease. More than 186,000 people were hospitalized, and 440 died.[1] The official case count exceeded those in the preceding and following years by more than threefold.[2] The burden was likely much higher than reported, given that many dengue infections do not present with clinical symptoms or symptoms resemble other diseases, and thus cases go unrecorded.[1] This dengue outbreak affected people in all parts of the country and greatly strained Sri Lanka's economy and public health system.[1]

In Sri Lanka, the risk of dengue is strongly seasonal, peaking several weeks after each of two monsoon seasons that occur annually.[1] In 2017, the country experienced severe flooding from very heavy monsoon rainfall and a tropical cyclone, which created widespread mosquito-breeding sites.[3] According to Gerhard Tauscher, operations manager in Sri Lanka for the International Federation of Red Cross and Red Crescent Societies, "Dengue tends to seek out the poor who live in densely populated places where sanitation is inadequate, rubbish piles up, water pools and mosquitoes thrive."[3] It is likely that climate-change-fueled extreme precipitation, high temperatures, and storms played a role in this and other dengue outbreaks.

Dengue is an example of a vector-borne disease. Humans have faced a burden of **vector-borne diseases (VBDs)** for thousands of years. VBDs are illnesses caused by **pathogens**, most often bacteria, viruses, protozoa, or worms, that are transmitted to humans by **vectors**, most often a blood-sucking insect (mosquito, tick, or fly). Most vectors become infected with a pathogen by biting an infected host (human or animal) and then spreading the pathogen by subsequently biting an uninfected host. **Hosts** are animals (including humans) that become infected by a pathogen. A host may be a **reservoir host**, meaning that the host can become infected and then spread the pathogen to uninfected vectors when bitten, perpetuating the cycle of disease. Some pathogens are also **parasites**, which means that they take nutrients from the host but confer no benefit to the host in return.

In modern times, many VBDs, particularly malaria and dengue, have caused enormous suffering and death and have greatly constrained human livelihoods and economic development. According to the World Health Organization (WHO), VBDs account for more than 17% of all infectious diseases, causing more than 700,000 deaths annually and threatening more than half the world's population.[4] The most deadly VBD

is malaria, but other VBDs that do not cause significant mortality still can have serious and debilitating outcomes.[5] Much of the current VBD burden is borne by low- and middle-income countries in tropical and subtropical regions.[5] For some VBDs, drug treatments and vaccines have been developed against the pathogens. For others, the emphasis is on vector-control measures because treatments and vaccines are lacking.

Characteristics of the world's major VBDs are summarized in **Table 7.1**. For most diseases and regions, large gaps in disease surveillance data make it difficult to accurately report disease incidence and mortality. For example, if a recent estimate of Zika virus infections in the Americas of more than 132 million cases between 2015 and 2018 is accurate, only about 1% of actual cases are being reported.[6] Certain countries and regions are **endemic** for one or more VBDs, meaning that disease cases consistently occur in that geographic area.

VBDs can be thought of as whole eco-systems, with vector–pathogen–host relationships linked to specific environmental conditions. Many VBDs show strong seasonality, indicating climate sensitivity, and changes in intensity and variability of temperature, precipitation, and humidity play an important role in disease transmission. For many VBDs, climate change may affect vector and host habitat suitability and geographic spread, behaviors and life cycles of vectors, pathogens, and hosts, and disease emergence. With future climate change, regional burdens of some VBDs are projected to become more severe, and others less severe. Current evidence suggests that shifts are already occurring and, in some cases, can be attributed in part to climate change. Disease surveillance is lacking, however, especially in endemic countries, and data are scarce on historical VBD burdens for comparison.

Table 7.1 Examples of Vector-Borne Diseases, Listed By Vector Type and Approximate Order of Global Incidence for Each Type. All incidence and mortality data, unless otherwise indicated, are from the 2019 Global Burden of Disease report compiled by the Institute for Health Metrics and Evaluation.

Disease	Pathogen(s)	Vector(s)	Endemic countries/regions	Global incidence	Global mortality
Mosquito-borne diseases					
Malaria	5 species of *Plasmodium* (protist)	41 species of *Anopheles* mosquito	Sub-Saharan Africa, South and Southeast Asia, Central and South America, Mexico, Haiti, Dominican Republic, Iran, Saudi Arabia, Yemen, North Korea, South Korea	231 million	643,000
Dengue	4 dengue viruses (DENV) (*Flaviviridae*)	*Aedes aegypti*, *Aedes albopictus* mosquitoes	South and Southeast Asia, Africa, parts of the Middle East, Mexico, the Caribbean, Central and South America	57 million	36,000
Zika virus fever	Zika virus (ZIKV) (*Flaviviridae*)	*Aedes aegypti*, *Aedes albopictus* mosquitoes	Americas, Africa, Asia, and Western Pacific	269,000	3
Chikungunya fever	Chikungunya virus (CHIKV) (*Flaviviridae*)	*Aedes aegypti*, *Aedes albopictus* mosquitoes	Americas, Africa, Asia, islands in the Indian Ocean, Western and South Pacific, and the Caribbean	184,789 in the Americas[a]	81 in the Americas[a]
Yellow fever	Yellow fever virus (*Flaviviridae*)	*Aedes aegypti*, *Aedes albopictus* mosquitoes	Africa, South America	111,000	4,300
Japanese encephalitis	Japanese encephalitis virus (*Flaviviridae*)	*Culex* mosquitoes	Asia	68,000[b]	13,600–20,400[b]
West Nile fever	West Nile virus (WNV) (*Flaviviridae*)	3 species of *Culex* mosquito	Africa, Europe, the Middle East, North America, and West Asia	>5,000 in North America and Europe[c]	167 in U.S.[c]

(continues)

Table 7.1 Examples of Vector-Borne Diseases, Listed By Vector Type and Approximate Order of Global Incidence for Each Type. All incidence and mortality data, unless otherwise indicated, are from the 2019 Global Burden of Disease report compiled by the Institute for Health Metrics and Evaluation. *(continued)*

Disease	Pathogen(s)	Vector(s)	Endemic countries/regions	Global incidence	Global mortality
Tick-borne diseases					
Lyme disease	4 species of *Borrelia* (bacteria)	4 species of *Ixodes* tick	U.S. and Canada, Europe, Asia	>100,000 in North America and Europe[d] *(estimate)*	*Rare*
Tick-borne encephalitis	Tick-borne encephalitis virus (TBEV) *(Flaviviridae)*	3 species of tick *(Ixodes* and *Haemaphysalis)*	Europe, China, Mongolia, Russia	10,000–20,000[b] *(estimate)*	100–4,000[b] *(estimate; case fatality varies by region)*
Other vector-borne diseases					
Schistosomiasis	5 species of *Schistosoma* (blood fluke)	4 species of freshwater snail	Africa, Southeast Asia, Brazil, Venezuela, Iraq, Yemen, China	140 million *(prevalence)*	11,500
Leishmaniasis	Over 20 species of *Leishmania* (protist)	Sand flies of the genera *Phlebotomus* and *Lutzomyia*	Northeastern Africa, Southern Europe, the Middle East, Southeastern Mexico, Central and South America	657,000	5,700
Chagas disease (American trypanosomiasis)	*Trypanosoma cruzi* (protist)	Many species of triatominae ("kissing bugs")	Southern U.S., Mexico, Central and South America	173,000	7,900

[a] Data from Pan American Health Organization. Reported cases and deaths in 2019. https://www3.paho.org/data/index.php/en/mnu-topics/chikv-en/551-chikv-subregions-en.html

[b] Data from World Health Organization. Japanese encephalitis. Published May 9, 2019. https://www.who.int/news-room/fact-sheets/detail/japanese-encephalitis

[c] Reported cases in 2018. Data from Centers for Disease Control and Prevention. West Nile virus disease cases and presumptive viremic blood donors by state – United States, 2018. https://www.cdc.gov/westnile /resources/pdfs/data/WNV-Disease-Cases-PVDs-by-State-2018-P.pdf; Government of Canada. Surveillance of West Nile virus. Updated February 1, 2021. https://www.canada.ca/en/public-health/services/diseases /west-nile-virus/surveillance-west-nile-virus.html#a1; European Centre for Disease Prevention and Control. Historical data by year – West Nile virus seasonal surveillance. Updated December 5, 2019. https://www.ecdc .europa.eu/en/west-nile-fever/surveillance-and-disease-data/historical

[d] Data from Centers for Disease Control and Prevention. Lyme disease: recent surveillance data. Updated April 29, 2021. Accessed June 24, 2021. https://www.cdc.gov/lyme/datasurveillance/recent-surveillance-data. html; Government of Canada. Surveillance of Lyme disease. Updated March 3, 2021. Accessed June 24, 2021. https://www.canada.ca/en/public-health/services/diseases/lyme-disease/surveillance-lyme-disease.html; Semenza JC, Suk JE. Vector-borne diseases and climate change: a European perspective. *FEMS Microbiol Lett.* 2018:365. doi:10.1093/femsle/fnx244

Data from Institute for Health Metrics and Evaluation. GBD Compare. Accessed October 15, 2020. https://vizhub.healthdata.org/gbd-compare

Many complex factors influence VBDs, including poverty, lack of adequate food, lack of improved water and sanitation systems, political instability, human migration, and inadequate health services. Climate change is compounding the effects of these health determinants. Other variables that affect VBD ecosystems and disease transmission are land use and agricultural practices, changes in biodiversity, demographic changes, globalization, and international travel and migration.[10]

Much of the climate sensitivity of VBDs relates to vector biology. Most vector species are cold-blooded and lack the ability to control body temperature. Thus, their internal temperature is modulated by external temperature, making vector reproduction, development, survival, and behavior temperature-sensitive. Mosquitoes, flies, ticks, and other vectors have optimal temperature ranges for survival. At higher temperatures (up to a maximum threshold), vector species undergo faster reproduction and have reduced mortality and increased biting activity.[11]

Vectors thrive in suitable micro-climates. In the short term, mosquitoes in particular respond to temperature and precipitation changes to create cycles of disease transmission in endemic areas.[12] In the long term, suitable habitat ranges may shift as increasing temperatures expand or contract the edges of geographic suitability for VBDs, resulting in shifting disease incidence.[13] Globally, climate change is predicted to lead to the poleward spread of disease vectors and certain pathogens and animal hosts, as well as a likely contraction of the equatorial boundaries of disease as temperatures warm to extremes.[5,12] This will have the effect of tropical and subtropical VBDs moving from areas that become too hot into currently temperate but warming zones as temperatures become well-suited for disease transmission. The abundance of vectors and pathogens in currently endemic areas may also increase. Establishment of vector, pathogen, and reservoir hosts in new locations must occur before new diseases cases result.

Vectors are also sensitive to precipitation and moisture levels. Mosquito vectors require rainfall to create standing water breeding sites that will not dry up or wash away for 9–12 days.[11] Mosquitoes lay eggs in standing water, which is also required for larval and pupal development and adult hydration. Rainfall creates standing water in both natural spaces, including marshes, mangrove swamps, rice paddies, ditches, and stream and river edges, and in human-made structures and containers that may be intentional or unintentional water collection sites. Too much rainfall or flooding may reduce vector populations if breeding habitat is diluted or washed out, and too little rain may create dry conditions that cause standing water to evaporate rapidly. If standing water is lacking, vectors will spend more time seeking adequate water and less time seeking a blood meal from a host, so disease transmission tends to decline.[12] Shifting precipitation with climate change alters local water cycles such that some areas are becoming wetter and some drier, although both wet and dry areas may experience extreme precipitation events that lead to disease outbreaks.[5]

Sensitivity to climatic conditions is reflected in two important features of vector and pathogen life cycles—vectorial capacity and extrinsic incubation period. **Vectorial capacity (VC)** is a measure of the ability of a vector to transmit disease. VC reflects the number of new cases or infected hosts per infectious vector and is based on vector and pathogen life cycles, vector density and survival, and biting rate, all of which may be temperature- and moisture-sensitive.[5]

$$VC = ma^2p^nb/{-}\ln(p)$$

- VC: vectorial capacity
- m: population density of vector relative to host
- a: human biting rate (probability that vector will feed on susceptible host species)
- p: probability of daily survival of infectious vector
- n: number of days after infection that vector can transmit pathogen (duration of extrinsic incubation period)
- b: vector competence (probability that vector acquires pathogen after feeding on infected host)

VC may decline if the vector population is low, if there is a low frequency of blood meals taken by the vector, or if the vector life span is shorter than the time it takes the pathogen to develop in the vector, which is called the **extrinsic incubation period (EIP)**. EIP is measured as the time interval from when a pathogen is ingested by the vector to when it can be transmitted to a host (the term "n" in the VC equation). EIP is highly temperature dependent. At warmer temperatures, EIP is shorter because the pathogen's life cycle in the vector speeds up. If temperatures are too high, vector mortality rates may increase faster than EIP shortens, which reduces disease transmission. At colder temperatures, EIP lengthens and may become longer than the life expectancy of the vector, which stops disease transmission.[12]

In addition to VBDs shifting because of changes in habitat suitability, emergence of VBD cases in new areas may be *travel-related* and occur when infected travelers or migrants from endemic regions enter nonendemic areas and become ill. New *local* transmission may occur when infected people transmit the pathogen to previously uninfected local vectors, which then transmit the pathogen to uninfected people. This is referred to as **autochthonous** transmission, in which the disease passes from an infected to an uninfected person in the same location.

Many VBDs pose a high risk of morbidity and mortality, and they may also exacerbate the burdens of other diseases and worsen or constrain livelihoods, health systems, and social safety nets. Climate change is likely to be a threat magnifier for VBDs, particularly for populations facing climate extremes in VBD-endemic areas. For example, in the drylands of sub-Saharan Africa, high VBD burdens can be explained in part by drought and resulting scarce water resources that concentrate people, vectors, livestock, and other animals in places where water remains. These locations may become localized hot spots for disease transmission.[4]

Research is ongoing to characterize the complex climate sensitivities of VBDs and project future disease risks. In this chapter, two mosquito-borne diseases (**malaria** and **dengue**), one tick-borne disease (**Lyme disease**), and one fly-borne disease (**leishmaniasis** [See **Box 7.1**]) are discussed in detail. The climate sensitivities of other mosquito- or tick-borne diseases are likely to follow patterns similar to dengue or Lyme disease, respectively.

Box 7.1 Leishmaniasis in Bihar, India

Hannah Richelieu

Many infectious diseases are classified as *neglected tropical diseases*. Even though they cause tremendous suffering around the world, they have not been prioritized like the "big three" infectious diseases (malaria, tuberculosis, and HIV/AIDS) or the "big five" mosquito-borne diseases (malaria, dengue, Zika, chikungunya, and yellow fever). **Leishmaniasis** is one of these neglected

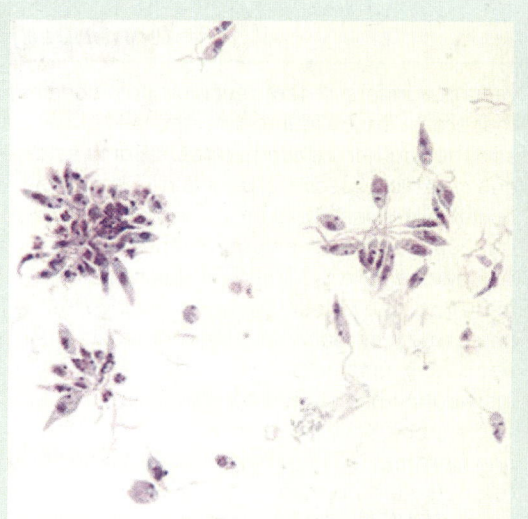 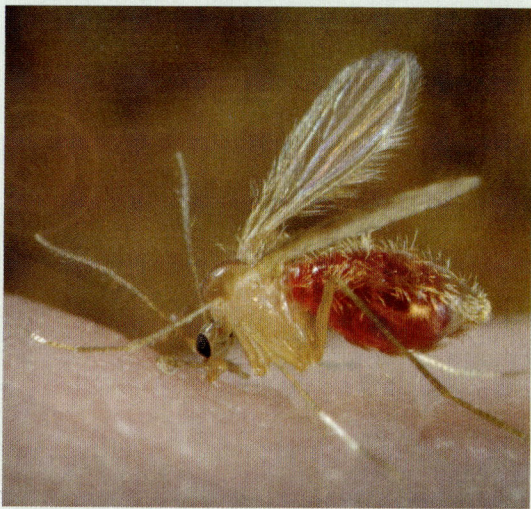

Figure 7.1 Left: *Leishmania* Parasite. Right: Phlebotomine Sand Fly Vector for Leishmaniasis.
Reproduced from CDC/Frank Collins, from the CDC's Public Health Image Library.

tropical diseases. It is caused by the protist ***Leishmania*** and transmitted to humans by the **phlebotomine sand fly** (**Figure 7.1**).[14] Visceral leishmaniasis (also called kala-azar) is the most serious form of the disease and can result in fever, weight loss, enlargement of the spleen or liver, and death.[14] The case fatality rate for visceral leishmaniasis is about 95% if untreated, and about 10% if effectively treated.[15]

The disease is highly climate sensitive. Temperature and precipitation have been found to affect the distribution and development of the *Leishmania* parasite and the sand fly.[16] According to WHO, the range of visceral leishmaniasis incidence has increased over the past 20 years. This is partially due to anthropogenic environmental changes such as urbanization, deforestation, and the building of dams, which have created new fly habitats and brought humans and flies in greater contact.[17]

More than 90% of visceral leishmaniasis cases occur in just six countries: India, Bangladesh, Brazil, Ethiopia, South Sudan, and Sudan.[14] Visceral leishmaniasis tends to be predominate in poor rural villages with minimal access to medical care facilities.[18]

In India, about 80% of visceral leishmaniasis cases occur in the state of Bihar. A recent study traced potentially symptomatic visceral leishmaniasis cases and interviewed patients or family members to obtain sociodemographic data and information about disease progression.[19] The study found that more than 80% of the visceral leishmaniasis patients in Bihar belonged to marginalized castes. In India, caste has a direct link to socioeconomic status, and *Musahars*, the lowest caste in Bihar, have an average daily income of less than USD $1.10.[20] Additionally, about 67% of visceral leishmaniasis patients in Bihar lived in mud or thatched houses.[19] Poor housing with mud floors and cracked walls or roofs provides a resting habitat for phlebotomine sand flies and thus increases the likelihood of human interaction with an infected vector.[18]

Low-income patients have a decreased chance of survival from leishmaniasis, and in Bihar, the proportion of patients from marginalized castes who died from visceral leishmaniasis was significantly higher than for nonmarginalized castes.[19] Treatment is very expensive, and impoverished

(continues)

> **Box 7.1** Leishmaniasis in Bihar, India   *(continued)*
>
> families often need to sell off their assets in order to afford it.[18] Using data from the medical humanitarian group Médecins Sans Frontières (Doctors Without Borders), researchers found that in Bihar, regardless of age, gender, and distance to a healthcare facility, patients in the lowest castes received diagnosis and treatment for visceral leishmaniasis significantly later than patients in higher castes. Reasons for this difference included low levels of education, various forms of social exclusion, and preferential use of traditional healers rather than medical personnel.[20] As a result, worse clinical presentations at the time of admission to a healthcare facility, specifically low hemoglobin levels and greatly enlarged spleens, were observed, contributing to higher death rates.[20]
>
> Models of future global leishmaniasis risk are uncertain, but at least one model predicted that conditions will become more favorable in North America for two sand fly vectors and four rodent reservoir host species that are currently found in northern Mexico and the southern United States.[21] Both vector and reservoir species are predicted to expand northward, possibly as far as parts of southern Canada, doubling the human population at risk of exposure by 2080.[21] For leishmaniasis in Europe, the phlebotomine sand fly currently has a much broader range than the parasite, and it is expanding northward.[10] Recently, fly distribution has shifted northward in Europe, and suitable vectors are now found in Belgium and Germany.[13] There has also been a north-to-south shift in vector habitat in temperate regions of Argentina that are experiencing warming.[13]
>
> It is likely that hot spots for leishmaniasis may persist in places like Bihar, where social and structural vulnerabilities are prevalent among marginalized groups.[16] Prevention and treatment of visceral leishmaniasis likely require attention to vulnerable communities and allocation of significant resources to control disease.

Malaria

Of all known human VBDs, **malaria** is the most deadly, and its climate sensitivity is perhaps the best characterized. Malaria results from infection by the parasitic protozoan **Plasmodium** and is transmitted through the bite of an **Anopheles mosquito** (the malaria vector) (**Figure 7.2**).[22] Symptoms of malaria are often flulike, including fever and chills. Without treatment, severe malaria may result, causing anemia and potentially deadly neurological and respiratory conditions.

Pathogen and Vector

More than 100 species of *Plasmodium* can infect animals, including mammals, birds, and reptiles. Four species of *Plasmodium* infect humans and cause malaria, and a fifth species infects macaques[23]:

- *Plasmodium falciparum*—most common malaria parasite, causing most malaria deaths; predominant in African countries
- *Plasmodium vivax*—causes milder disease but can infect the liver for several years and cause malaria relapses; predominant in Asian and South American countries
- *Plasmodium ovale*—uncommon, and infection is mostly asymptomatic; usually found in West African countries
- *Plasmodium malariae*—uncommon and causes milder disease; widespread throughout countries in sub-Saharan Africa, Southeast Asia, Western Pacific, and the Amazon Basin

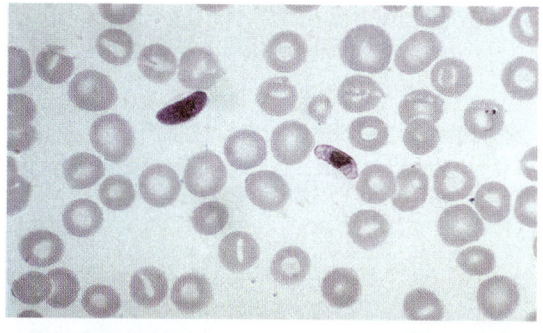
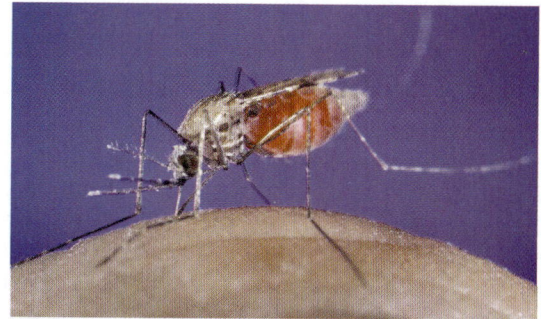
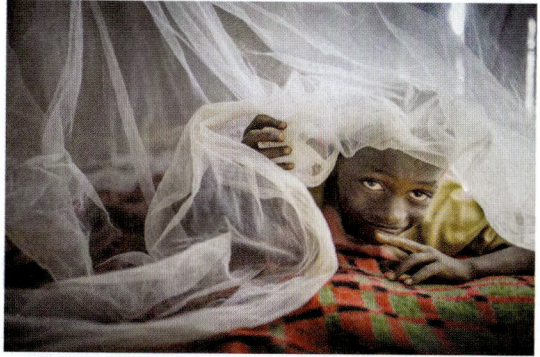
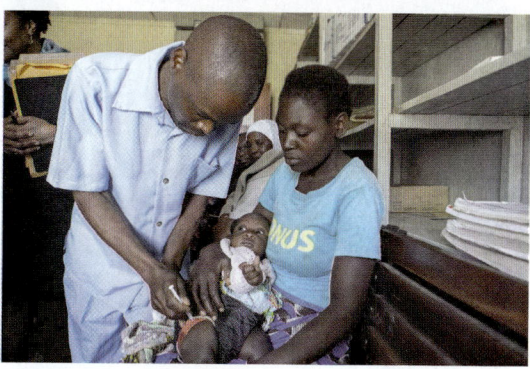

Figure 7.2 Upper Left: *Plasmodium falciparum* and Target Red Blood Cells. Upper Right: Female *Anopheles gambiae* Mosquito Taking a Blood Meal from a Human Host. Lower Left: A Boy in Zambia Sleeps under a Mosquito Net. Lower Right: A Baby Is One of the First to Receive a Malaria Vaccine.

Upper Left: Reproduced from CDC/Dr. Mae Melvin from the CDC's Public Health Image Library. https://phil.cdc.gov; Upper Right: Reproduced from CDC/Jim Gathany from the CDC's Public Health Image Library; Lower Left: Reproduced from World Vision/Jon Warren. https://www.worldvision.org/health-news-stories/what-is-malaria-facts; Lower Right: Courtesy of WHO/M. Nieuwenhof.

- *Plasmodium knowlesi*—infects macaques and may be able to cause *zoonotic* malaria in humans (*zoonotic* refers to human diseases originating in animals); found in Southeast Asian countries

Malaria can be transmitted by approximately 40 species of *Anopheles* mosquitoes with varying geographical ranges. In Africa, the predominant vector is *An. gambiae*, which thrives in rural settings. *An. darlingi* predominates in South America, and *An. pseudopunctipennis* and *An. albimanis* are found in Central America and the Caribbean. Numerous species are found across Asia.[24] Vector species that transmit malaria in endemic areas are also found in nonendemic areas, so disease expansion to new locations is a persistent risk if vectors encounter pathogens. Much of the climate sensitivity of malaria relates to shifting mosquito habitat suitability.

Transmission

Malaria disease transmission occurs when humans are bitten by an infected female *Anopheles* mosquito. Only female mosquitoes carry the *Plasmodium* parasite because they need a blood meal to produce eggs, and if taken from an infected host, mosquitoes may acquire pathogen. An infected mosquito can then transmit the parasite to the next person it bites. Effective malaria vectors must be sufficiently abundant that mosquitoes encounter infected hosts, acquire the malaria parasite, and survive long enough after becoming infected to allow the parasite to undergo the entire mosquito portion of its life cycle and be transmitted to uninfected hosts. Malaria in humans requires vectors inhabiting areas with human populations, coming in frequent contact with humans, and carrying a high load of parasite.[23] The biting behavior of *Anopheles* mosquitoes occurs mostly between dusk and dawn.

The life cycle of the *Plasmodium* pathogen occurs partly in the mosquito (**vector-dependent** phase) and partly in the human (**host-dependent** phase) (**Figure 7.3**).[23] When *Plasmodium* is ingested

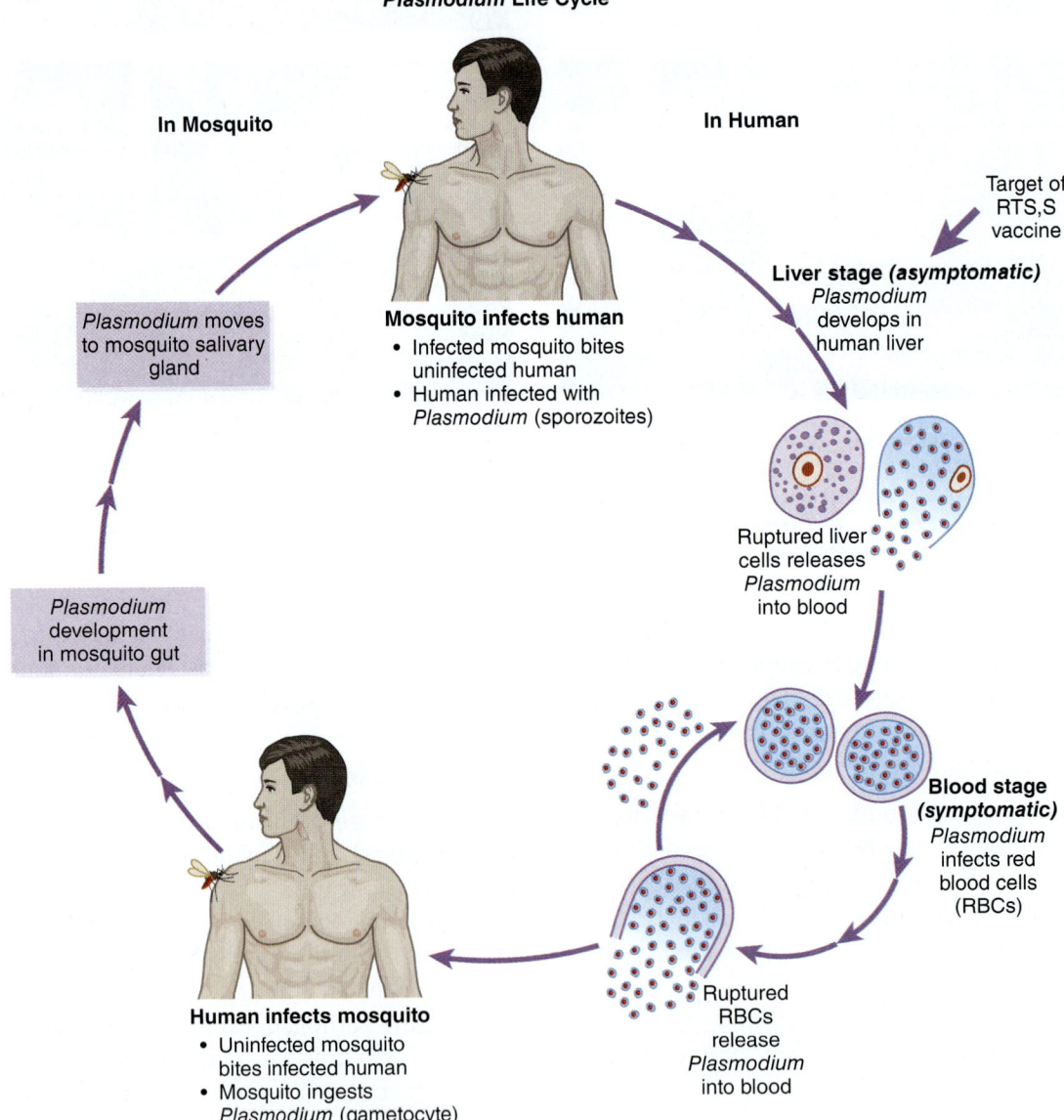

Figure 7.3 Plasmodium Life Cycle in the Mosquito Vector and Human Host.

Data from Centers for Disease Control and Prevention. About malaria. Accessed August 24, 2020. https://www.cdc.gov/malaria/about

by the mosquito, it moves from the mosquito's gut to its salivary gland in a time period (EIP) dependent on environmental conditions and vector survival. When the infected mosquito bites another host, the pathogen can be transmitted to the host, where it grows and develops first in liver cells (dormant phase) and then in red blood cells (symptomatic phase). Infected red blood cells rupture, releasing more pathogen to infect more red blood cells in a cycle.

Symptoms

Malaria is a multiorgan disease. The severity of illness and its clinical manifestations and outcomes depend on many factors, including the specific *Plasmodium* strain, the parasite load in the blood, and the patient's age, malaria-specific immunity, overall immunity, and nutritional status.[25] Malaria cases may be symptomatic or asymptomatic. Symptoms are generally flulike, with fever, chills, aches, and fatigue common. Children are at particular risk of malaria. Symptoms may mimic other common childhood illnesses, such as gastroenteritis, meningitis, and pneumonia, making malaria difficult to track in places where laboratory diagnostic testing is lacking. Fever is the key symptom, and children are more likely than adults to have very high fevers of >40°C (>104°F).[25] Headache, nausea, and vomiting are also common (especially for *P. falciparum* infection) and may hamper treatment with oral antimalaria drugs. Pneumonia and acute diarrhea are malaria **comorbidities** (diseases or conditions simultaneously present with another disease in the same patient) that strongly predict mortality.[25]

Clinical symptoms of **severe malaria** in children are fever, seizures, vomiting, respiratory distress, hypoglycemia, and impaired consciousness, and **cerebral malaria (CM)** may result.[25] CM is considered "unrousable coma" in a malaria patient without other neurological conditions[25] and is characterized in children by abnormal behaviors and body postures, hallucinations, seizures, lowered consciousness, and coma. Many CM patients recover, but 20% die, and 10% have persistent neurological problems, including cognitive impairment or epilepsy.[25] In severe malaria, *Plasmodium*-infected cells adhere to other cells and clog small blood vessels in various organs, including the brain in cases of CM.[26]

Symptoms of malaria arise in the blood stage of *Plasmodium* infection, when red blood cells (RBCs) are invaded and destroyed. RBCs carry iron-rich hemoglobin, which serves a key function to bind to oxygen in the blood and deliver it to organs and tissues so that aerobic metabolism can occur. Functional RBC levels decrease in patients with malaria, resulting in reduced oxygen flow to organs and the symptoms of **malarial anemia**—a condition in a malaria patient marked by RBC or hemoglobin deficiency—which include fatigue, pallor of the skin, dizziness, shortness of breath, rapid heartbeat, and neurological problems.

Plasmodium-infected pregnant women and children are at high risk of malarial anemia, and malaria is a common cause of childhood anemia in endemic areas of sub-Saharan African countries. Other causes are malnutrition, worm infections, and HIV/AIDS.[22] Anemia can be fatal, cause cognitive impairment in children, and increase the risk of death from cerebral malaria. Pregnant women, people who are malnourished, and those with a compromised immune system or no spleen are also at increased mortality risk.[23] The spleen plays an important role in malaria, filtering out old or damaged red blood cells, including those parasitized by

Plasmodium. **Congenital malaria** can also occur, in which an infected pregnant woman passes the infection to her fetus via the placenta. This condition is most common with *P. vivax* infection and can lead to premature birth and low birth weight. Infected infants may experience fever, irritability, vomiting, diarrhea, and impaired eating.[25]

People can acquire immunity from surviving prior *Plasmodium* infections. This immunity is at least semiprotective against subsequent malaria. In newborns, immunity is acquired from the mother, but children between the ages of 6 months and 5 years are at highest risk, particularly for severe malaria, because they lack immunity.[25] Treatments for malaria include anti-*Plasmodium* drugs, the most effective being artemisinin-based combination therapy (ACT). In certain areas where people take antimalarial medication prophylactically, disease transmission may be disrupted, or symptoms may be mild.[25]

Incidence and Mortality

More than 40% of the world's population (3 billion people) is at risk of malaria in 108 endemic countries (**Figure 7.4**).[25] In children living in highly endemic countries in western Africa, the malaria burden is extraordinarily high, and children may experience more than one infection per year. Malaria accounts for a significant portion of all clinical visits for children and is a major reason for school absenteeism.[25] Malaria is the 6th leading cause of death in children under 5 globally, and the 3rd leading cause in sub-Saharan African countries. The high burden of malaria constrains the ability of members of malaria-affected households to earn livelihoods, as well as economic development at the country level.[27]

In 2019, WHO reported the *incidence* of malaria to be an estimated 229 million cases in 87 endemic countries, down 4% since 2000.[22] WHO reported malaria *mortality* in

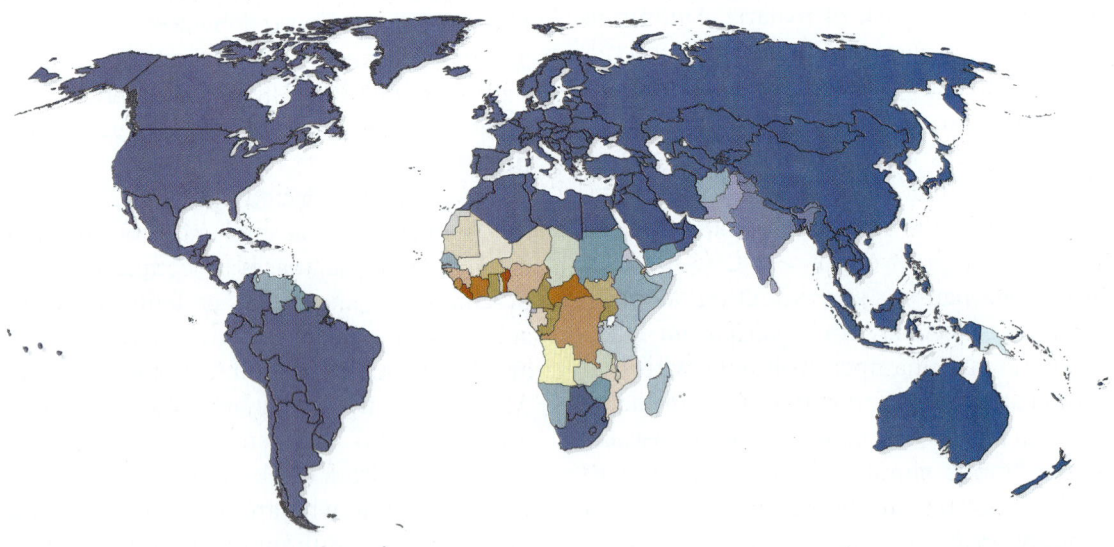

Figure 7.4 Map of Malaria Incidence Rates, 2019. Higher rates were in countries shaded red, orange, and tan.

Reproduced from Institute for Health Metrics and Evaluation. GBD Compare. Accessed October 15, 2020. https://vizhub.healthdata.org/gbd-compare

2019 to be 409,000 deaths, a decrease of 44% since 2000.[22] Progress has stalled in recent years, and both incidence and mortality have increased slightly. The Global Burden of Disease (GBD) report compiled by the Institute for Health Metrics and Evaluation describes similar trends for malaria cases and deaths worldwide.[7] In 2019, the GBD report cites 231 million malaria cases and 643,000 deaths.

Most malaria cases in 2019 (91%) occurred in sub-Saharan African countries, with 25% in Nigeria alone and half of all cases in just five countries (Nigeria, Democratic Republic of the Congo [DR Congo], Uganda, India, and Mozambique).[7] Children of *all ages* accounted for three-quarters of global malaria cases in 2019.[7] Children *under* 5 accounted for 37% of all cases in 2019 (85 million), and 95% of these cases occurred in sub-Saharan African countries.[7]

Malaria has the highest global age-standardized **incidence rate** among VBDs (3,247 new cases per 100,000 people in 2019), more than 4 times higher than for dengue.[7] This rate declined 20% since 1990 but has leveled off recently (**Figure 7.5**).[7] The global incidence rate in children under 5 in 2019 was **over 4 times higher** than in people of all ages (12,852 new cases per 100,000 children under 5).[7] This rate declined 39% since 1990 but is now leveling off.[7]

Malaria is also the deadliest VBD, with a death rate 19 times higher than that for dengue.[7] Nigeria alone accounted for 30% of all malaria deaths in 2019, and over half occurred in just five countries (Nigeria, DR Congo, India, Côte d'Ivoire, and Burkina Faso).[7] This global mortality was down 33% from the peak in 2004, at 961,000 deaths.[7] Total malaria deaths in children under 5 peaked in 2003, at 631,500 (two-thirds of all global malaria deaths that year).[7] By 2019,

deaths in children under 5 had declined 44%, to 356,000 deaths, which was still over half of all malaria deaths.[7] Nearly all malaria deaths in children under 5 in 2019 occurred in countries in sub-Saharan Africa.[7]

Malaria **mortality rates** have declined about 40% since 1990 (Figure 7.5). In 2019, the malaria death rate in people of all ages was 9 deaths per 100,000, and in children under 5, it was 54 deaths per 100,000.[7] Since 1990, malaria death rates in children under 5 have consistently been **more than 6 times higher** than overall rates.[7] As with malaria incidence, declines in mortality have stalled in recent years.

The top 10 countries with the highest malaria *incidence* rates in 2019 were all in western sub-Saharan Africa except for Mozambique (**Table 7.2**). In 2019, the highest incidence rate was in Benin, **12 times higher** than the global average. All top 10 countries saw a decline in the malaria incidence rate from 1990 to 2019, ranging from 5% in Benin to 38% in Mozambique.[7] Still, these incidence rates are some of the highest for any human disease and indicate enormous risks in these countries.

One of the major success stories for reduction of malaria incidence over the past 20 years is São Tomé and Principe. This small island nation off the western coast of Africa had the highest malaria incidence rate in 2000 but reduced it 93% by 2019.[7] This impressive decline was due to an aggressive United Nations-supported malaria control program that focused on disease detection, treatment, and vector control. In contrast, Venezuela is one of very few countries with a growing malaria burden. Its malaria incidence rate increased nearly 16-fold since 2010,[7] due mainly to extreme political and economic instability and a crumbling public health system.

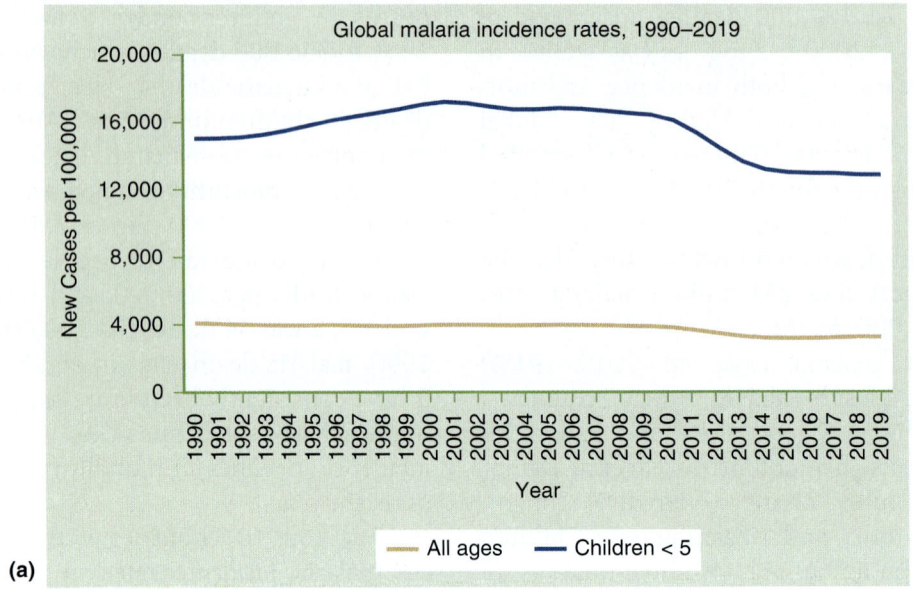

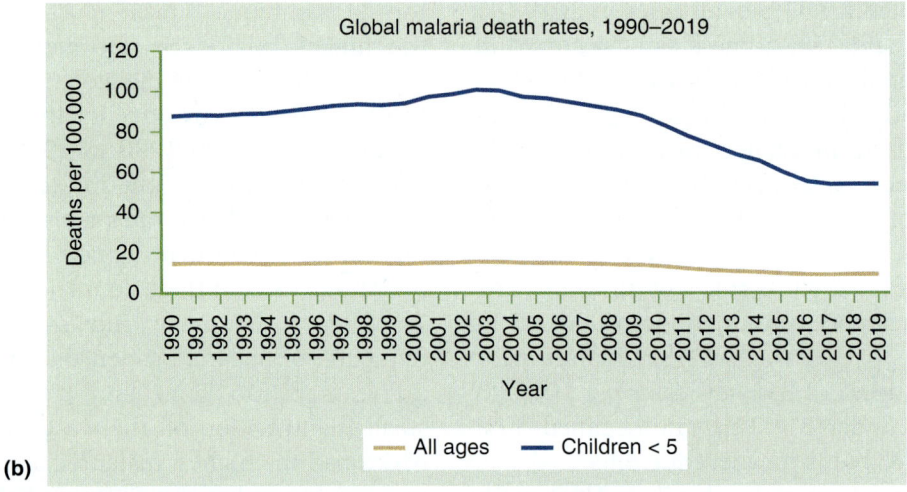

Figure 7.5 Global Malaria Incidence Rates **(a)** and Death Rates **(b)** for All Ages (Age-Standardized; Gray) and Children Under 5 (Blue), 1990–2019.

Data from Institute for Health Metrics and Evaluation. GBD Compare. Accessed October 15, 2020. https://vizhub.healthdata.org/gbd-compare

Nine of the top 10 countries with the highest malaria *incidence* rates among people of all ages also had the highest rates in children under 5 (Table 7.2).[7] The top two countries, Benin and Liberia, had incidence rates **nearly 10 times higher** than the global average for children under 5, and 35 times higher than the global rate in people of all ages. These incidence rates are extraordinarily high and suggest that on average, each young child in these top countries suffers one or more malaria infections every 1–2 years. All these countries saw the malaria incidence rate in children under 5 decline since

Table 7.2 Ten Countries with Highest Malaria Incidence Rates for All Ages (Age-Standardized) and in Children Under 5 Years of Age, 2019.

Rank	Country	Incident Cases per 100,000, All Ages, 2019	Country	Incident Cases per 100,000, Children Under 5, 2019
1	Benin	27,623	Benin	105,374
2	Liberia	27,428	Liberia	104,895
3	Central African Republic	25,301	Central African Republic	95,589
4	Côte d'Ivoire	24,842	Côte d'Ivoire	92,610
5	Sierra Leone	24,123	Sierra Leone	90,370
6	Democratic Republic of the Congo	23,514	Democratic Republic of the Congo	87,914
7	Gabon	22,046	Equatorial Guinea	81,003
8	Mozambique	21,491	Gabon	77,771
9	Togo	21,389	Mozambique	77,443
10	Nigeria	20,460	Togo	73,533
	World	*3,247*	*World*	*12,852*

Data from Institute for Health Metrics and Evaluation. GBD Compare. Accessed October 15, 2020. https://vizhub.healthdata.org/gbd-compare

2000, ranging from 0.4% in Benin to 41% in Mozambique and Togo.[7]

The top 10 countries for malaria *mortality* rates in 2019 were also all in western sub-Saharan Africa except for Mozambique (**Table 7.3**). Sierra Leone had the highest death rate—**18 times higher** than the global rate. In all these countries, death rates have decreased since 1990, ranging from 8% in Liberia to 50% in Burkina Faso.[7]

Eight of these 10 countries also had the highest malaria death rates in children under 5 in 2019 (Table 7.3). In these countries, death rates in children under 5 were **three- to fourfold higher** than in people of all ages.[7] The death rate in Sierra Leone, where malaria contributes to 20% of child mortality,[28] is 10 times higher than the global average for children under 5. Death rates in all these countries declined from 2000 to 2019, ranging from 18% in Benin to 62% in DR Congo.[7]

In Europe and North America, malaria is much less prevalent, and most cases are not transmitted locally, but brought to the region by international travelers. About 2,000 cases of malaria are diagnosed in the United States each year, nearly all in travelers or immigrants entering the country from countries where malaria transmission occurs.[23] Very few malaria deaths are recorded in the United States.

Prevention

Since 2000, the international community has set ambitious targets to combat malaria in developing countries, outlined in the United

Table 7.3 Ten Countries with Highest Malaria Death Rates for People of All Ages (Age-Standardized) and Children Under 5, 2019.

Rank	Country	Deaths per 100,000, All Ages, 2019	Country	Deaths per 100,000, Children Under 5, 2019
1	Sierra Leone	163	Sierra Leone	524
2	Côte d'Ivoire	144	Burkina Faso	405
3	Liberia	131	Guinea	382
4	Burkina Faso	125	Côte d'Ivoire	349
5	Benin	122	Mali	347
6	Nigeria	112	Benin	344
7	Guinea	105	Niger	341
8	Niger	103	Democratic Republic of the Congo	323
9	Togo	92	Liberia	293
10	Mozambique	88	Nigeria	285
	World	*9*	*World*	*54*

Data from Institute for Health Metrics and Evaluation. GBD Compare. Accessed October 15, 2020. https://vizhub.healthdata.org/gbd-compare

Nations' 2000–2015 Millennium Development Goals (MDGs) and 2015–2030 Sustainable Development Goals (SDGs). Declines in malaria cases and deaths over the past two decades are due in large part to this concerted international effort.

MDG 6 (*Combat HIV/AIDS, malaria and other diseases*) set a target to "**have halted by 2015 and begun to reverse the incidence of malaria and other major diseases**."[29] This target was achieved for malaria, as incidence declined globally and in most countries. WHO reported a 37% decline in global malaria incidence between 2000 and 2015 and prevention of millions of malaria deaths.[30] This was due primarily to greatly increased funding and technical support for malaria prevention, mostly in sub-Saharan African countries. Prevention emphasized effective low-tech vector-control strategies,

including using insecticide-treated mosquito nets (ITNs) when sleeping, using window screens and mosquito repellents, covering skin with clothing, and conducting indoor residual spraying with insecticides.[22] Another key strategy is to eliminate mosquito breeding habitats by removing standing water pools and water-filled containers that could serve as egg-laying habitats, as well as covering domestic water storage containers kept outdoors. Mosquito larvicides are sometimes used to treat water in outdoor storage containers. Community education programs raise awareness about the risks of mosquito-borne diseases and how they can be prevented.

The percentage of pregnant women and children sleeping under ITNs in sub-Saharan African countries increased from 2–3% in 2000 to 52% in 2019.[22] This coverage varies widely from country to country, however, and large gaps remain. Malaria diagnosis has become more frequent, and treatment with anti-*Plasmodium* ACT has been scaled up. Prophylactic use of ACT in pregnant women and children to prevent malaria is common in endemic regions with high malaria burdens.[22] One major threat to continuing progress in malaria control is the emergence of drug-resistant *Plasmodium* and insecticide-resistant *Anopheles* mosquitoes.[23]

As a follow-up to MDG 6, **SDG 3** (*Ensure healthy lives and promote well-being for all at all ages*) sets a target to, "**by 2030, end the epidemics of AIDS, tuberculosis, malaria and neglected tropical diseases and combat hepatitis, water-borne diseases and other communicable diseases.**"[31] Specifically, WHO's Global Technical Strategy for Malaria 2016–2030 aims to reduce global malaria incidence and mortality rates by at least 90% by 2030.[22]

In 2019, an estimated $3 billion was spent on malaria control, with nearly 70% of this funding coming from international sources and the remaining provided by malaria-endemic countries.[22] Nearly three-quarters of this funding benefited the WHO African Region. The largest donor countries were the United States ($1.1 billion), the United Kingdom ($0.2 billion), and France, Japan, Germany, and Canada. The **Bill and Melinda Gates Foundation** is also a major donor of global health aid for infectious disease control, including malaria. Much of this malaria funding moves through **The Global Fund to Fight AIDS, Tuberculosis and Malaria**, established in 2002 to accelerate action on malaria and other deadly infectious diseases.[32] The Global Fund is a partnership among governments, technical agencies, the private sector, civil society, and impacted communities. Without this funding, it is estimated that global malaria incidence and

mortality would have been 62–65% higher, rather than declining.[32] Nevertheless, at the current pace, it is unlikely that the SDG target to "end" malaria will be met, and focusing on global targets rather than region-specific targets ignores disparate malaria burdens among endemic countries.

After decades of research and development, the first viable vaccine against malaria is now available. This vaccine, Mosquirix, also known by its scientific name, **RTS,S**, was developed by the British pharmaceutical company GlaxoSmithKline in collaboration with the U.S.-based Walter Reed Army Institute of Research. This vaccine blocks the development of *P. falciparum* in the human liver stage before symptoms appear. It does not protect against *P. vivax* or other *Plasmodium* species.[33] Results of clinical trials concluded in 2014 showed that the vaccine reduced a child's risk of malaria by 40% and severe malaria by 30%. Side effects included higher risks of meningitis and cerebral malaria, as well as higher female mortality among children receiving the vaccine, and the efficacy wore off 4 years after the final dose.[33]

Starting in 2019, WHO initiated a four-year pilot program in Ghana, Kenya, and Malawi to immunize children with four doses of RTS,S between 6 and 24 months of age in order to establish the safety of the vaccine and feasibility of successfully administering the full course of the vaccine in local settings. Approximately 360,000 children in the three countries will be vaccinated as part of their routine immunization program.[33] The ministries of health of Ghana, Kenya, and Malawi are full partners in this pilot program. Funding is being provided by the Global Fund, **Gavi, the Vaccine Alliance** (a public–private partnership to expand vaccine development and coverage), and Unitaid, an international partnership created in 2006 by the governments of Brazil, Chile, France, Norway, and the United Kingdom to provide better medicines and technologies in the global response to HIV/AIDS, tuberculosis, and malaria. GlaxoSmithKline will donate up to 10 million doses of the vaccine for the pilot program, and the UN Children's agency UNICEF is playing a key role in implementation.[33] The goal is for the vaccine to be approved by WHO for use in all malaria-endemic countries. If a broad rollout is approved, it is not known what the cost of the vaccine will be or who will pay for it.

One location for the vaccine pilot program is western Kenya (**Figure 7.6**). Over the past 25 years, malaria prevalence has declined 88% in Kenya, but in regions of the country surrounding Lake Victoria and bordering Uganda and Tanzania, malaria risk and the need for an effective vaccine are high. In a 24-month period between 2015 and 2018, this region of Kenya admitted 5,766 malaria patients 1 month to 15 years of age in four hospitals (more than 38% of all hospital admissions in this age group), with a median age of 36 months. More than 40% of cases were severe malaria, with symptoms including severe anemia, respiratory distress, and impaired consciousness.[34]

Climate Change Effects

Much of the climate sensitivity of malaria is attributed to habitat suitability for mosquito vector species. Mosquitoes go through four life cycle stages: egg, larva, pupa, and adult. Standing water is required for laying eggs and larval and pupal growth, and is highly dependent on precipitation. Mosquito development from egg to adult is strongly influenced by ambient temperature and ranges from approximately 7 to 14 days in tropical climates to longer durations in cold climates.[23] Pathogen development in the vector is also temperature dependent, requiring a minimum air temperature of 19–20°C (66–68°F) for *Plasmodium falciparum* and 15–16°C (59-61°F) for *Plasmodium vivax*.[11] The upper limit for vector and pathogen survival, and thus for malaria transmission, is 35–37°C (95–99°F).[11]

Compelling evidence for *current* climate impacts on malaria transmission comes from reports that vector species' ranges are expanding to higher altitudes and that *P. falciparum* infections are spreading to tropical highland regions in East Africa, Nepal, and Colombia.[11,35] For example, warmer

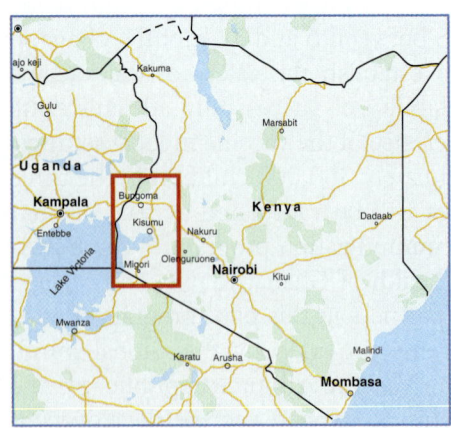

Figure 7.6 Western Kenya Locations Where the Malaria Vaccine Pilot Program Is Being Administered, and a Kenyan child.

Left: Map data © 2019 Google; Right: © himarkley/E+/Getty Images.

temperatures have increased the distribution of the *Anopheles arabiensis* vector on the slopes of Mount Kilimanjaro in Kenya. In Uganda, extreme flooding in 2013 was followed by a 30% increase in malaria, and in Zambia in 2008–2010, increased malaria was linked to unusually high rainfalls.[13] In Rwanda, an increase in mean temperature explains 80% of the monthly malaria variance in high-altitude regions.[13] People living in non-endemic areas that are vulnerable to new malaria transmission attributable to climate change may lack the protective immunity that comes from prior infections and thus may be more vulnerable to severe disease.[11] With changing conditions, malaria may also reemerge in endemic areas where cases have declined as a result of control measures. This has been seen in the temperate Anhui province in China, where malaria cases have increased dramatically, linked to changing rainfall patterns.[11]

Projecting *future* shifts in malaria attributed to climate change is difficult given data gaps, uncertain climatic shifts, and changing disease trends over the past two decades because of malaria interventions. Nevertheless, many climate–disease models project that the range of malaria transmission may decrease in tropical regions and increase in certain temperate and highland areas.[5] In China, a very large and climatically diverse country, increasing minimum temperatures is predicted to increase malaria in the north, and increasing maximum temperatures may decrease malaria in the south.[13] In Ethiopia, a 1°C rise in temperature may lead to an additional 3 million cases in children under 15 years of age in that country alone.[32] Models suggest that future malaria transmission could occur in Europe, but vector control, land use management, and a high degree of socioeconomic development is likely to counteract this effect.[10] Malaria was

endemic in the United States until the early 1950s but was eliminated through aggressive vector-control measures that drained wetlands and involved heavy use of synthetic insecticides, such as DDT.[5]

The net effect of the shifting malaria burden globally may be small, but it may be significant in places in or near endemic regions. For populations that continue to bear a high malaria burden, it will be imperative to understand the climatic and nonclimatic drivers of malaria and take necessary climate and public health actions to minimize this burden in the future.

Dengue

Along with malaria, **dengue** is one of the most serious mosquito-borne diseases, and the global burden has increased in recent years. Dengue is one of several viral diseases, including Zika, chikungunya, and yellow fever, that are spread by the same mosquito vector.

Pathogen and Vector

The dengue pathogen is the dengue virus, an **arbovirus** transmitted via an insect vector (**ar**thropod-**bo**rne **virus**). Dengue virus has four **serotypes**, and in many countries all four are endemic.[36] Once infected, immunity to that serotype is lifelong, but people are still susceptible to infection from the other serotypes. The vectors are species of **Aedes mosquito**, including *Aedes aegypti* (yellow fever mosquito) and *Aedes albopictus* (Asian tiger mosquito) (**Figure 7.7**). *Aedes* mosquitoes are found mainly in urban habitats and are daytime feeders.

Transmission

Dengue transmission starts when a female mosquito feeds on a dengue-infected person, who may or may not be symptomatic.

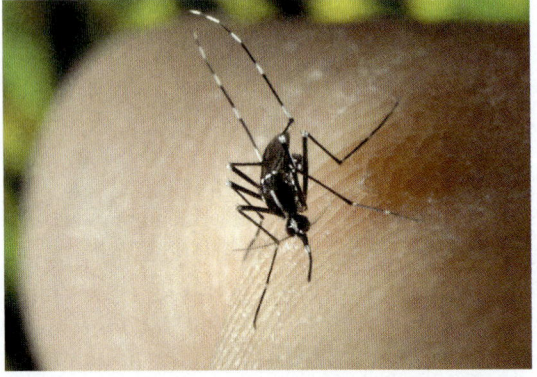 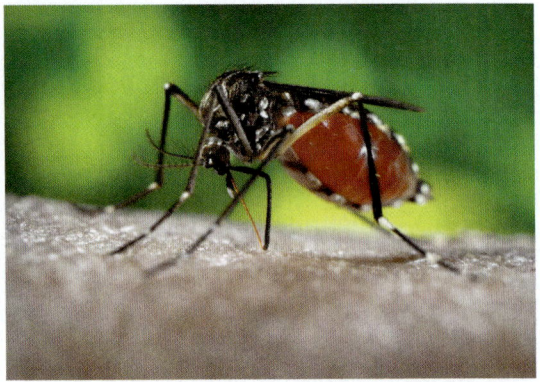

 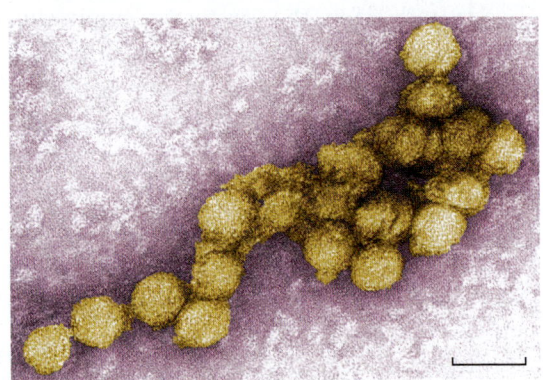

Figure 7.7 Upper Left: *Aedes albopictus*. Upper Right: *Aedes aegypti*. Lower Left: Mosquito Breeding Habitat in Standing Water. Lower Right: Micrograph of *Flavivirus* (Dengue Virus Family).

The virus replicates in the mosquito gut and then disseminates to other tissues, including the salivary glands. Once infectious, the mosquito is capable of transmitting the virus to uninfected humans for the remainder of its life cycle.[36] As with malaria, pregnant women may be able to pass dengue virus to their fetus.

Symptoms

Cases are classified as dengue or **severe dengue**. Clinically, dengue is suspected when a high fever (40°C/104°F) is accompanied by at least two of the following symptoms: severe headache, pain behind the eyes, muscle and joint pain, nausea, vomiting,

swollen glands, or a rash.[36] In symptomatic cases, people may think they have the flu or other illness characterized by fever, but not necessarily dengue, so the actual global incidence vastly exceeds the number of reported cases. In addition, cases may be asymptomatic, so many people may not know they are infected. Severe dengue is usually a result of subsequent infection by dengue virus after a person has had a prior infection. Severe dengue can be deadly, causing severe bleeding, plasma leakage from blood vessels, respiratory distress, and organ failure. Clinical warning signs include severe abdominal pain, persistent vomiting, blood in vomit, rapid breathing, bleeding gums, fatigue, and restlessness.[36] No specific treatments exist for dengue, but fever-reducing medications are prescribed to target fever and muscle pains. Medical care by experienced physicians and nurses is required for severe dengue, mostly to maintain patients' body fluid volume.[36]

Incidence and Mortality

The number of dengue cases reported to WHO increased over eightfold in the last two decades, from just over 500,000 cases in 2000 to 5.2 million in 2019, the largest number of global cases ever reported.[36] WHO acknowledges that it may miss most dengue cases and estimates that there may be as many as 390 million infections each year.[33] Nearly 4 billion people in 129 countries are at risk, with the largest burdens in South Asia, Southeast Asia, and the Pacific region.[33] Dengue cases are also reported in nonendemic regions because of international travel. Recently, WHO has reported that autochthonous cases of dengue have occurred in some Southern European countries.[33] Dengue incidence since the 1950s

has been driven primarily by increasing urbanization, globalization, and ineffective vector control in tropical and subtropical areas. In recent decades, climate change may also be a contributing factor to this rise. The increase in dengue case numbers is also partly explained by changes in surveillance and reporting practices over time.

According to the GBD report, the global dengue incidence rate increased steadily between 1990 and 2010, then declined, and has now leveled off (**Figure 7.8**).[7] Of the nearly 57 million dengue cases counted in the GBD report in 2019, half were in India, 11% were in China, and 5% were in Indonesia.[7]

The countries with the highest dengue *incidence rates* in 2019 are all island nations in the South Pacific (**Table 7.4**). This is likely the result of a number of factors, including improved disease surveillance, introduction of dengue from international travelers, and susceptibility to mosquito vector outbreaks due to increasingly frequent extreme storms with significant rainfall that creates breeding habitats. The top two countries, Niue and the Northern Mariana Islands, had incidence rates **12 times higher** than the global average in 2019.[7]

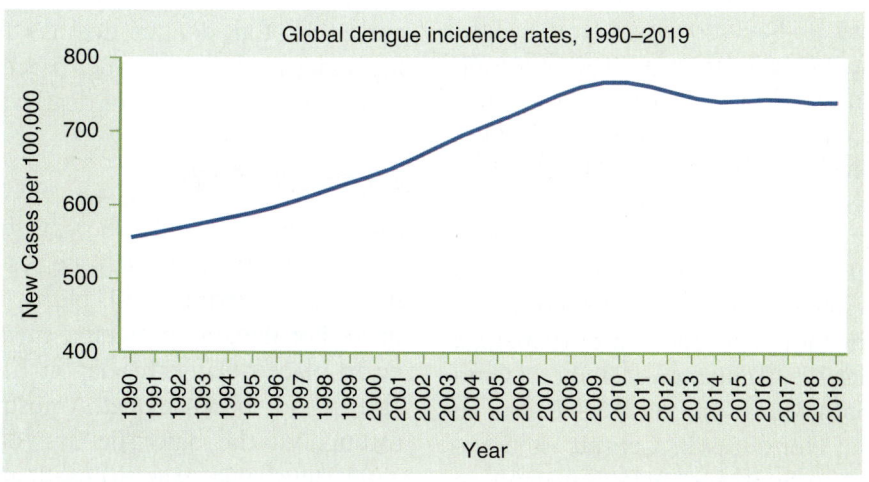

Figure 7.8 Global Dengue Incidence Rates (Age-Standardized) from 1990 to 2019.

Data from Institute for Health Metrics and Evaluation. GBD Compare. Accessed October 15, 2020. https://vizhub.healthdata.org/gbd-compare

Table 7.4 Ten Countries with the Highest Dengue Incidence Rates and Death Rates (Age-Standardized), 2019.

Rank	Country	Incident Cases per 100,000, 2019	Country	Deaths per 100,000, 2019
1	Niue	8,750	Solomon Islands	12.3
2	Northern Mariana Islands	8,687	Indonesia	4.7
3	Kiribati	8,051	Tonga	3.6
4	Palau	7,536	Maldives	2.5
5	Seychelles	7,530	Honduras	2.0
6	Nauru	5,703	Philippines	1.9
7	Tokelau	5,422	India	1.7
8	Cook Islands	5,247	Sri Lanka	1.4
9	Tuvalu	4,678	Suriname	1.3
10	Samoa	4,444	Oman	1.0
	World	*740*	*World*	*0.5*

Data from Institute for Health Metrics and Evaluation. GBD Compare. Accessed October 15, 2020. https://vizhub.healthdata.org/gbd-compare

Dengue is less fatal than malaria but still caused 36,000 deaths in 2019.[7] More than half of all global dengue deaths in 2019 occurred in India (53%), and 28% occurred in Indonesia.[7] The countries with the highest dengue *mortality rates* are the Solomon Islands and Indonesia, with death rates **26 times** and **10 times higher**, respectively, than the global average (Table 7.4).[7] Fortunately, Indonesia's death rate is down 42% since 1990, and the Solomon Islands rate is down 13%, more than the global decline of just 4%. Other countries with high mortality rates are dispersed geographically: Tonga, a Pacific island nation; Maldives, India, and Sri Lanka in South Asia; Honduras in Central America; Philippines in Southeast Asia; Suriname in tropical South America; and Oman in the Middle East.[7]

The proportion of all dengue cases that occur in children under 5 is lower than for malaria. In 2019, children under 5 accounted for 5.4% of all global dengue cases and 18% of all dengue deaths.[7] In contrast, for malaria, children under 5 accounted for 37% of cases and 54% of deaths.[7]

Prevention

As with malaria and other mosquito-borne diseases, vector-control strategies to prevent mosquito breeding and biting are prioritized. For dengue, protective measures must be in place during the day at home, work, and school because *Aedes* mosquitoes bite throughout the day. The first dengue vaccine, Dengvaxia, was licensed in late 2015 and has so far been approved for use in 20 countries.[37] Because people who have never been exposed to dengue are at higher risk of severe dengue after receiving the vaccine, it is currently recommended only for persons living in endemic areas, who are between 9 and 45 years of age, and who have had at least one documented dengue virus infection previously.[36]

Because of its very high dengue burden, Indonesia was the site of a novel vector-control experiment with engineered mosquitoes. A three-year trial concluded in 2020, in which *Wolbachia*-infected *Aedes aegypti* mosquitoes were released in the dengue-endemic city of Yogyakarta on the island of Java. *Wolbachia* is a bacteria that commonly infect insects, and in *Aedes* mosquitoes, *Wolbachia* effectively competes with viruses, resulting in less disease transmission. The study found that in areas where the infected mosquitoes were released, the number of cases of dengue decreased 77% compared with areas where they were not released.[38]

Climate Change Effects

Like malaria, dengue transmission is sensitive to climatic conditions. Dengue cases tend to increase in rainy seasons as more water collects, providing mosquito breeding habitats.[36] The VC of *Aedes* mosquitoes has increased approximately 10% since 1980.[39] In the laboratory, *Aedes aegypti* survival is strongly temperature dependent, with no survival at 15°C (59°F), 90% survival at 20°C (68°F), and 60% survival at 35°C (95°F). Development times from egg to adult vary, from 60 days at 15°C, to 12 days at 20°C, to 6 days at 27–34°C (81–93°F). Mosquito-feeding behavior is also temperature dependent, with the percentage of mosquitoes that complete a blood meal within 30 minutes of host availability varying from 50% at 22–28°C to 0% at 33°C.[5] As with the malaria pathogen, the EIP for dengue virus is temperature sensitive.[36]

Because of shifting rainfall patterns and temperature increases, global habitat suitability for *Aedes aegypti*, which predominates in tropical and subtropical regions, has increased 1–2% per decade since the 1950s and is projected to increase 3–5% per decade between now and 2050.[39] The global range of *Aedes albopictus*, considered the world's most invasive mosquito, has expanded in many areas, including at least 32 U.S. states and 25 European countries.[36] *Aedes albopictus* is thought to be more tolerant of colder temperatures, including mild winters, than *Aedes aegypti*. In addition to urbanization and shifts in climatic conditions, the spread of *Aedes* mosquitoes has been due to international trade in tires (which collect water and create a breeding habitat) and the lucky bamboo houseplant (on which mosquitoes often lay eggs).[36] Many parts of the world have suitable *Aedes* mosquito habitats, but the vector is not yet established there, or the vector is established but the disease pathogen is not. These areas may be vulnerable in the future.

Evidence is strong for *current* climate change impacts on dengue transmission.[13] In Singapore, dengue incidence has risen significantly over the past 40 years, and 14% of this risk is attributed to increased temperatures. Dengue cases are on the rise in Nepal, where increasing minimum temperatures are expanding dengue transmission at higher altitudes. From 1992 to 2011 in Puerto Rico, a 1°C increase in sea surface temperature correlated to a 3.4-fold increase in dengue transmission. In Vietnam, each 50 mm rainfall increase and 1% humidity increase leads to a 1% increase in dengue outbreak risk. In Australia, decreased rainfall and more drought conditions are increasing *Aedes aegypti* mosquito densities because of higher use of water storage containers that become breeding habitats. In China, tropical cyclone activity is linked to increased dengue burden in several parts of the country.

Locally acquired cases of VBDs transmitted by *Aedes* mosquitoes are on the rise. For example, the first locally transmitted case of dengue in Japan was reported in 2014, and local chikungunya incidence was recorded in recent epidemics in the Caribbean, Central America, and South America.[11] Local transmission of dengue has occurred in the Florida Keys, Texas, and Hawaii in recent years, as well as Zika in Texas and Florida. Local transmission of malaria, dengue, and chikungunya have occurred recently in parts of Europe.[11] In 2014, autochthonous chikungunya transmission in France was linked to heavy rainfall, leading to spikes in mosquito populations.

Future increased suitability for dengue is predicted for southern Europe in the summer months and for chikungunya in Western Europe.[10] Models suggest that human movement will explain much of the spread of dengue over the next 5–15 years, but that after that, climate-change-driven temperature increases and urbanization will be the major explanatory variables.[5] In some areas, *Aedes* mosquitoes and the diseases they transmit may decline with climate change. For example, in Ecuador, *Aedes* mosquito range is projected to decline 40–70% between 2030 and 2100, and human populations at risk for *Aedes*-borne diseases may decline 40–80%.[40] In addition, four of five *Anopheles* vectors for malaria will decline in this region, with remaining vectors limited to highland areas.[40]

Lyme Disease

Lyme disease, also called Lyme borreliosis, is a bacterial tick-borne disease caused by a bacterial pathogen. Lyme disease is the most common VBD affecting humans in the northern hemisphere[11] and is one of the most prevalent tick-borne diseases in the world.

Pathogen and Vector

Lyme disease is caused by infection by any of three genospecies of *Borrelia burgdorferi* and transmitted by the bite of infected female *Ixodes* (blacklegged) ticks (**Figure 7.9**).[8] Numerous species of the *Ixodes* tick are widespread in much of the world, and Lyme disease is prevalent in northern climates, particularly in the eastern half of the United States, in southern Canada, in much of Europe, and in parts of Russia, China, and Mongolia.[41] *Ixodes* ticks are vectors for other diseases as well, including *Ixodes scapularis* ticks that transmit anaplasmosis, babesiosis, and Powassan encephalitis in the United States. Powassan encephalitis in particular is less prevalent than Lyme disease but much more deadly.[42]

Lyme disease was first described in 1976 when a group of children living near Lyme, Connecticut, were believed to have developed juvenile rheumatoid arthritis. They were eventually diagnosed with late symptoms of an apparent tick-borne disease that had been recognized elsewhere. In 1981, the bacterial pathogen *Borrelia burgdorferi* was identified and found in an *Ixodes scapularis* tick.[43] It is now known that the pathogen has been endemic for a long time in much of the United States, but disease outbreaks in humans began only in the later decades of the 20th century. These outbreaks were linked primarily to forest fragmentation, suburban development, and deer population increases that created optimal conditions for the growth and spread of ticks as well as human exposures to ticks.[11]

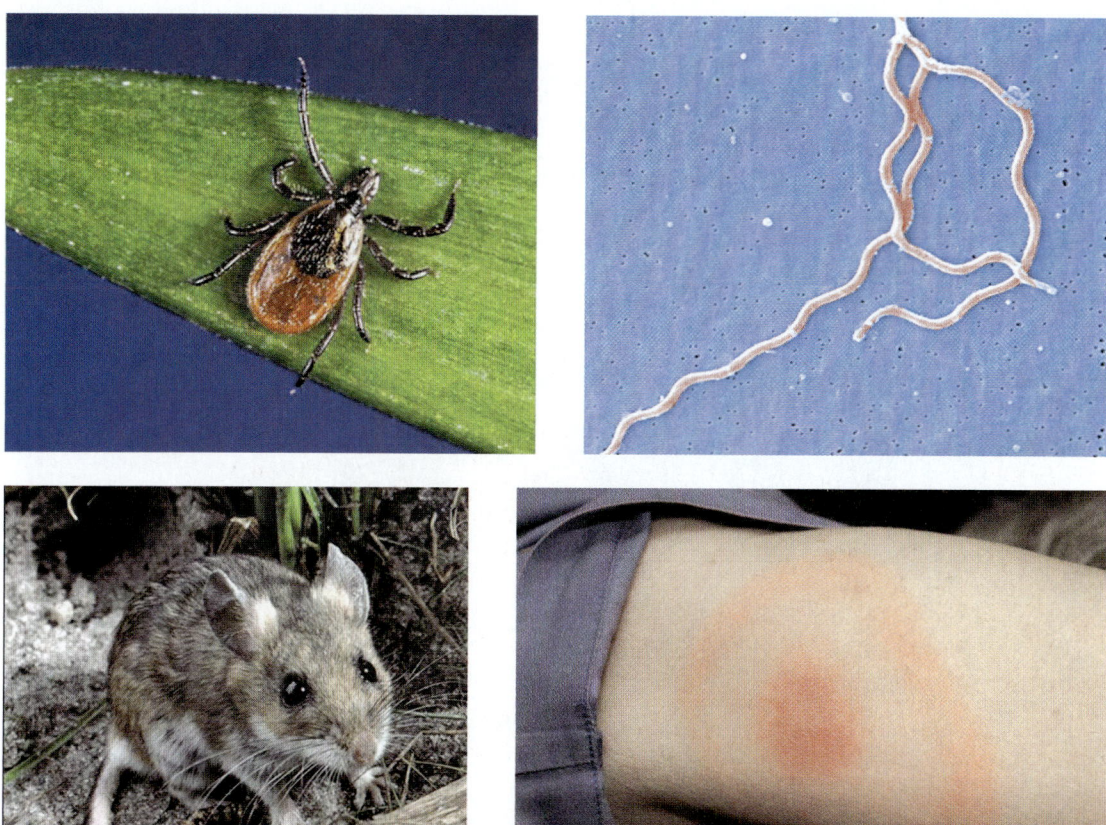

Figure 7.9 Lyme Disease Vector, Pathogen, Reservoir Host, and Classic Symptom. Upper Left: Lyme disease vector *Ixodes scapularis* (blacklegged tick). Upper Right: Pathogen *Borrelia burgdorferi*. Lower Left: Reservoir host *Peromyscus leucopus* (white-footed mouse). Lower Right: Classic "bull's eye" rash (*erythema migrans*), which is a common early symptom of Lyme disease.

Upper Left: Reproduced from CDC/Michael L. Levin, PhD from the CDC's Public Health Image Library; Upper Right, Lower Left, and Lower Right: Courtesy of the Centers for Disease Control and Prevention.

Transmission

Unlike mosquito vectors, *Ixodes* ticks have a multiyear life cycle (**Figure 7.10**), and the pathogen develops more slowly in the vector than do mosquito-borne pathogens, on the order of months rather than days.[12] The primary reservoir host is the white-footed mouse, though other competent hosts in the eastern United States include chipmunks, shrews, eastern gray squirrels, and certain birds.[44] Tick larvae that take a blood meal from a host infected with the *Borrelia* pathogen become infected and stay infected through the nymph stage. Nymphs require a blood meal to develop into adults and, if infected, can transmit the pathogen to an uninfected host. Most human infections occur through the bites of nymph ticks. In the United States, nymph feeding activity is highest in May and June, which coincides with human activities outdoors and thus human exposures to ticks.[10] Adult ticks feed mainly on large mammals such as deer, which, along

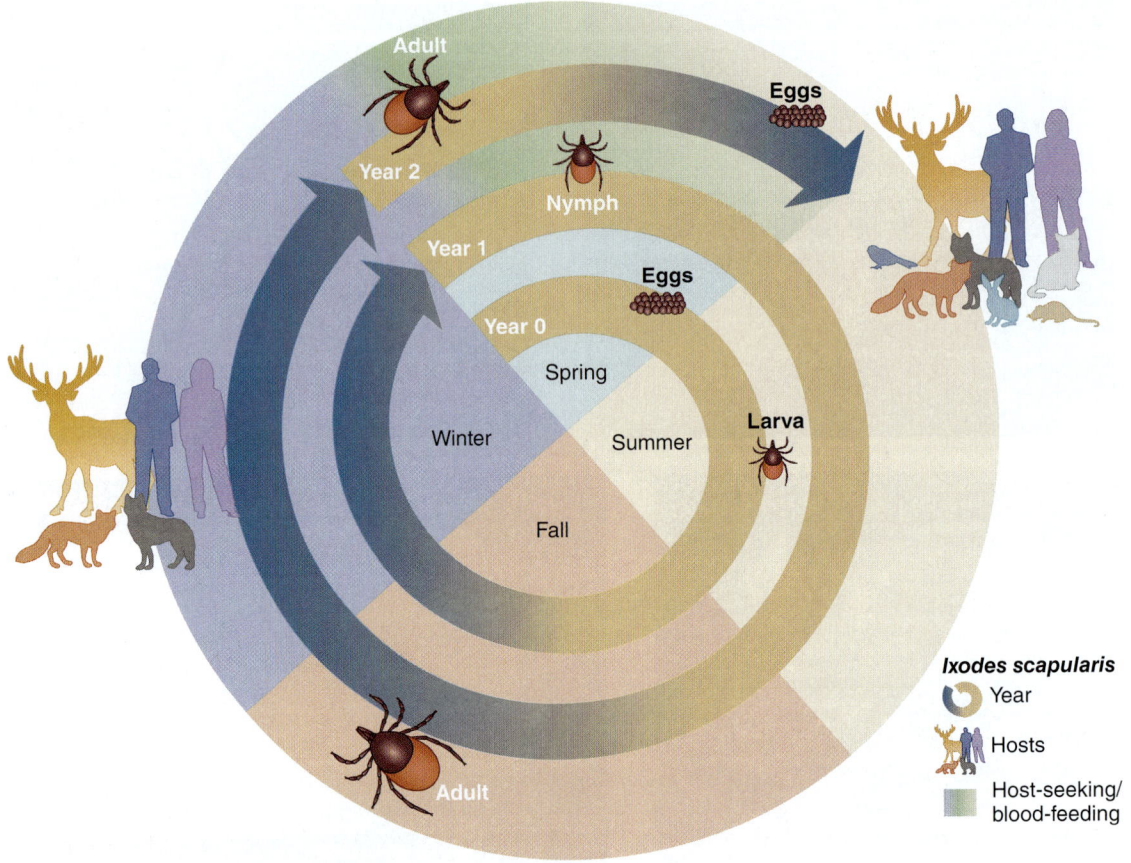

Figure 7.10 Life Cycle of *Ixodes scapularis* Tick.
Reproduced from Centers for Disease Control and Prevention. How ticks spread disease: how ticks survive. https://www.cdc.gov/ticks/life_cycle_and_hosts.html

with humans, are not competent reservoir hosts. Nymph ticks are very small (less than 2 mm) and thus are difficult to spot on the body, whereas adult ticks are larger and more easily discovered and removed.[8]

Symptoms

Symptoms of Lyme disease include fever, headache, fatigue, joint and muscle pain, and a characteristic "bull's eye" skin rash called *erythema migrans* at the site of the tick bite (Figure 7.9).[8] Treatment with antibiotics is effective if the disease is diagnosed in an early stage.[8] If left untreated, infection can lead to arthritis in one or more joints (often the knee), facial palsies (muscle paralysis), meningitis (inflammation of the meninges, membranes that cover the brain and spinal cord), and "Lyme carditis" (inflammation of the heart), which in rare cases leads to sudden cardiac death.[8] Inflammatory pathways in the body triggered by *Borrelia* infection are responsible for most disease symptoms.[45]

Incidence and Mortality

Most cases of Lyme disease in the United States occur in the northeastern and upper midwestern states (**Figure 7.11**). Certain U.S.

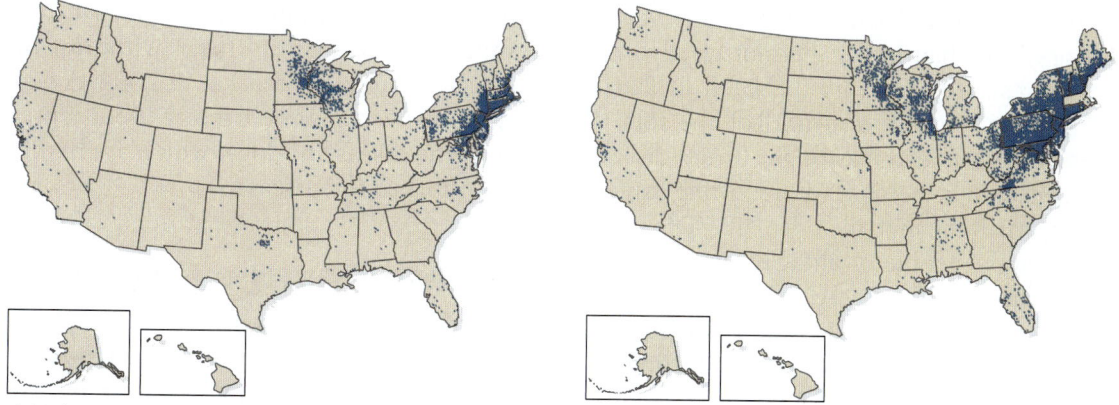

1 dot placed randomly within county of resistance for each reported case　　1 dot placed randomly within county of resistance for each confirmed case

Figure 7.11 Lyme Disease Incidence in the U.S. in 2001 and 2019.

Reproduced from Centers for Disease Control and Prevention. Lyme disease maps: historical data. Accessed June 24, 2021. https://www.cdc.gov/lyme/stats/maps.html

states have very high incidence rates for Lyme disease (**Table 7.5**). The top state, Maine, has a 3-year incidence rate for Lyme disease for 2017–2019 of 104 cases per 100,000 people, more than 13 times higher than the national average for the 43 states that reported Lyme disease cases to the U.S. Centers for Disease Control and Prevention (CDC).[8] The year 2019 was very active for Lyme disease, and Maine reported a 56% increase

Table 7.5 U.S. States with the Highest Lyme Disease Incidence Rates per 100,000 (3-Year Average, 2017–2019), According to the CDC.

Rank	State	Incidence Rate	Confirmed Cases in 2019
1	Maine	104	1,629
2	Vermont	91	706
3	New Hampshire	74	1,106
4	Pennsylvania	62	6,763
5	Rhode Island	56	527
6	Delaware	53	619
7	New Jersey	33	2,400
8	West Virginia	32	703
9	Connecticut	32	795
10	Wisconsin	24	1,219
	Entire U.S.	*7.8*	*23,453*

Data from U.S. Centers for Disease Control and Prevention. Lyme disease. Accessed September 30, 2020. https://www.cdc.gov/lyme/index.html

in cases compared with the previous year.[46] Pennsylvania had by far the most total Lyme disease cases counted by the CDC in 2019: 6,763 cases (nearly one-third of the total).[8] Massachusetts also has high Lyme disease incidence, but because of reporting differences, cases are not included in CDC data. Importantly, people of all ages are at risk of contracting Lyme disease. In 2019, the age groups with the most cases were 55–69, but children ages 5–9 also had many infections.[8]

CDC surveillance data likely underestimate Lyme disease incidence, given that data are not available for all states and that disease cases are not always diagnosed accurately or reported to the CDC. In fact, less than 10% of the estimated yearly cases in the United States are believed to be reflected in CDC data. A 2021 CDC analysis of health insurance claims data revealed that 476,000 Americans were diagnosed and treated for Lyme disease annually during 2010–2018.[47]

Prevention

Critical preventive strategies for Lyme disease include limiting human exposure to ticks via vector-control measures, such as using insect repellants, covering exposed skin with clothing, removing underbrush and litter debris in wooded areas, and checking for ticks on the skin after outdoor activities.[8] Two Lyme disease vaccines, both engineered against an outer surface protein of *Borrelia burgdorferi*, were developed in the mid-1990s. One received approval from the U.S. Food and Drug Administration in 1998 after clinical trials showed a 76% reduction in Lyme disease incidence among vaccinated individuals.[48] The vaccine was available for about 3 years, and physicians reported fewer Lyme disease cases. However, public concern and negative publicity about the vaccine's side effects, including a possible link to an autoimmune form of arthritis, resulted in the manufacturer withdrawing the vaccine from the market in 2002. A version of this vaccine is currently used to prevent Lyme disease in dogs, which are highly susceptible to the disease.

Climate Change Effects

Tick life cycles are sensitive to meteorological conditions, particularly humidity, precipitation, and temperature. Lyme disease incidence is linked to warmer winters, increased summer temperatures, and low seasonal temperature variation, and the range of *Ixodes* ticks is constrained at very low temperatures.[10] Ticks spend most of their life cycle in surface soil or leaf litter in wooded areas and, as such, are somewhat insulated from changes in ambient temperature. Ticks do not require standing water but need high humidity, which is achieved when moisture is trapped in the litter layer.[11] Long-term drought lowers tick numbers and prevents tick invasions into new areas.[12]

In the United States since 1996, *Ixodes* ticks have moved northward. Both ticks and cases of Lyme disease have expanded into southern Canada, with most cases in Quebec, Ontario, and Nova Scotia.[11] Fewer than 100 cases per year were reported in Canada from 2002 to 2006,[12] but a peak of 2,636 cases was recorded in 2019.[9] Warming temperatures are believed to be an important factor in *Ixodes* tick establishment in southern Canada. Higher temperatures, higher humidity, and lower cumulative precipitation are linked to an earlier start to Lyme disease transmission in the spring,[5] and climate models predict a longer season and earlier onset of cases.[49] In Europe, tick-borne disease incidence has increased 400% over the past 30 years, attributed to enhanced surveillance and diagnosis as well as climate change.[10]

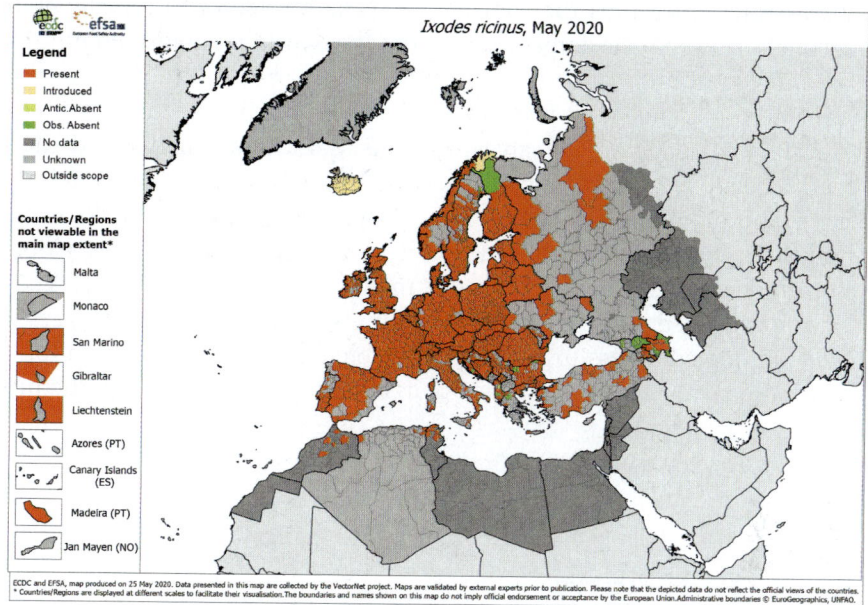

Figure 7.12 VectorNet Tick Surveillance in Europe. The range of *Ixodes ricinus* ticks in 2020 is indicated in red.

Reproduced from European Centre for Disease Prevention and Control and European Food Safety Authority. Tick maps: Ixodes ricinus—current known distribution: March 2021 [internet]. Accessed June 29, 2021. https://ecdc .europa.eu/en/disease-vectors/surveillance-and-disease-data/tick-maps

The current burdens include approximately 65,000 cases of Lyme disease and more than 2,000 cases of tick-borne encephalitis each year.[10] Tick-borne encephalitis incidence increased 50-fold in northern Russia between the 1980s and the first decade of the 2000s, which was linked to increasing mean annual air temperature from 1990 to 2009.[11]

Changes in weather and climate that impact the densities of competent reservoir hosts like the white-footed mouse are particularly important for influencing tick-borne disease transmission.[12] Unlike mosquito vectors, which can fly and thus rapidly expand into new areas, ticks are dispersed by host species, mostly in woodland habitats. Milder and shorter winters in Quebec are linked to a northward spread of the white-footed mouse.[5] Other factors, such as land use changes, establishment of new forested areas, wildlife management practices, and expansion of towns and suburban areas into previously undeveloped and primarily wooded areas, have helped fuel the increase in Lyme disease.[11]

The European Centre for Disease Prevention and Control and the European Food Safety Authority run VectorNet, which conducts VBD surveillance in Europe.[10] VectorNet publishes vector range maps, including for the *Ixodes* tick that transmits Lyme disease and tick-borne encephalitis (**Figure 7.12**). These tools are important to predict and monitor disease outbreaks and to record shifting trends in vector locations, which may be linked to climate change.

Adaptations

Limiting the climate change impacts on VBD burdens requires sustained action to mitigate greenhouse gas emissions along with strengthened public health adaptive capacity.

Continued VBD control measures that have been prioritized globally via the targets set by the MDGs and SDGs and implemented in VBD-endemic regions will likely need to be ramped up significantly in many areas. In addition, early diagnosis and treatment of diseases, vaccination programs for diseases that have a safe and effective vaccine, and improved water and sanitation management for vector control will need to be strengthened, along with improved access to healthcare services.

The burden of VBDs falls predominantly on the world's poorest communities, those with high disease burdens generally, and those that lack access to adequate public health systems, disease treatments, and effective vector control. Given the likelihood that climate change will worsen the risks of many VBDs, it is imperative to redouble global efforts to minimize climatic drivers of VBD transmission, prevent and treat VBD cases in currently endemic areas, and pay attention to those at risk of disease expansion in the future.

Discussion Questions

1. What are the relative burdens of various vector-borne diseases in humans?
2. Describe the *Plasmodium* life cycle and which stage causes symptomatic malaria.
3. Describe how malaria affects children compared with adults.
4. What are the trends in malaria, dengue, and Lyme disease over the past two decades? Which regions and countries have the highest burdens? Why?
5. What is the difference between dengue and severe dengue?
6. What are MDG6 and SDG3? How have they contributed to international efforts to reduce malaria burdens in endemic countries?
7. Describe the life cycle of *Ixodes* ticks and how Lyme disease transmission occurs.
8. What does leishmaniasis incidence and mortality reveal about public health, economic, and social disparities and inequities?
9. Describe how mosquito- and tick-borne diseases are climate sensitive and how their incidence may shift with future climate change.

References

1. Tissera H, Jayamanne BDW, Raut R, et al. Severe dengue epidemic, Sri Lanka, 2017. *Emerg Infect Dis J.* 2020;26:682. doi:10.3201/eid2604.190435
2. Sri Lanka Ministry of Health. Disease surveillance: trends. March 14, 2018. Accessed October 7, 2020. http://www.epid.gov.lk/web/index.php?option=com_casesanddeaths&Itemid=448&lang=en
3. McKirdy E. "Unprecedented" outbreak of dengue fever plagues Sri Lanka. *CNN.* July 25, 2017. Accessed October 7, 2020. https://www.cnn.com

/2017/07/25/health/sri-lanka-dengue-fever/index
.html

4. Ramirez B. Support for research towards understanding the population health vulnerabilities to vector-borne diseases: increasing resilience under climate change conditions in Africa. *Infect Dis Poverty.* 2017;6:164. doi:10.1186/s40249-017 -0378-z

5. Rocklöv J, Dubrow R. Climate change: an enduring challenge for vector-borne disease prevention and control. *Nature Immunol.* 2020;21:479–483. doi:10.1038/s41590-020-0648-y

6. Moore SM, Oidtman RJ, Soda KJ, et al. Leveraging multiple data types to estimate the size of the Zika epidemic in the Americas. *PLOS Negl Trop Dis.* 2020;14:e0008640. doi:10.1371/journal.pntd .0008640

7. Institute for Health Metrics and Evaluation. GBD Compare. October 15, 2020. Accessed October 15, 2020. https://vizhub.healthdata.org/gbd-compare

8. Centers for Disease Control and Prevention. Lyme disease: recent surveillance data. Accessed June 24, 2021. https://www.cdc.gov/lyme/datasurveillance /recent-surveillance-data.html

9. Government of Canada. Surveillance of Lyme disease. Updated March 3, 2021. Accessed June 24, 2021. https://www.canada.ca/en/public-health/services /diseases/lyme-disease/surveillance-lyme-disease .html

10. Semenza JC, Suk JE. Vector-borne diseases and climate change: a European perspective. *FEMS Microbiol Lett.* 2018;365. doi:10.1093/femsle/fnx244

11. Caminade C, McIntyre KM, Jones AE. Impact of recent and future climate change on vector-borne diseases. *Ann N Y Acad Sci.* 2019;1436:157–173. doi:10.1111/nyas.13950

12. Ogden NH, Lindsay LR. Effects of climate and climate change on vectors and vector-borne diseases: ticks are different. *Trends Parasitol.* 2016;32:646–656. doi:10.1016/j.pt.2016.04.015

13. Fouque F, Reeder JC. Impact of past and on-going changes on climate and weather on vector-borne diseases transmission: a look at the evidence. *Infect Dis Poverty.* 2019;8:51. doi:10.1186/s40249-019 -0565-1

14. Pigott DM, Bhatt S, Golding N, et al. Global distribution maps of the leishmaniases. *eLife.* 2014;3:e02851. doi:10.7554/eLife.02851

15. Jervis S, Chapman LAC, Dwivedi S, et al. Variations in visceral leishmaniasis burden, mortality and the pathway to care within Bihar, India. *Parasit Vectors.* 2017;10:601. doi:10.1186/s13071-017-2530-9

16. Purse BV, Masante D, Golding N, et al. How will climate change pathways and mitigation options alter incidence of vector-borne diseases? A framework for leishmaniasis in South and Meso-America. *PLoS One.* 2017;12:e0183583. doi:10.1371/journal.pone .0183583

17. World Health Organization. A global brief on vector-borne diseases. 2014. Accessed August 24, 2020. https://apps.who.int/iris/handle/10665/111008

18. Okwor I, Uzonna J. Social and economic burden of human leishmaniasis. *Am J Trop Med Hyg.* 2016;94:489–493. doi:10.4269/ajtmh.15-0408

19. Das A, Karthick M, Dwivedi S, et al. Epidemiologic correlates of mortality among symptomatic visceral leishmaniasis cases: findings from situation assessment in high endemic foci in India. *PLoS Negl Trop Dis.* 2016;10:e0005150-e. doi:10.1371/journal .pntd.0005150

20. Pascual Martínez F, Picado A, Roddy P, et al. Low castes have poor access to visceral leishmaniasis treatment in Bihar, India. *Trop Med Int Health* 2012;17: 666–673. doi:10.1111/j.1365-3156.2012.02960.x

21. González C, Wang O, Strutz SE, et al. Climate change and risk of leishmaniasis in North America: predictions from ecological niche models of vector and reservoir species. *PLoS Negl Trop Dis.* 2010;4:e585. doi:10.1371/journal.pntd.0000585

22. World Health Organization. World malaria report 2020. November 30, 2020. Accessed March 2, 2021. https://www.who.int/teams/global-malaria-programme /reports/world-malaria-report-2020

23. Centers for Disease Control and Prevention. About malaria. Accessed August 24, 2020. https://www .cdc.gov/malaria/about

24. Sinka ME, Bangs MJ, Manguin S, et al. A global map of dominant malaria vectors. *Parasit Vectors.* 2012;5:69. doi:10.1186/1756-3305-5-69

25. Schumacher R-F, Spinelli E. Malaria in children. *Mediterr J Hematol Infect Dis.* 2012;4:e2012073-e. doi:10.4084/MJHID.2012.073

26. Rasti N, Wahlgren M, Chen Q. Molecular aspects of malaria pathogenesis. *FEMS Immunol Med Microbiol.* 2004;41:9–26. doi:10.1016/j.femsim .2004.01.010

27. Sarma N, Patouillard E, Cibulskis RE, et al. The economic burden of malaria: revisiting the evidence. *Am J Trop Med Hyg.* 2019;101:1405–1415. doi:10.4269/ajtmh.19-0386

28. World Health Organization. Tackling malaria in Sierra Leone. December 13, 2016. Accessed October 9, 2020. https://www.afro.who.int/news /tackling-malaria-sierra-leone

29. United Nations. News on Millennium Development Goals. Accessed August 11, 2020. https://www.un.org/millenniumgoals/

30. World Health Organization. World malaria report 2015. December 11, 2015. Accessed March 2, 2021. https://www.who.int/publications/i/item/9789241565158

31. United Nations. The 17 goals. Accessed August 11, 2020. https://sdgs.un.org/goals

32. The Global Fund. Malaria. Accessed June 24, 2021. https://www.theglobalfund.org/en/malaria/

33. World Health Organization. Malaria: the malaria vaccine implementation programme (MVIP). March 2, 2020. Accessed August 23, 2020. https://www.who.int/malaria/media/malaria-vaccine-implementation-qa/en/

34. Akech S, Chepkirui M, Ogero M, et al. The clinical profile of severe pediatric malaria in an area targeted for routine RTS,S/AS01 malaria vaccination in western Kenya. *Clin Inf Dis.* 2019;71:372–380. doi:10.1093/cid/ciz844

35. Siraj AS, Santos-Vega M, Bouma MJ, et al. Altitudinal changes in malaria incidence in highlands of Ethiopia and Colombia. *Science.* 2014;343:1154–1158. doi:10.1126/science.1244325

36. World Health Organization. Dengue and severe dengue. May 19, 2021. Accessed June 23, 2021. https://www.who.int/news-room/fact-sheets/detail/dengue-and-severe-dengue

37. Centers for Disease Control and Prevention. Dengue vaccine. September 23, 2019. Accessed August 23, 2020. https://www.cdc.gov/dengue/prevention/dengue-vaccine.html

38. Callaway E. The mosquito strategy that could eliminate dengue. *Nature.* August 27, 2020. doi:10.1038/d41586-020-02492-1

39. Iwamura T, Guzman-Holst A, Murray KA. Accelerating invasion potential of disease vector *Aedes aegypti* under climate change. *Nature Comm.* 2020;11:2130. doi:10.1038/s41467-020-16010-4

40. Escobar LE, Romero-Alvarez D, Leon R, et al. Declining prevalence of disease vectors under climate change. *Sci Rep.* 2016;6:39150. doi:10.1038/srep39150

41. Steere AC, Strle F, Wormser GP, et al. Lyme borreliosis. *Nat Rev Dis Primers* 2016;2:16090. doi:10.1038/nrdp.2016.90

42. Centers for Disease Control and Prevention. Powassan virus. July 17, 2019. Accessed October 2, 2020. https://www.cdc.gov/powassan/index.html

43. Steere AC, Coburn J, Glickstein L. The emergence of Lyme disease. *J Clin Invest.* 2004;113:1093–1101. doi:10.1172/JCI21681

44. Levi T, Keesing F, Oggenfuss K, et al. Accelerated phenology of blacklegged ticks under climate warming. *Philos Trans R Soc Lond B Biol Sci.* 2015;370:20130556. doi:10.1098/rstb.2013.0556

45. Singh SK, Girschick HJ. Lyme borreliosis: from infection to autoimmunity. *Clin Microbiol Infect.* 2004;10:598–614. doi:10.1111/j.1469-0691.2004.00895.x

46. Maine Center for Disease Control and Prevention. Lyme disease. Accessed October 2, 2020. https://www.maine.gov/dhhs/mecdc/infectious-disease/epi/vector-borne/lyme/index.shtml

47. Kugeler K, Schwartz A, Delorey M, et al. Estimating the frequency of Lyme disease diagnoses, United States, 2010–2018. *Emerg Inf Dis.* 2021;27:616. doi:10.3201/eid2702.202731

48. Nigrovic LE, Thompson KM. The Lyme vaccine: a cautionary tale. *Epidemiol Infect.* 2007;135:1–8. doi:10.1017/S0950268806007096

49. United States Global Change Research Program. *The Impacts of Climate Change on Human Health in the United States: A Scientific Assessment.* 2016. Accessed January 27, 2021. https://health2016.globalchange.gov

CHAPTER 8

Water

KEY TERMS

WASH
Water
Sanitation
Hygiene
Fecal–oral transmission
Waterborne diseases
Water-washed diseases

Water-based diseases
Water-related diseases
Diarrheal diseases
Leptospirosis
Legionellosis
Schistosomiasis
Onchocerciasis

Soil-transmitted helminth
  infections
Vibriosis
Cholera
Harmful algal blooms (HABs)
Shellfish poisoning
Water scarcity

LEARNING OBJECTIVES

- Define water, sanitation, and hygiene (the three components of WASH).
- Describe how unsafe WASH leads to human diseases.
- Understand the global burden of diarrheal diseases, particularly in children under 5.
- Understand transmission and burdens of other water-related diseases, particularly vibriosis and worm infections.
- Understand the concept and examples of water scarcity.
- Describe how climate change threatens water quantity and quality.

In July 2020, Bangladesh, Bhutan, India, and Nepal endured weeks of torrential monsoon rains, deadly flooding, and landslides. UNICEF, the United Nations (UN) children's organization, estimated that over 4 million children were impacted and in need of lifesaving services, such as clean water and hygiene supplies, to prevent the spread of disease. Hundreds of people were reported to have gone missing or died, including many children presumed drowned, and millions more were displaced. Common disease outbreaks as a result of extreme floods include waterborne illnesses, such as cholera and leptospirosis, and water-related mosquito-borne diseases, such as dengue and malaria. Said Jean Gough, UNICEF regional director for South Asia, "Even for a region that is all-too-familiar with the devastating impact of extreme weather, the recent heavy monsoon rains, rising floods and continued landslides are creating a perfect storm for children and families affected."[1]

Humans face many water-related health issues, most of which may be exacerbated by climate change. Water is a basic need for human survival and well-being, and access to adequate safe water is a fundamental human right. Water quality and quantity are threatened by climate change in many ways, and with increasing frequency and intensity of precipitation, storms, and storm surges, people are more frequently exposed to stormwaters and floodwaters, and are more susceptible to waterborne diseases, water-related injuries, and other adverse health impacts. Billions of people lack access to high-quality water for drinking, improved sanitation systems, and hygiene facilities. Pressure on water systems is constantly growing around the world, as water usage expands by 1% each year.[2] As many as 4 billion people worldwide experience severe water scarcity for at least one month per year, exacerbated by warming temperatures, erratic rainfall, and drought.[2] Without climate-resilient and sustainable water management, many aspects of human life are threatened, including health, physical safety, food security, ecosystem services, energy systems, healthy cities, and equitable socioeconomic development.[2]

Water, Sanitation, and Hygiene

A lack of access to safe and adequate water, sanitation, and hygiene (**WASH**) is a leading environmental determinant of ill health. **Water** refers to improved sources of freshwater for human use—including drinking, bathing, food preparation, and irrigation—that are protected against contamination, especially from disease pathogens and toxic chemicals. Common examples of improved water systems that

allow for the provision of safe water include public water suppliers, piped household water connections, public water pipes, boreholes for groundwater extraction, protected dug wells, protected natural springs, and rainwater collection.[3] In contrast, unimproved drinking water sources include unprotected dug wells, unprotected natural springs, untreated surface water, water from a tanker truck or vendor's cart, and bottled water.

Sanitation refers to improved systems that physically separate humans from human excreta. Access to basic sanitation means that people are using improved sanitation facilities, such as toilets and latrines with sewer or septic connections, or latrines with other protective measures to prevent human exposure to excreta. Unimproved sanitation facilities do not ensure a hygienic separation of humans and their excreta and include open-pit latrines, bucket latrines, open defecation on the ground or in water bodies, and systems in which excreta is mixed with other forms of solid waste.[3] **Hygiene** refers to access to facilities for handwashing with improved water and soap.

Improved WASH interventions are necessary to break the transmission of waterborne, foodborne, and certain airborne diseases that are spread via **fecal–oral transmission**. This pathway allows disease-causing pathogens in contaminated feces to pass from infected to uninfected people, either directly or via contaminated water, soil, food, or flies (**Figure 8.1**).

According to the World Health Organization (WHO), 5.3 billion people (71% of the world's population) used *safely managed drinking water* in 2017, defined as drinking water from an improved source located on premises, available when needed and free from contamination.[4] However, many gaps remain.[2,4]

- Nearly 2.2 billion people lack *safely managed drinking water*.

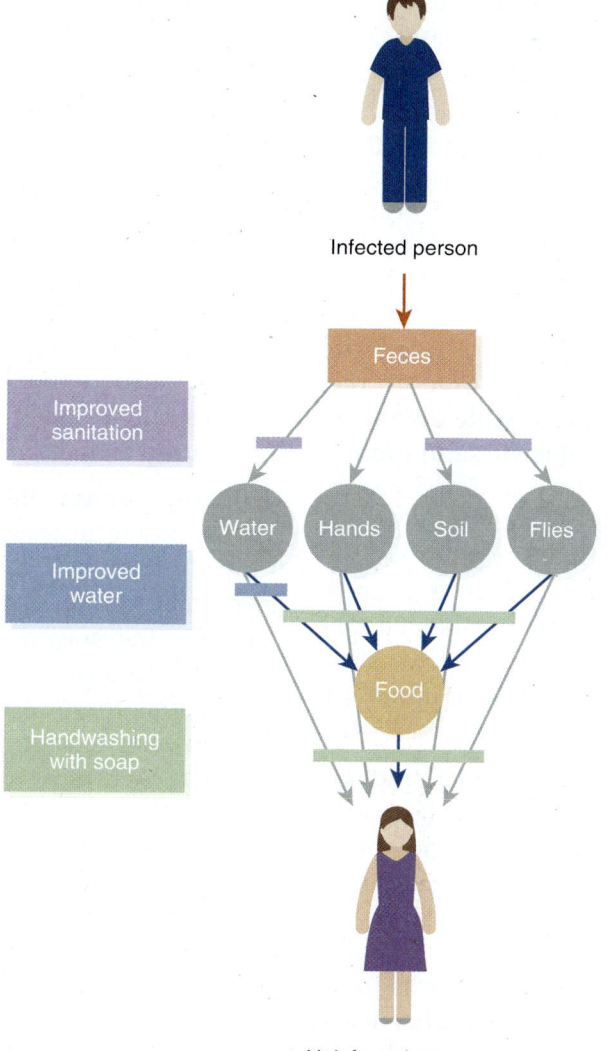

Figure 8.1 Fecal–Oral Route of Disease Transmission, and Barriers to That Transmission from Improved Water (Blue), Sanitation (Purple), and Handwashing (Green).

Data from CEO Water Mandate. Exploring the business case for sanitation: health for productivity. Accessed July 21, 2021. https://ceowatermandate.org/sanitation/impacts/productivity

- Nearly 1 billion people lack even *basic drinking water*, defined as drinking water that can be collected from a protected source in less than 30 minutes.
- Approximately 144 million people collect *untreated surface water* for drinking.
- Approximately 1.6 billion people face *economic water shortage*, meaning that

they live in an area without adequate investment to satisfy the demand for water, even if water is abundant.

In addition, major gaps persist in access to improved sanitation and hygiene. An estimated 4.2 billion people lack *safely managed sanitation*, defined as using hygienic toilets that are not shared with other households and from which wastes are treated and disposed of safely. An estimated 3 billion people lack *basic handwashing facilities* with soap and water in the home.

In spite of these persistent gaps, progress has been made in recent years to improve WASH access, largely catalyzed by international efforts to achieve the Millennium Development Goals (MDGs) and Sustainable Development Goals (SDGs). MDG 7 (*Ensure environmental sustainability*) set a target "**to halve, by 2015, the proportion of the population without sustainable access to safe drinking water and basic sanitation.**"[5] This target was achieved by 2015. Follow-up targets for 2015–2030 are outlined in the SDGs, particularly **SDG 3** (*Ensure healthy lives and promote well-being for all at all ages*) and **SDG 6** (*Ensure availability and sustainable management of water and sanitation for all*).[6] Specific SDG targets related to improved WASH include:[6]

- Target 3.3: "**By 2030, end the epidemics of AIDS, tuberculosis, malaria and neglected tropical diseases and combat hepatitis, waterborne diseases and other communicable diseases**"
- Target 3.9: "**By 2030, substantially reduce the number of deaths and illnesses from hazardous chemicals and air, water and soil pollution and contamination**"
- Target 6.1: "**By 2030, achieve universal and equitable access to safe and affordable drinking water for all**"
- Target 6.2: "**By 2030, achieve access to adequate and equitable sanitation and hygiene for all and end open defecation, paying special attention to the needs of women and girls and those in vulnerable situations**"
- Target 6.3: "**By 2030, improve water quality by reducing pollution, eliminating**

dumping and minimizing release of hazardous chemicals and materials, halving the proportion of untreated wastewater and substantially increasing recycling and safe reuse globally"
- Target 6.a: "**By 2030, expand international cooperation and capacity-building support to developing countries in water- and sanitation-related activities and programmes**"
- Target 6.b: "**Support and strengthen the participation of local communities in improving water and sanitation management**"

Many human diseases are associated with exposure to water of poor quality and to pathogens that require water for disease transmission. **Waterborne diseases** are those transmitted via direct exposure to water contaminated with pathogens. Often, these pathogens are spread via the fecal–oral transmission route, including most diarrheal diseases. **Water-washed diseases** result from improper sanitation or lack of adequate handwashing facilities that prevent removal of pathogens from the skin or from foods or other surfaces, including numerous soil-transmitted worm infections. **Water-based diseases** are caused by pathogens that spend all or most of their life cycles in water or depend on aquatic organisms for part of their life cycles. An example of a water-based disease is Legionnaires' disease, a respiratory illness that is caused by infection by the bacterium *Legionella pneumophila*, which lives in freshwater, stored water sources, and water pipes. **Water-related diseases** include those transmitted by insects that breed in water or live near water, including mosquito-borne malaria and dengue.

Table 8.1 lists the leading water-related diseases in the United States that are likely to cause significant illness or death and for

Table 8.1 Water-Related Disease Characteristics and Surveillance by the U.S. Centers for Disease Control and Prevention. Diseases are categorized by type and listed alphabetically within each category.

Disease	Pathogen	Symptoms	U.S. Case Incidence in 2014 (*estimated*)	% Cases That Are Waterborne
Diarrheal diseases				
Campylobacteriosis	*Campylobacter jejuni* [bacterium that causes 90% of cases]	Diarrhea, dysentery, fever, cramps, nausea, vomiting	1,540,000	13%
Cryptosporidiosis	*Cryptosporidium hominus* and *C. parvum* [parasitic protist]	Watery diarrhea, cramps, fever, dehydration, nausea, vomiting, weight loss	823,000	43%
Giardiasis	*Giardia intestinalis, G. lamblia, G. duodenalis* [parasitic protist]	Diarrhea, gas, cramps, nausea, vomiting, dehydration; in persistent cases, weight loss, micronutrient deficiencies	1,070,000	44%
Norovirus infection	Multiple strains of norovirus	Diarrhea, stomach pain, nausea, vomiting, fever, headache, body aches	21,800,000	6%
Salmonellosis, nontyphoidal[a]	*Salmonella typhimurium* [bacterium]	Diarrhea, cramps, fever	1,350,000	6%
Shiga toxin–producing *Escherichia coli* [STEC] infection, serotype O157	*Escherichia coli* serotype O157 [bacterum]	Bloody or watery diarrhea, cramps, vomiting; in severe cases hemolytic uremic syndrome [kidney damage]	64,200	5%
STEC infection, non-O157 serotype	*Escherichia coli* non-O157 serotype [bacterium]		219,000	6%
Shigellosis	*Shigella sonnei, S. flexneri* [bacteria] [Species common in U.S.]	Diarrhea [may be bloody], fever, pain, feeling the need to pass stool even when bowels are empty; in rare cases, postinfectious arthritis	449,000	4%

[continues]

Table 8.1 Water-Related Disease Characteristics and Surveillance by the U.S. Centers for Disease Control and Prevention. Diseases are categorized by type and listed alphabetically within each category.   *(continued)*

Disease	Pathogen	Symptoms	U.S. Case Incidence in 2014 (*estimated*)	% Cases That Are Waterborne
Diarrheal diseases				
Vibriosis	*Vibrio alginolyticus* (bacterium)	Watery diarrhea, cramping, nausea, vomiting, fever, chills; open wound infection	36,700	37%
	V. parahaemolyticus		92,400	24%
	V. vulnificus		249	77%
	Other *Vibrio*		42,600	2%
Diseases that are primarily respiratory				
Legionnaires' disease	*Legionella pneumophila* (bacterium that causes 90% of cases)	Pneumonia-like symptoms, including cough, shortness of breath, fever, headache, muscle aches; may also cause diarrhea, nausea, confusion	11,400	97%
Nontuberculous (NTM) infection	Mycobacteria other than *M. tuberculosis* (which causes TB) and *M. leprae* (leprosy)	Fever, weight loss, night sweats, decreased appetite, loss of energy, cough, shortness of breath, rash	97,000	72%
Pseudomonas pneumonia	*Pseudomonas aeruginosa* (bacterium that causes most cases)	Lung pneumonia, coughing, congestion	31,700	51%
Pseudomonas septicemia		Blood infection, high fever, chills, confusion, shock; in rare cases, diarrhea	26,100	22%
Other diseases				
Acute otitis externa (swimmer's ear)	*Pseudomonas aeruginosa* or *Staphylococcus aureus* (bacteria)	Ear canal redness, itch, discomfort, fluid drainage, decreased hearing; in severe cases, pain radiating to the face, neck, head, ear canal blockage, swollen lymph nodes, fever	5,980,000	79%
TOTAL			**>33,600,000**	

[a] *Salmonella typhi* and *paratyphi* infections cause the life-threatening diarrheal diseases typhoid fever and paratyphoid fever. Most cases reported in the United States come from infections acquired while traveling in other countries.
Data from Collier S, Deng L, Adam E, et al. Estimate of burden and direct healthcare cost of infectious waterborne disease in the United States. *Emerg Infect Dis.* 2021;27(1):140–149. doi:10.3201/eid2701.190676

which waterborne transmission is plausible.[7] Most are **diarrheal diseases** caused by a range of infectious agents. There are three clinical types of diarrhea:

- Acute watery diarrhea, characterized by frequent loose, watery stools
- Acute bloody diarrhea (dysentery), a serious condition characterized by intestinal inflammation, diarrhea with blood in stools, and vomiting
- Persistent diarrhea, which lasts 14 days or longer

In addition to the infections listed in Table 8.1, rotavirus infection, which is vaccine-preventable, is another major global diarrheal disease. In all cases, the most dangerous symptom is severe dehydration, which can be fatal if not treated promptly with rehydration solutions containing electrolytes (salts). Dehydration leads to a range of symptoms. Children may exhibit dry mouth, sunken eyes and cheeks, listlessness or irritability, no tears when crying, or no wet diapers for at least 3 hours. Adults may experience extreme thirst, less frequent urination, fatigue, dizziness, and confusion.

Each year in the United States, waterborne diseases lead to more than 600,000 hospital emergency department visits and approximately 118,000 hospitalizations, 6,630 deaths and $3.33 billion in direct healthcare costs.[7] Norovirus infection and acute otitis externa (swimmer's ear) are the most common illnesses, and most hospitalizations and deaths are caused by nontuberculous mycobacteria (NTM), *Pseudomonas*, and *Legionella* infections. Many disease cases go undiagnosed or unreported each year, so actual incidence is much higher. Children and people with compromised immune systems are at higher risk of these pathogenic infections. In the United States, waterborne diseases common a century ago, including typhoid fever (caused by *Salmonella typhi* infection) and cholera (caused by *Vibrio cholerae* infection), are rare today, but they persist in other regions.[7] Many of these pathogens may also be present in foods and cause disease via foodborne transmission.

Diarrheal Disease Burdens

Globally, unsafe water, unsafe sanitation, and lack of access to handwashing facilities are the most important risk factors associated with diarrheal disease transmission.[8] **Table 8.2** lists the global disease burdens associated with unsafe WASH among people of all ages and children under 5 in 2019, according to the Global Burden of Disease (GBD) report compiled by the Institute for Health Metrics and Evaluation.[8] Enteric diseases—those that cause an intestinal infection—are attributed to all three risk factors, and certain respiratory diseases are attributed to unsafe hygiene. More deaths and disability-adjusted life years (DALYs) are attributed to unsafe water than to unsafe sanitation and unsafe hygiene. In total, more than 2.6 million deaths and 140 million DALYs are attributed to unsafe WASH, with 36% of deaths and 61% of DALYs occurring in children under 5. Fortunately, death and DALYs rates attributed to unimproved WASH declined 50–59% in people of all ages and 69–70% in children under 5 between 1990 and 2019.[8]

Diarrheal diseases cause a significant proportion of global morbidity and mortality. In 2019, diarrheal diseases caused more than 1.5 million deaths (3% of all global deaths) and 81 million DALYs.[8] In children under 5, over half a million deaths (10% of all deaths in children under 5) and nearly 46 million DALYs were due to diarrheal diseases.[8] As a group, diarrheal diseases were

Table 8.2 Global Burden of Total Deaths and DALYs from Enteric or Respiratory Diseases Attributed to Unsafe Water, Sanitation, or Hygiene in People of All Ages and in Children Under 5, 2019.

Risk Factor	Cause	Total Deaths, 2019	Total DALYs (millions), 2019
All ages			
Unsafe water source	Enteric infections	1,230,000	65.1
Unsafe sanitation	Enteric infections	757,000	41.4
Lack of handwashing	Enteric infections	358,000	20.0
Lack of handwashing	Respiratory infections/TB	270,000	14.0
	TOTAL	**2,615,000**	**140.5 million**
Children under 5			
Unsafe water source	Enteric infections	419,000	37.9
Unsafe sanitation	Enteric infections	280,000	25.2
Lack of handwashing	Enteric infections	141,000	12.6
Lack of handwashing	Respiratory infections/TB	114,000	10.1
	TOTAL	**954,000**	**85.8 million**

Data from Institute for Health Metrics and Evaluation. GBD Compare. Accessed October 15, 2020. https://vizhub.healthdata.org/gbd-compare

the fourth leading cause of global deaths and DALYs in children under 5 in 2019, and in sub-Saharan African countries, the leading cause of DALYs and second leading cause of deaths among children under 5.[8] Since 1990, the global burden of diarrheal diseases has decreased significantly, with deaths down nearly 50% and DALYs down nearly 60% in people of all ages, and down 70% in children under 5.[8]

This declining burden of diarrheal diseases is due almost entirely to reductions in years of life lost due to premature deaths (YLLs) in children. YLLs are a major component of DALYs along with YLDs, years of healthy life lost to disability. The YLLs value will be greater for children than adults because if they die prematurely, many more years of life are lost. The YLLs rate is a form of mortality rate, but instead of counting the *number* of deaths in a population, YLLs sum the *total years of life lost for every premature death* (the difference between the ideal life expectancy and the age at premature death). Thus, the YLLs value gives more weight to diseases that are deadly to children and younger adults.

The YLLs rate for diarrheal diseases declined 71% in children under 5 between 1990 and 2019 (**Figure 8.2**).[8] The impressive decline is attributed primarily to fewer children being exposed to unsafe sanitation, a reduction in severe child undernutrition, improved rates of infant breastfeeding, higher use of oral rehydration solutions as a treatment for diarrheal diseases in children, and vaccination against rotaviruses, the most common global cause of severe diarrheal

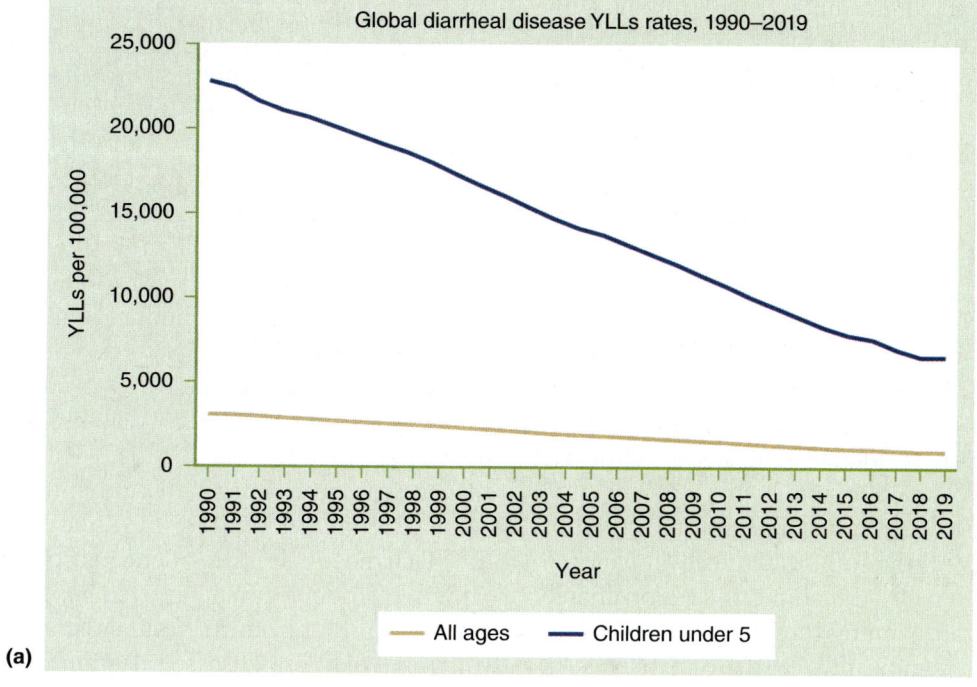

(a)

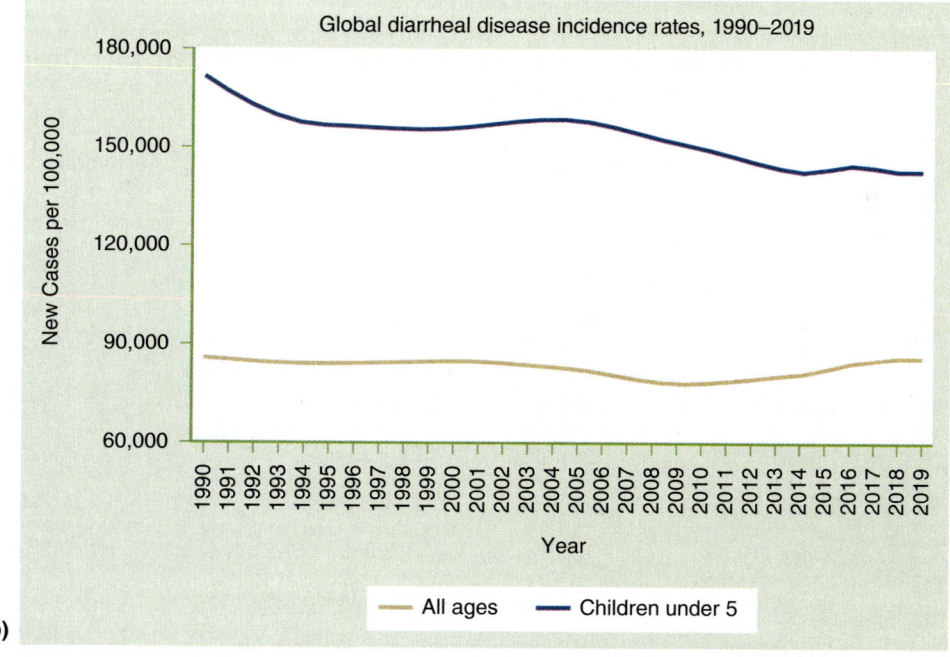

(b)

Figure 8.2 Trends in Global Burden of Diarrheal Diseases in People of All Ages (Age-Standardized) and in Children Under 5, 1990–2019. **(a)** Years of life lost to premature death (YLLs) per 100,000 people; **(b)** Incident cases per 100,000 people.

Data from Institute for Health Metrics and Evaluation. GBD Compare. Accessed October 15, 2020. https://vizhub.healthdata.org/gbd-compare

diseases in young children.[9] During this same period, the *incidence* of diarrheal diseases in children under 5 declined only 17%, indicating that children are still getting sick, but fewer are dying of these diseases.[8] Global incidence in people of all ages remained level, although older adults experienced an *increase* during this period.

Diarrheal disease burdens vary greatly by region and country.[8] The Solomon Islands, Chad, and Niger had the highest *incidence* rates in 2019, **more than twice** the global average. Most of the top countries saw an *increase* in incidence rate since 1990. The Central African Republic had by far the highest *death* rate, **8 times higher** than the global average. All the top 10 countries for mortality were in sub-Saharan Africa. Fortunately, deaths rates among these top countries have declined significantly (24–73%) since 1990.

Table 8.3 lists the 10 countries with the highest diarrheal disease burden in children under 5 (seven countries in sub-Saharan Africa plus Yemen, Haiti, and Afghanistan).[8] Sudan, Chad, and Yemen had the highest *incidence* rates in 2019, **approximately double** the global average. These rates did not vary much from 1990 values. Four of these countries were also in the top 10 for diarrheal disease *death* rates in children under 5 in 2019, including Chad at #1, with a death rate **10 times higher** than the global average.[8] All the top 10 countries for diarrheal disease mortality were in sub-Saharan Africa. Unlike incidence rates, death rates in these countries have declined significantly (30–75%) since 1990.

Diarrheal disease incidence is so high in children under 5 that, on average, children in the top 10 countries suffer from more than one diarrheal disease infection per year.

Table 8.3 Ten Countries With Highest Diarrheal Disease Incidence Rates and Death Rates in 2019 in Children under 5.

Rank in 2019	Country	Incident Cases per 100,000, 2019	Country	Death Rate per 100,000, 2019
1	Sudan	292,406	Chad	752
2	Chad	285,783	Central African Republic	643
3	Yemen	277,725	Niger	473
4	Mauritania	275,469	Nigeria	395
5	Senegal	272,561	Togo	376
6	Burundi	261,676	Cameroon	320
7	Haiti	259,991	Madagascar	278
8	Niger	255,994	Burundi	272
9	Cameroon	252,845	Somalia	255
10	Afghanistan	245,884	South Sudan	228
	World	*142,971*	*World*	*75*

Data from Institute for Health Metrics and Evaluation. GBD Compare. Accessed October 15, 2020. https://vizhub.healthdata.org/gbd-compare

The incidence rate in children under 5 is **1.7 times higher** than in people of all ages, and the death rate is **3.6 times higher**. India accounts for the most *total* diarrheal disease deaths and DALYs among people of all ages, and Nigeria has the most *total* deaths and DALYs in children under 5.[8]

Other WASH-Related Diseases

Other serious WASH-related diseases include leptospirosis, Legionnaires' disease, and worm infections. **Leptospirosis** is a zoonotic disease, originating in livestock and dogs and spreading to humans, that is caused by five pathogenic bacteria of the genus *Leptospira* (**Figure 8.3**).[10] Humans become infected when they are exposed to water, food, or soil contaminated with the urine of infected animals. Bacteria enter the body through skin abrasions or the mucous membranes of the mouth, eyes, and nose. *Leptospira* bacteria are present worldwide but are seasonally endemic in tropical and subtropical areas, especially those with poor sanitation and in agricultural areas with livestock operations

or rodent infestations. Outbreaks often occur after storms, heavy precipitation, or floods. Transmission from mammalian host to environment and back is aided by extensive rainfall and movement of water across the land. Use of canals and irrigation channels in agricultural settings also raises the risk of *Leptospira* transport.[10]

Symptoms of leptospirosis include sudden fever, chills, headache, muscle pain, red itchy eyes, skin rash, difficulty urinating, nausea, vomiting, and diarrhea. Prompt treatment with antibiotics prevents a more severe illness, known as Weil's disease, which occurs in about 10% of patients and is characterized by kidney or liver failure, jaundice, heart failure, meningitis (inflammation of the brain), respiratory failure, and death.[10]

Legionellosis, commonly known as Legionnaires' disease (LD), results from infection by *Legionella*, bacteria found in natural and human-made freshwater systems. *Legionella* are transmitted via small aerosols (airborne water droplets) that can be inhaled and cause lung infections (Figure 8.3).[11] The first reported LD cases were associated with an outbreak at an American Legion convention in Philadelphia

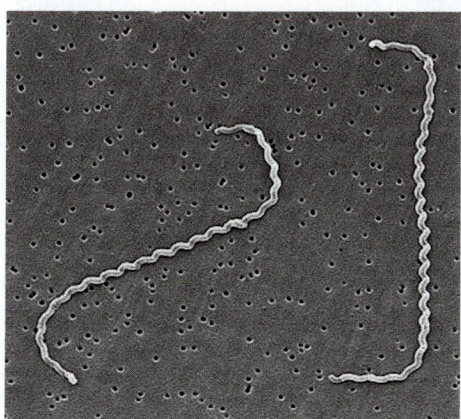 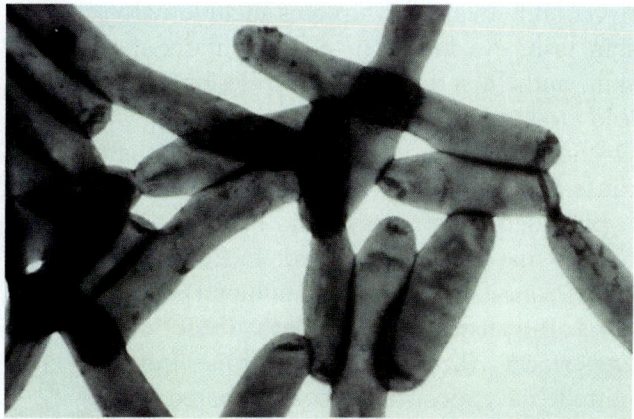

Figure 8.3 Micrographs of *Leptospira* Bacteria and *Legionella* Bacteria.
Left: Content Provider: Rob Weyant/Janice Haney Carr/CDC; Right: Reproduced from CDC/Dr. Francis Chandler from the CDC's Public Health Image Library.

in 1976, where 182 people became ill, and 29 died. *Legionella* thrives in biofilms, layers of microbial cells that adhere to each other and to inner surfaces of corroded pipes, and in stored water sources that warm as air temperatures increase. LD is seasonal, peaking in warm summer months, and sensitive to humidity and rainfall.[11]

Water-related worm infections are common in certain countries around the world but tend to be uncommon in the United States. **Schistosomiasis** (also known as bilharzia or snail fever) is caused by several species of a parasitic flatworm (*Schistosoma*) that grow in certain freshwater snails and are transmitted when humans come in skin contact with water containing these snails. Thus, schistosomiasis is also a vector-borne disease (VBD). Symptoms include rash, itchy skin, fever, chills, cough, and muscle aches. If left untreated, schistosomiasis can lead to organ damage, including inflammation or scarring of the intestine, liver, or bladder. In 2019, schistosomiasis caused 11,500 global deaths and more than 1.6 million DALYs.[8]

Onchocerciasis (also known as river blindness) is caused by the parasitic worm *Onchocerca volvulus*. It is a VBD transmitted by blackflies that live and breed near fast-flowing rivers and streams. Symptoms include itchy skin rash, eye infection, nodules under the skin, and, in serious cases, eye lesions that lead to visual impairment or blindness. Onchocerciasis is the second leading global cause of blindness after eye cataracts and in hyperendemic areas may blind half the population that becomes infected. In 2019, the disease caused more than 1.2 million DALYs.[8]

Soil-transmitted helminth infections are worm (helminth) infections transmitted via exposure to contaminated soil that often result from a lack of sanitation and cause significant health burdens (**Figure 8.4**).[12] As many as a billion people

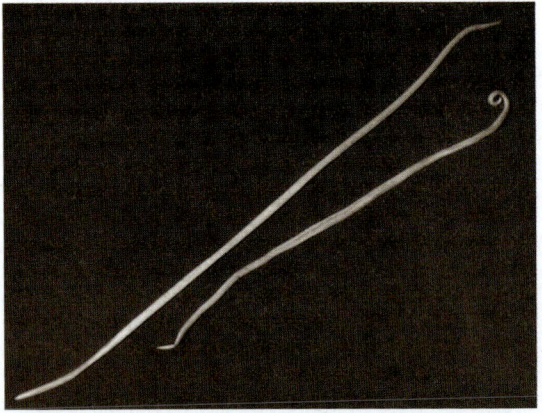

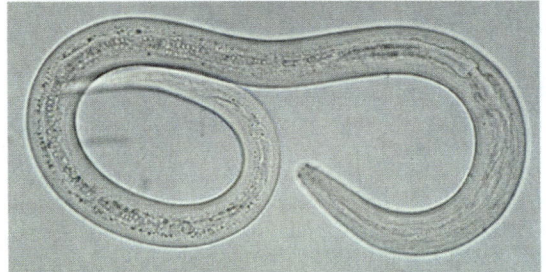

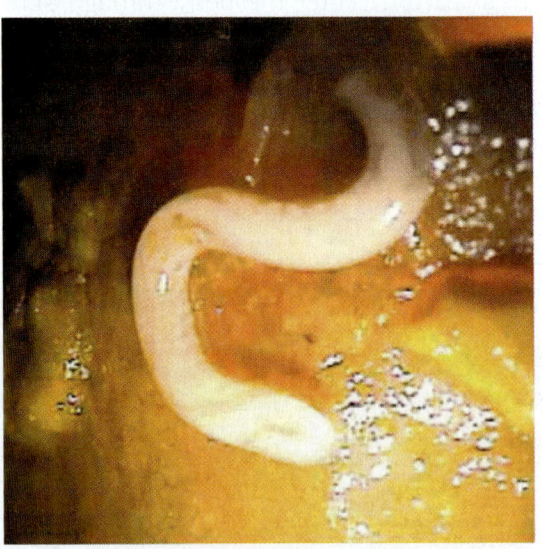

Figure 8.4 Soil-Transmitted Helminths That Cause Human Diseases. Top to bottom: *Ascaris lumbricoides*, hookworm larva, and *Trichuris trichiura* in a human intestine.

Top: Courtesy of Centers for Disease Control and Prevention; Middle: Reproduced from Hookworm filariform larvae. Centers for Disease Control and Prevention; 2019. Accessed September 17, 2019. https://www.cdc .gov/dpdx/hookworm/index.html; Bottom: Reproduced from Posterior end of an adult *T. trichiura*, taken during a colonoscopy. Image courtesy of Duke University Medical Center. Centers for Disease Control and Prevention; 2017. Accessed December 19, 2017. https://www.cdc.gov/dpdx/trichuriasis/index.html

worldwide suffer from **ascariasis** due to *Ascaris lumbricoides* (roundworm) infection. The major symptom of ascariasis is abdominal pain. **Hookworm disease**, caused primarily by *Ancylostoma duodenale* or *Necator americanus* infections, also affects about 700 million people worldwide (See **Box 8.1**). Symptoms in cases of severe hookworm infection include abdominal pain, diarrhea, loss of appetite, weight loss, fatigue, and anemia. Infected children may experience diminished physical and cognitive growth as a result of nutrient deficiencies caused by worm infections. As many as 700 million people suffer from trichuriasis due to *Trichuris trichiura* (whipworm) infection, which can cause diarrhea and, in children, nutrient deficiencies.

These three worm diseases caused 2 million global DALYs and over 2,000 deaths in 2019. Fortunately, deaths have declined 74% and DALYs have declined 63% since 1990.[8] Improvements are due primarily to global efforts to improve access to safe WASH and prevention of worm infections in children in endemic countries by administering antihelminthic medications prophylactically as part of routine pediatric care.

Mold infections are another water-related health risk. The risk of mold growth in homes and other structures increases with more frequent and intense rainfall, storms, and flooding fueled by climate change. Exposure to several mold species causes short- and long-term health problems, including invasive and sometimes deadly mold infections,

Box 8.1 Soil-Transmitted Helminth Infections in Lowndes County, Alabama.

Soil-transmitted helminth infections are thought to primarily affect people in developing regions where improved WASH is lacking, so it surprised many when a recent study uncovered high rates of hookworm infection among residents of Lowndes County, Alabama.[13] This location was chosen to monitor because of its high poverty rate and lack of improved sanitation systems, with nearly three-quarters of residents exposed to raw sewage as a result of faulty septic systems or waste pipes that deposit human waste from homes into backyards (**Figure 8.5**). Exposure risk increases further when these minimal waste systems are overwashed in heavy rainfall, resulting in human waste backing up into household sinks and bathtubs.[14]

Figure 8.5 A straight line sewage pipe from a home in rural Lowndes County, Alabama, dumps untreated sewage directly into the environment.
© Mickey Welsh – USA TODAY NETWORK.

If human feces infested with hookworms are deposited on the ground, the parasite grows in the soil and can penetrate the skin, particularly of the feet, of people who come into direct contact with the ground. Worms grow and reproduce in the small intestine, and the cycle of disease transmission continues if the infected host's feces are introduced into the environment and infect others. Hookworms thrive in warm, moist climates

(continues)

Box 8.1 Soil-Transmitted Helminth Infections in Lowndes County, Alabama. *(continued)*

where people have limited access to improved WASH, and rural areas that lack paved surfaces and improved sanitation systems are at higher risk.[15] In the past, infections were known to be prevalent in this area of the southern United States, with as many as 40% of the population infected in the 1950s, but in recent decades, monitoring declined.[15]

People infected in Lowndes County were treated with anti-helminthic medications. However, unlike in poor countries, where the drug costs pennies, in the United States, it may cost up to $500 per pill, and many of the affected families lack health insurance. Because of the prevalence of hookworm infection in this area, a public–private

collaboration has been set up to provide drugs at subsidized costs.[15]

The U.S. CDC is funding an ongoing follow-up study of children in Lowndes County and two other counties in Alabama. Researchers suspect that hookworm may also be a problem in parts of Mississippi, Louisiana, Texas, and South Carolina that have similar conditions and poor communities at risk.[15] In the meantime, the Alabama Center for Rural Enterprise filed a lawsuit in 2019 alleging that the Alabama Department of Public Health and the Lowndes County Health Department discriminated against the predominantly Black community included in the study by not taking action to improve their sanitation systems.[16]

---

allergies, and respiratory illnesses. One type of mold, *Aspergillus*, produces the carcinogen aflatoxin B1. After many of the major hurricanes that have hit the United States in recent years, homes that were flooded had significant mold growth.[17]

Women and girls face unique risks associated with lack of access to improved WASH.[18] Cooking and washing with contaminated water exposes females to deadly pathogens on a regular basis. Women are the primary caregivers in the household and may face difficult burdens keeping their family healthy without safe water, a toilet, or soap. Women and girls are responsible for water collection in 8 out of 10 households that use water from off-premises sources, including unsafe sources such as rivers, streams, and unprotected holes in the ground.[18] Fetching water is time-consuming and difficult work, and it places females at risk of extreme energy expenditure, injury, harassment, and physical and sexual violence.

Long collection times can result in women and girls being late for or unable to attend school or work. A study in Tanzania showed that reducing water collection times from 30 to 15 minutes per day increased girls' school attendance by 12%.[18] In some cases, people may resort to using contaminated water sources closer to home rather than traveling increasingly far to gather improved water.

In some situations, females risk attack while collecting water or using open defecation sites or toilets or latrines not in or near the home. They also have specific hygiene needs during menstruation, pregnancy, and child-rearing. Managing menstruation without adequate facilities causes girls to miss school and women to stay home from work. One million deaths each year are associated with childbirth under unhygienic conditions, and WASH-associated infections account for 26% of neonatal deaths and 11% of maternal deaths.[18] Addressing the WASH needs of females is key to achieving gender equity, yet women and girls are often excluded from WASH decision-making because of traditional gender roles and power structures.

WASH-Related Climate Change Impacts

Even though major progress has been made to reduce WASH-related diseases, burdens remain high, and increased stress on WASH systems from climate change threatens this progress. Climate-related impacts will vary locally and regionally, and human exposures and responses will depend largely on baseline water conditions, health status of the population, public health infrastructure, and dependence on rain-fed and irrigated agriculture. Increasing temperatures, erratic rainfall, and loss of ecosystem services that naturally purify and protect water supplies will reduce water quality and increase the threat of pathogens and other contaminants.[19] For example, a major outbreak of gastroenteritis in New Zealand was linked to contamination of a local water supply following heavy rainfall. An estimated 5,500 people fell ill, resulting in 45 hospitalizations and three deaths.[20] Flooding moves contaminants through the environment, including from sewage and wastewater, and drought conditions concentrate pollutants in ever-shrinking water sources.[21] Certain bacteria—specifically fecal coliforms, *E. coli*, and enterococci—that tend to thrive at higher temperatures are commonly monitored in marine and freshwater environments as indicators of water quality and safety for human exposures.

Public water supplies provide improved treated drinking water and dramatically reduce disease burdens, but problems may still occur because of climate change, aging infrastructure, and regulatory failures. Climate-change-fueled outbreaks of pathogenic water contamination threaten both public and private water supplies. Water capture and reuse may become common as water supplies dry up, which is a beneficial strategy to conserve resources but may result in more human exposure to stored water at risk of contamination.

Climate-change-induced human displacement may remove people from reliable WASH services and lead to people living in crowded emergency shelters or camps without adequate WASH. This often increases the risk of infectious disease outbreaks, including diarrheal diseases, along with mental stress and other health problems.[19] In addition, in some parts of the world, healthcare facilities lack improved WASH services, a situation that may worsen as climate change increases human health burdens. One-third of healthcare facilities have no handwashing supplies, 1 in 4 facilities lack basic water services, and 1 in 10 have no sanitation system.[22] This means that 1.8 billion people use healthcare facilities that lack basic water services, and 800 million use facilities with no toilets. In the world's least developed countries, the problem is even greater, with half of healthcare facilities lacking basic water services.[22]

Vibriosis

Vibriosis is a collection of infectious diseases sensitive to climate change that are caused by several species of *Vibrio* bacteria that inhabit marine and estuarine waters all over the world. More than 100 species of *Vibrio* exist, and 12 are known to infect humans, though most infections are by four species: *V. cholera, V. vulnificus, V. parahaemolyticus,* and *V. alginolyticus. Vibrio* growth is seasonal and temperature-sensitive. In general, *Vibrio* thrives in warm, low-salinity waters, and most cases of vibriosis occur in summer months when temperatures are warmer.

Cholera is the deadliest form of vibriosis, caused by *V. cholerae* infection. Cholera incidence occurs mostly during warm months, and transmission is via the fecal–oral route. *V. cholerae* serotypes O1 and O139 account for most disease cases, although more than 200 serotypes have been identified. About 10% of people with cholera experience severe symptoms, which include profuse watery diarrhea, vomiting, thirst, leg cramps, restlessness, or irritability. Without prompt treatment, dehydration sets in, indicated by a rapid heart rate, dry mucous membranes, low blood pressure, and loss of skin elasticity. Severe dehydration can lead to kidney failure, shock, coma, and death.[23]

It is estimated that nearly 3 million cases of cholera and 95,000 deaths occur globally each year, mostly in low-income countries.[24] In 2021, several countries had cholera outbreaks (**Figure 8.6**). Yemen had the most cases in 2020 (over 200,000 cases). An ongoing civil war in Yemen has destroyed public health infrastructure, reduced access to improved WASH, and blocked international humanitarian aid, leading to severe health threats and the largest cholera outbreak on record. Bangladesh had the second highest number of cases in 2020 (over 100,000).[25] Both countries reported tens of thousands of cholera cases in the first part of 2021.[25]

Treatment for cholera is with intravenous fluids or oral rehydration therapy with salts to combat dehydration and electrolyte loss, as well as antibiotics to eliminate *Vibrio*, although drug resistance is common. Cholera vaccines are available for tourists traveling to endemic areas and for populations at risk during disease outbreaks. As for all diseases transmitted by the fecal–oral route, access

- • 1
- • 100
- 🔴 100,000
- 🟩 Affected countries

Figure 8.6 Regions of the World with Cholera Cases in the First Half of 2021.

to improved WASH greatly reduces cholera incidence.

Vibrio cholerae thrives at very warm sea surface temperatures—around 30°C (86°F)—and low water salinity, and growth is also associated with intense monsoon rains and the El Niño phase of the El Niño Southern Oscillation (ENSO) climate cycle.[21] Most of the variation in cholera incidence in Bangladesh from 1980 to 2000 was attributed to El Niño,[26] which causes warmer sea surface temperatures, increased rainfall, flooding, and nutrient runoff from the land. This creates suitable conditions for *Vibrio* and for tiny zooplankton called copepods that serve as reservoirs for *Vibrio*. Copepods are the most abundant animal in the ocean, critically important in ocean food webs, and many bacteria, including *Vibrio cholerae*, reside in their guts and on their shells. One single copepod is able to carry an infective dose of *Vibrio cholera*, and copepod abundance is positively associated with cholera cases.[24] In the future, climate change is projected to improve conditions in many parts of the world for copepod and *Vibrio* growth, and thus for cholera risk.

Other forms of vibriosis are more common in humans than cholera. Infection by pathogenic strains of *Vibrio parahaemolyticus* is the most common cause of gastroenteritis from seafood consumption, which is increasing in incidence worldwide, particularly in parts of Asia.[21] Gastroenteritis is inflammation of the stomach and intestines that causes vomiting and diarrhea, usually due to viral infection or the action of bacterial toxins, as is the case with *Vibrio* species. *V. parahaemolyticus* thrives in very warm water and low salinity, notably in coastal brackish waters, and may infect more than half of harvested oysters and clams in many parts of the world.[27] These bacterial infections are treated with antibiotics, but *V. parahaemolyticus* is reported to be highly multidrug resistant.[28]

In the United States, an estimated 80,000 cases of vibriosis occur each year, mostly in immunocompromised people, and are characterized by watery diarrhea or wound infections if skin is exposed to contaminated water.[29] In addition, approximately 100 deaths occur each year. Most cases are due to consuming raw or undercooked oysters or other seafood that are contaminated. *V. parahaemolyticus* is the most common bacterium detected in seafood, but *V. vulnificus* causes the most fatalities linked to seafood consumption. In 2014, 18% of *V. vulnificus* patients in the United States died, compared with 1% for *V. parahaemolyticus*.[30] Infections may be underreported because in most cases, symptoms resolve within a few days without treatment, so affected people may not seek medical care or even know they have shellfish poisoning.

Most cases occur from May through October, when water temperatures are warmer. *Vibrio* grows very quickly and can respond rapidly to changing environmental conditions. Higher water temperature increases *Vibrio* motility, pathogenicity, and antimicrobial resistance.[21] Locally, sea level rise and flooding can spread *Vibrio* into new water sources and soils. In some cases, extreme precipitation reduces sea water salinity, which may allow bacteria like *Vibrio* to thrive.

Research suggests that vibriosis is increasing worldwide, including in places where the disease was not previously known. Noncholera *Vibrio* is monitored as an important indicator of climate change in marine systems.[29] Between 1996 and 2005, the annual incidence of vibriosis increased 41% in the United States, and recent vibriosis outbreaks have occurred in more temperate regions, including the Baltic Sea, northern Spain, Chile, Alaska, and the coastal northeastern United States.[29] *Vibrio* abundance has increased in the North Sea with warming sea surface temperatures, particularly above 18°C (64°F). Cases have even been reported less than 100 miles from the Arctic Circle. Rising temperatures likely explain part of this increasing incidence and geographic shifts, although improved surveillance, diagnosis, and reporting, rising coastal populations, and increased shellfish consumption likely also play a role.

Harmful Algal Blooms

Another risk to human health posed by waterborne organisms is shellfish poisoning due to **harmful algal blooms (HABs)** in marine or freshwater ecosystems. The U.S. Global Change Research Program has identified marine and freshwater HABs as a critical water-associated health impact of climate change in the United States, predicting that higher water temperatures will shift seasonality and geographic range of freshwater and marine harmful algae and *Vibrio* bacteria, which may increase human exposures and disease incidence.[31] This increased risk is also attributed to extreme precipitation causing more runoff, which leads to water contamination, and coastal storms and storm surges that damage water infrastructure and increase exposures to harmful algal species.

Marine HABs

The upper ocean is inhabited by phytoplankton, also known as **microalgae**. Two main classes of microalgae, *dinoflagellates* and *diatoms*, produce **biotoxins** to defend against predators. These biotoxins are also harmful to humans and other species when consumed (**Figure 8.7**). Microalgae form the base of aquatic food webs, and when ocean conditions are optimal—specifically temperature, sunlight, and nutrient levels—explosive phytoplankton growth occurs, known as a *bloom*. Blooms may cover very large areas of water and last for weeks.

Bivalve shellfish such as oysters, clams, mussels, and scallops filter large amounts of sea water and can accumulate biotoxins produced by harmful microalgae (**Figure 8.8**).[32] Human exposure to biotoxins through the consumption of contaminated shellfish may lead to toxic **shellfish poisoning**, including **paralytic shellfish poisoning (PSP)**, **amnesic shellfish poisoning (ASP)**, **diarrhetic shellfish poisoning (DSP)**, and **neurotoxic shellfish poisoning (NSP)**.[33] In all cases, the biotoxins have poisonous effects on the brain and other organs. In addition, human exposure and illness may occur via ingestion of biotoxin-contaminated drinking water or inhalation of airborne biotoxins, as with red tide.

PSP is caused by saxitoxin, which is produced by several species of marine dinoflagellates, including *Alexandrium fundyense* (New England red tide), and freshwater cyanobacteria.[32] Human PSP cases have been reported in Europe, Asia, New Zealand, Australia, and South Africa, and in Alaska, West Coast states, and New England in the United States. In PSP cases, saxitoxin blocks the flow of sodium ions in the brain and causes muscle paralysis, difficulty breathing, and, in extreme cases, death by respiratory failure. In nonfatal cases, paralysis resolves quickly, but persistent muscle weakness may occur. Standard cooking methods for shellfish do not eliminate the toxin. The mortality rate for PSP is 2%—higher than for other forms of shellfish poisoning.[32]

ASP is caused by domoic acid produced by marine diatoms, such as *Pseudo-nitzschia*. In ASP, domoic acid damages the brain and kidneys and may cause memory loss (amnesia), seizures, unstable blood pressure, heart rhythm problems, and coma.

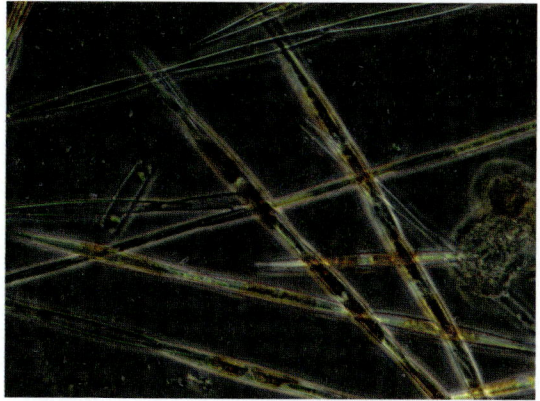
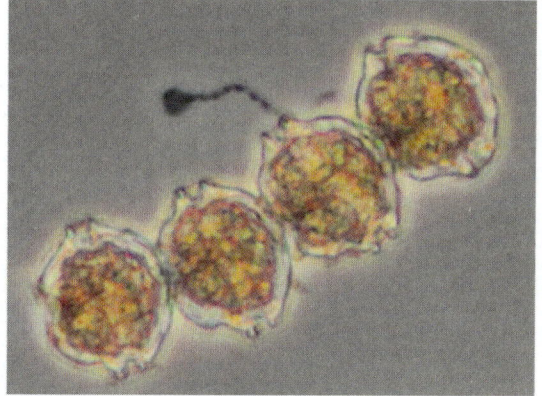

Figure 8.7 Marine Algae Species Associated with Harmful Algal Blooms. Top Left: Marine diatom *Pseudo-nitzschia*, which produces the neurotoxin domoic acid that causes amnesic shellfish poisoning. Top Right: Chain of *Alexandrium catenella* cells, a marine dinoflagellate that produces the neurotoxin saxitoxin, exposure to which causes paralytic shellfish poisoning. Bottom: Red tide in Whangarei, Northland, New Zealand. Red tides are caused by explosive growth of any of a number of dinoflagellate species.

Approximately 10% of patients develop long-term amnesia, as well as damage to peripheral nerves that leads to persistent tingling and prickling sensations.[32] The first reported ASP cases in humans occurred in 1987 in Prince Edward Island, Canada, when 143 people who had eaten domoic acid-contaminated blue mussels became ill, and 3 died. Cases have also been documented in Japan, but no cases of ASP have been recorded in the United States to date. In 2015, a bloom of domoic acid-producing *Pseudo-nitzschia* stretched from Alaska to Mexico,[34] and in recent years in the waters off coastal Maine, domoic acid toxicity has exceeded regulatory limits each year.[35]

DSP is caused by okadaic acid produced by marine dinoflagellates. In DSP, okadaic acid disrupts sodium signaling in the brain. Symptoms include nausea, vomiting, diarrhea, and abdominal pain, which are often mistaken for symptoms of other diarrheal diseases. Outbreaks of DSP have occurred in the Netherlands and Japan, and okadaic-acid-producing

Figure 8.8 Shellfish Consumed by Humans Who Are Susceptible to Biotoxin Infection from Harmful Algal Blooms. Clockwise from Upper Left: oysters, clams, mussels, scallops.

Top Left: © Artur Begel/Shutterstock; Top Right: © 1Roman Makedonsky/Shutterstock; Bottom Left: © billnoll/E+/Getty Images; Bottom Right: © Floortje/E+/Getty Images.

organisms have recently been found along the Gulf Coast of Texas.[32]

NSP is caused by brevetoxins produced by the dinoflagellate *Karenia brevis* (red tide), which bind to sodium channels in nerve and muscle cells, causing them to fire repetitively.[32] Brevetoxins can cause gastrointestinal symptoms and nerve tingling, headache, muscle pain, loss of control of bodily movements, vertigo, reversal of hot and cold temperature sensation, slow heart rate, and dilation of eye pupils. Brevetoxins can aerosolize in sea spray, and if inhaled, they can cause respiratory illness with severe bronchoconstriction, mucosal irritation, and cough.[32] NSP due to *Karenia brevis* exposure is most common in Florida and other Gulf of Mexico regions.[32]

Physical, chemical, and biological conditions in marine environments affect HAB species' growth rates, photosynthesis, motility, and nutrient acquisition rates.[36] Both bloom events and shellfish poisoning are predicted to worsen with climate change. Algal growth rates increase at higher temperatures, with higher CO_2 dissolved in the ocean, and, in some cases, with greater ocean acidification.[34,37] *Pseudonitzschia* blooms and resulting domoic acid levels in shellfish are associated with warmer ocean temperatures, and shellfish closures are more common in warmer years.[36] In recent years, HABs have emerged along the east coast of Tasmania, resulting in permanent public warnings since 2017.[36] In Chesapeake Bay in the eastern United States, warmer and wetter spring seasons and increased nitrogen nutrient inputs from runoff are linked to increased risk of HABs.[37]

Regions vulnerable to HABs may expand in some areas and contract in others, depending on how climate change affects local conditions and how ocean temperatures become more or less suitable for algae growth. Evidence suggests that marine HABs are moving poleward as warming progresses.[34] In European waters, suitable habitats for HABs are projected to shift northward for some species and southward for others, particularly in response to ocean temperature but also salinity and water dynamics.[37] Harmful algal expansions are projected for northeastern Asia with changing temperature, salinity, and nutrient loads, but little change is expected in Southeast Asia.[37]

The state of Maine provides a good example of government-led initiatives to protect people from exposure to HABs and minimize the risks of shellfish poisoning.[35] Shellfish poisoning is a "notifiable condition," which means that all diagnosed cases must be reported to the Maine CDC. Even though few cases are reported, the state presumes shellfish poisoning occurs even if people do not go to the doctor or if they end up being misdiagnosed. In addition, it is estimated that many cases occur in summer tourists who consume contaminated seafood but leave the state before symptoms appear. Maine has enacted shellfish closures regularly and aggressively as HABs have become more common. The state's Department of Marine Resources conducts regular surveillance and testing for HAB species and dictates these shellfish closures to minimize both health risks and economic losses for fishing communities. HABs themselves also cause significant mortality in fish, marine mammals, and seabirds.

The HABs outreach campaign by the Maine CDC makes the following recommendations to the public[33]:

- Consult the Department of Marine Resources on current closures for shellfish harvesting due to algal blooms and red tide notices.
- Do not eat finfish or shellfish sold as bait. Bait products do not need to meet the same food safety regulations as seafood for human consumption.
- Do not harvest shellfish from floating containers or from open ocean. Only harvest shellfish from approved locations.
- Consume shellfish from certified dealers only.
- Cooking or freezing contaminated shellfish does not kill toxins.

The risk of HABs is increasing along the Maine coast because the Gulf of Maine is warming faster than most oceans in the world. *Alexandrium catenella* (New England red tide), which produces PSP-causing saxitoxin, has become prevalent in the Gulf of Maine, which now sees annual blooms and shellfish closures.[35] In 2019, researchers installed four underwater robotic sampling devices in the eastern Gulf of Maine to detect New England red tide.

Freshwater HABs

HABs also occur in freshwater systems, mostly due to explosive growth of cyanobacteria fed by water warming and nutrient overloading. Toxic cyanobacteria (also called blue-green algae) include *Microcystis* (**Figure 8.9**), which produces the biotoxin microcystin. Upon exposure, microcystin causes a range of symptoms, including diarrhea, headaches, rashes, chest pain, and respiratory symptoms.[38] Microcystin is also a likely carcinogen. Exposure to freshwater HABs occurs through direct water contact

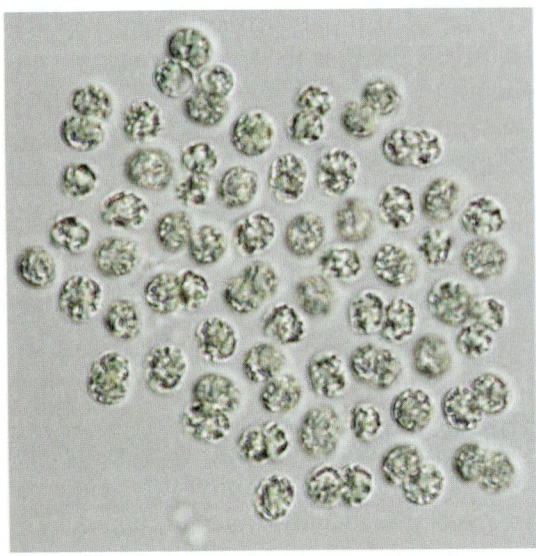 

Figure 8.9 Left: Cyanobacterium *Microcystis aeruginosa*. Right: *Microcystis* Bloom in Lake Dora, Florida, 2010.

Left: Courtesy of Barry H. Rosen, USGS; Right: Courtesy of NaraSouza, Enviroscience Inc.

(swimming, wading, or playing in the water), drinking contaminated water, or consuming contaminated fish.

Many bodies of surface freshwater provide drinking water for millions of people yet have recurring toxic blooms of harmful cyanobacteria. In China, for example, more than 60% of lakes have HABs.[2] Lake Erie, the drinking water source for millions of people in the United States and Canada, is persistently plagued with cyanobacteria blooms, which have increased in the past two decades.[38]

Water Scarcity

In addition to worsening water quality, climate change is altering water supplies and seasonal availability of water, a major problem in water-stressed parts of the world (**Figure 8.10**). **Water scarcity** refers to a long-term imbalance between water demand and water availability in a specific region, and major drivers are economic development, increased consumption levels, and population pressures. Climate change is already fueling loss of water resources in many places, notably in drought-prone regions.[2] For example, in June 2021, all but one of the major water reservoirs in California were down to 31–63% of total capacity.[39] Warming surface temperatures and more erratic rainfall, combined with increasing water usage, are creating severe drought conditions in many parts of the world (**Figure 8.11**). Over a 20-year period, from 1995 to 2015, 5% of natural disasters were attributed to drought, affecting 1.1 billion people, leading to 22,000 deaths, and causing $100 billion in damage.[2] Drought increases food scarcity, food insecurity, undernutrition, and famine for many people. Water scarcity will increase in the future, with more than half of the world's population living in water-stressed regions by 2050, and drylands around the world expanding significantly.[2]

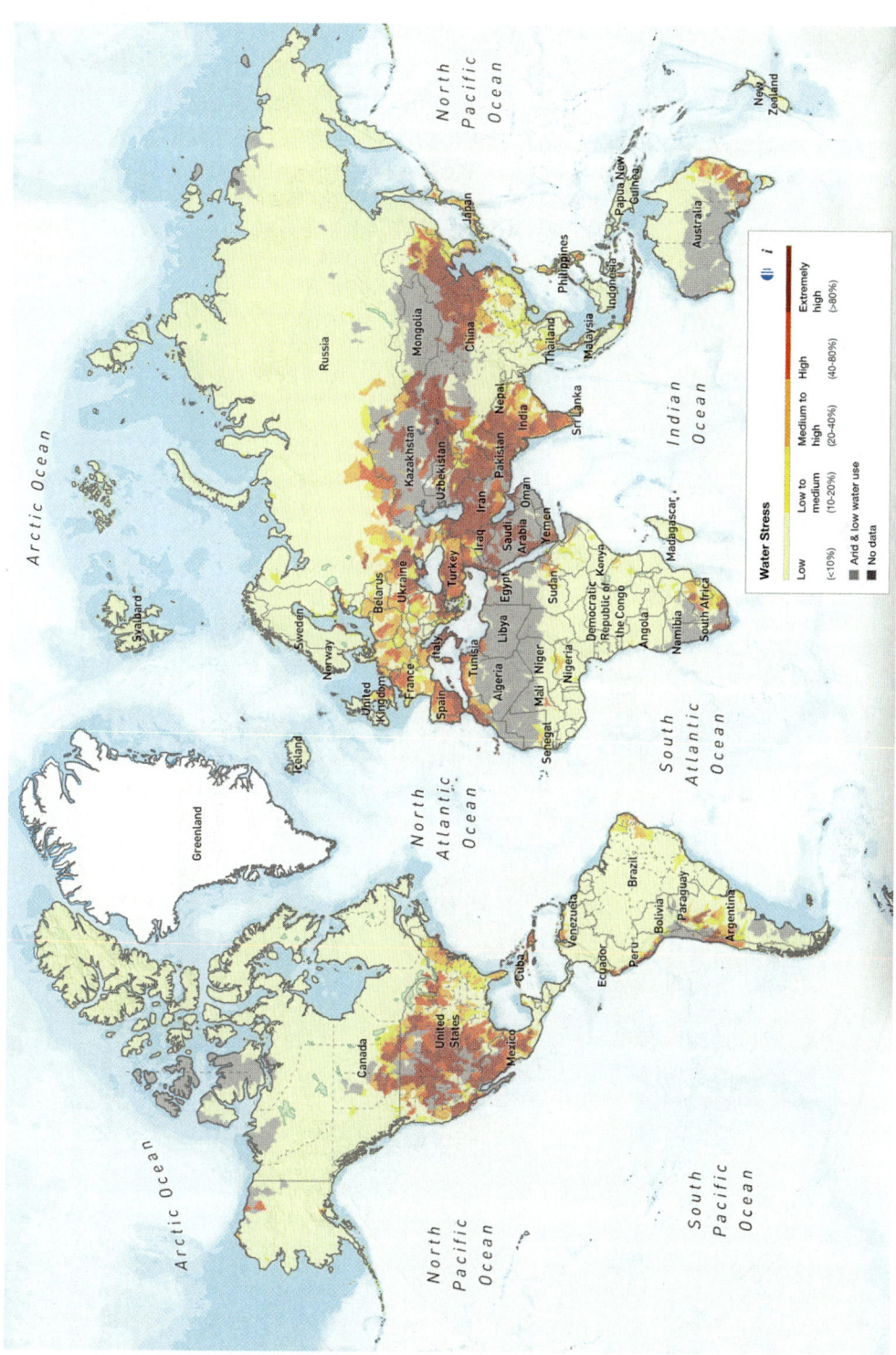

Figure 8.10 Map of Projected Water Stress in 2030. Regions in dark red face the greatest water stress.

Courtesy of WRI – World Resource Institute.

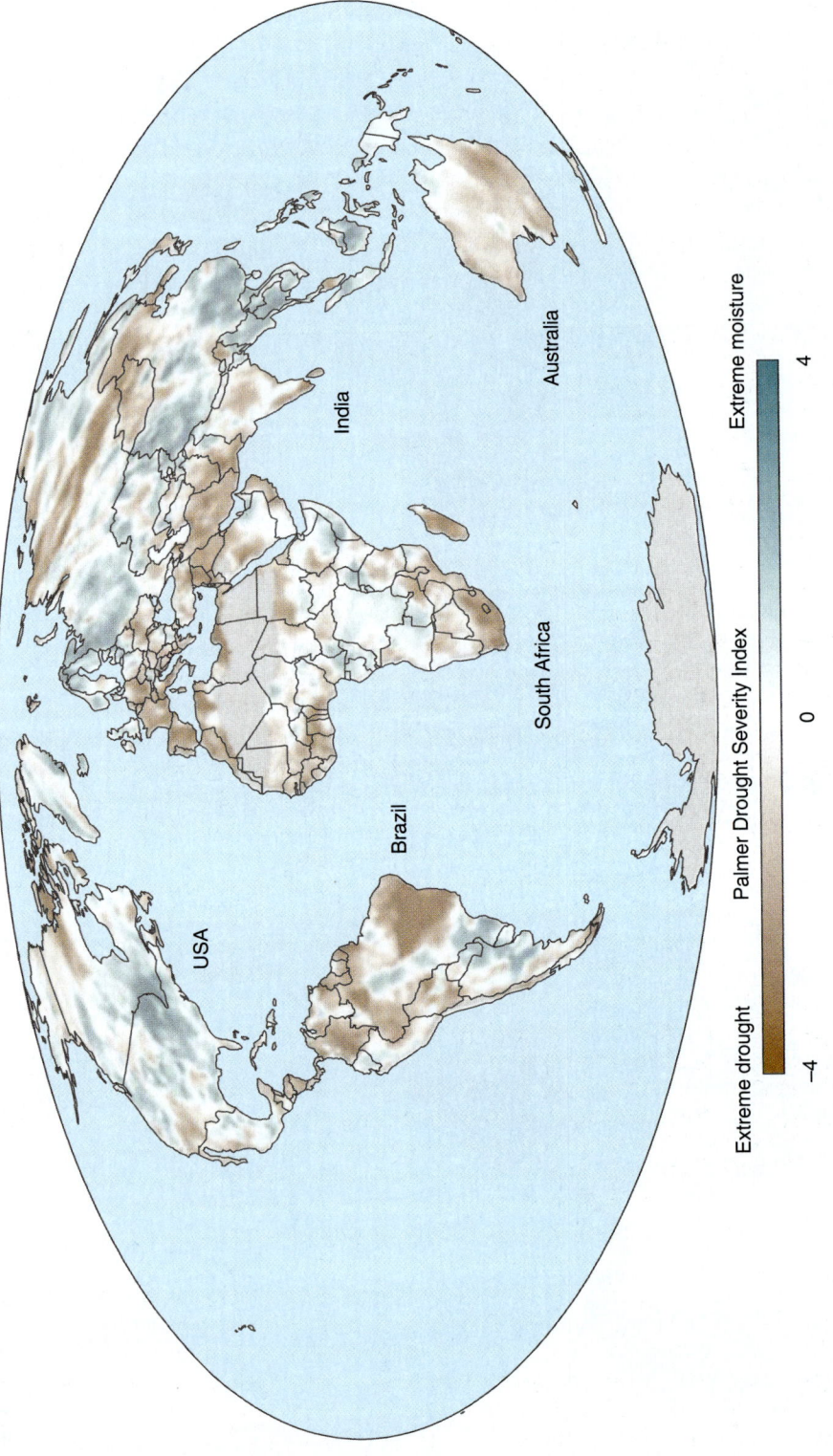

Figure 8.11 Map of Drought Severity in 2017.

These maps show a high degree of regional variation in vulnerability to water stressors. By 2050, nearly 700 million people living in hundreds of cities around the world will face declining freshwater availability because of climate change. Some cities, such as Santiago, Chile; Cape Town, South Africa; Amman, Jordan; and Melbourne, Australia, may face 30–50% declines in water quantity.[2] In places where the water supply is severely limited, people without access to alternative improved sources may resort to collecting water from unsafe sources, traveling farther to find water, and competing with others for scarce water resources, which may create conflict.

Water scarcity may also cause people to leave rural areas and migrate to cities. For example, in Ghana, increasing water stress due to more erratic rainfall, along with poor water management, is driving young farmers to abandon agriculture and seek new livelihoods in cities. Increased access to reliable water sources is an important climate adaptation for farmers that will improve agricultural productivity and reduce food instability, and it may lead to economic and social revitalization in rural areas.[40]

Water resources in mountain regions and their adjacent lowlands are threatened by accelerated melting of glaciers.[2] For example, several of the world's major rivers flow out of the Himalayan mountains and provide freshwater for more than a billion people in Tibet, Nepal, India, Pakistan, and Bangladesh. This water is replenished by annual monsoon rains and glacial melt. Glaciers are considered "water towers" because they store winter snowfall as ice high in the mountains, and they surrender it as meltwater in spring and summer. In this way, glaciers provide a steady flow of water that nourishes humans and ecosystems. Downstream, in the plains of Pakistan and northern India, the world's most extensive system of irrigated agriculture depends on the Indus River. Climate change is causing the rapid melting of the world's glaciers, and these critical rivers are expected to have diminished flow by 2050.[2]

Variable snow cover will impact flow in snow-dominated rivers, which has been observed in Eurasian and North American rivers. Decreased flow volumes and seasonal water availability have been observed in rivers in western Africa, southwestern Australia, the U.S. Pacific Northwest, and China's Yellow River basin.[2] Such decreases affect water withdrawal for agriculture, industry, and domestic supplies, as well as power generation, navigation, fisheries, recreation, and biodiversity and ecosystem health.[2] The Yellow River, the second largest river in China and vulnerable to water supply fluctuations, flows across nine provinces and autonomous regions and supplies water to approximately 9% of China's population. It also supplies irrigation water for 17% of the country's agricultural lands.[41] These threats to water availability are linked to adverse human health impacts.

Small island developing states (SIDS) are particularly vulnerable to increased water stress. Most SIDS have scarce freshwater, and the supply they do have is projected to decrease as a result of lower and erratic precipitation in the future and an increase in water demand due to population growth and tourism.[2] Most SIDS residents live near coastlines, and intrusion from sea level rise and more extreme storms may contaminate drinking water sources and coastal farmlands with saltwater. A lack of resilient water infrastructure and ineffective regulations compound the problems, and less robust economies and high debt loads prevent many SIDS from investing significantly in water systems.

In addition to WASH concerns, the SDGs have several targets that address water scarcity:[6]

- Target 6.4: "**By 2030, substantially increase water-use efficiency across all sectors and ensure sustainable withdrawals and supply of freshwater to address water scarcity and substantially reduce the number of people suffering from water scarcity**"
- Target 6.5: "**By 2030, implement integrated water resources management at all levels, including through transboundary cooperation as appropriate**"
- Target 6.6: "**By 2020, protect and restore water-related ecosystems, including mountains, forests, wetlands, rivers, aquifers and lakes**"

As with many other SDG targets, some progress is ongoing but may not be sustained with climate change impacts.

Adaptations

Minimizing the effects of climate change on water systems requires robust action to mitigate greenhouse gas emissions. Research on the impacts of climate change on water, especially due to rising land and water temperatures and changing precipitation patterns, will inform our ability to adapt and create climate-resilient water management systems. Prioritizing expanded access to improved WASH for the billions who lack it is critical, as well as protecting currently available safe water sources and sanitation and hygiene systems that may be threatened by climate change. Understanding water-related disease burdens, how they are shifting, and which populations are most at risk will help focus efforts to implement public

health interventions and climate adaptation strategies that will improve human health, survival, and well-being in the face of serious threats to water.

Climate action includes concerted international efforts to achieve the water-related SDG targets. Climate change is likely to slow or undermine progress on the SDGs, particularly if improvements to water systems are not climate-resilient. Much uncertainty persists in predicting changing water dynamics in the future, so more research and monitoring are needed. In particular, monitoring and protection of water quantity and quality in cities will become more important as urbanization continues worldwide.

Protecting water supplies from contamination requires strong drinking water regulations, which have been enacted in some countries around the world. WHO publishes Guidelines for Drinking Water Quality, which countries can choose to use to set rules and standards to ensure safe drinking water. A recent WHO analysis of the drinking water regulations in 104 countries and territories showed that protective standards for well-known hazards, such as heavy metals in drinking water, are widely implemented, but fewer protections exist for other chemical contaminants and pathogenic microbes.[42] Under the Safe Drinking Water Act in the United States, legally enforceable maximum levels of more than 90 biological, chemical, and radiological contaminants have been set for public water systems—although they do not apply to private water supplies or bottled water, which serve as drinking water sources for many Americans.[43] Countries in the European Union are protected by the Drinking Water Directive that regulates the quality of water intended for human consumption.[44]

Climate adaptation plans related to water will likely include both natural and engineered solutions, and affected communities will need to help direct planning, implementation, and assessment processes. Communicating the risks of water-related diseases and other impacts to the public, as well as engaging and empowering those at high risk, will be critical for success.

Discussion Questions

1. Discuss examples of water-related diseases, the pathogens that cause them, and their burdens, treatments, and prevention.
2. Discuss the ways that improved WASH prevents diseases, particularly those transmitted by the fecal–oral route.
3. How are the Sustainable Development Goals related to water and human health?
4. Discuss the importance of the concept of YLLs (years of life lost due to premature death) in the context of diarrheal disease burdens.
5. Discuss the ways that women and girls are disproportionately burdened by problems with water quality and quantity.
6. Discuss strategies to prevent harmful algal blooms and *Vibrio* outbreaks.
7. Discuss the threat of water scarcity today and in the face of future climate change.

References

1. UNICEF. Millions of children affected by devastating flooding in South Asia, with many more at risk as COVID-19 brings further challenges. July 23, 2020. Accessed December 1, 2020. https://www.unicef.org/press-releases/millions-children-affected-devastating-flooding-south-asia-many-more-risk-covid-19
2. UNESCO, UN-Water. *United Nations World Water Development Report 2020: Water and Climate Change*. 2020. Accessed December 1, 2020. https://en.unesco.org/themes/water-security/wwap/wwdr/2020
3. World Health Organization. *Water, sanitation, hygiene and health: A primer for health professionals*. 2019. Accessed December 1, 2020. https://www.who.int/publications/i/item/WHO-CED-PHE-WSH-19.149
4. World Health Organization, UNICEF. *Progress on Household Drinking Water, Sanitation and Hygiene 2000–2017: Special Focus on Inequalities*. 2019. Accessed December 1, 2020. https://www.unicef.org/reports/progress-on-drinking-water-sanitation-and-hygiene-2019
5. United Nations. News on Millennium Development Goals. Accessed August 11, 2020. https://www.un.org/millenniumgoals/
6. Sustainable Development Goals. United Nations; ©(2021) United Nations. Reprinted with the permission of the United Nations. https://www.undp.org/sustainable-development-goals
7. Collier S, Deng L, Adam E, et al. Estimate of burden and direct healthcare cost of infectious waterborne disease in the United States. *Emerg Infect Dis*. 2021; 27(1):140–149. doi:10.3201/eid2701.190676
8. Institute for Health Metrics and Evaluation. GBD Compare. Accessed October 15, 2020. https://vizhub.healthdata.org/gbd-compare
9. Troeger CE, Khalil IA, Blacker BF, et al. Quantifying risks and interventions that have affected the burden of diarrhoea among children younger than 5 years: an analysis of the Global Burden of Disease

Study 2017. *Lancet Infect Dis.* 2020;20(1):37–59. doi:10.1016/s1473-3099(19)30401-3

10. Walsh M. Leptospirosis. June 25, 2013. Accessed December 1, 2020. http://www.infectionlandscapes .org/2013/06/leptospirosis.html

11. Walker JT. The influence of climate change on waterborne disease and Legionella: a review. *Perspect Public Health.* 2018;138(5):282–286. doi:10.1177 /1757913918791198

12. Centers for Disease Control and Prevention. Parasites: soil-transmitted helminths. October 27, 2020. Accessed December 2, 2020. https://www .cdc.gov/parasites/sth/index.html

13. McKenna ML, McAtee S, Bryan PE, et al. Human intestinal parasite burden and poor sanitation in rural Alabama. *Am J Trop Med Hyg.* 2017;97(5): 1623–1628. doi:10.4269/ajtmh.17-0396

14. Pilkington E. Hookworm, a disease of extreme poverty, is thriving in the U.S. South. Why? *The Guardian.* September 5, 2017. Accessed December 10, 2020. https://www.theguardian .com/us-news/2017/sep/05/hookworm-lowndes -county-alabama-water-waste-treatment-poverty

15. Brink S. Why it can be harder to fight hookworms in Alabama than in Argentina. *NPR.* January 22, 2021. Accessed March 3, 2021. https://www.npr.org /sections/goatsandsoda/2021/01/22/959204833 /why-it-can-be-harder-to-fight-hookworms-in -alabama-than-in-argentina

16. Pillion D. Alabama failed Black community plagued by hookworm, complaint states. *AL.com.* Updated January 29, 2019. Accessed March 3, 2021. https:// www.al.com/news/2018/09/alabama_health _agencies_face_c.html

17. Hamblin J. The looming consequences of breathing mold. *The Atlantic.* August 30, 2017. Accessed December 10, 2020. https://www.theatlantic.com /health/archive/2017/08/mold-city/538224/

18. United Nations, UN-Water. Water and gender. Accessed December 3, 2020. https://www.unwater .org/water-facts/gender/

19. Veenema TG, Thornton CP, Lavin RP, et al. Climate change-related water disasters' impact on population health. *J Nurs Scholarsh.* 2017;49:625–634. doi:10 .1111/jnu.12328

20. Hales S. Climate change, extreme rainfall events, drinking water and enteric disease. *Rev Environ Health.* 2019;34:1–3. doi:10.1515/reveh-2019-2001

21. Hernroth BE, Baden SP. Alteration of host-pathogen interactions in the wake of climate change– increasing risk for shellfish associated infections?

Environ Res. 2018;161:425–438. doi:10.1016/j .envres.2017.11.032

22. World Health Organization. *Global Progress Report on WASH in Health Care Facilities: Fundamentals First.* December 14, 2020. Accessed December 14, 2020. https://www.who.int/publications/i/item /9789240017542

23. Centers for Disease Control and Prevention. Cholera - Vibrio cholerae infection. October 2, 2020. Accessed December 13, 2020. https://www .cdc.gov/cholera/illness.html

24. Christaki E, Dimitriou P, Pantavou K, et al. The impact of climate change on cholera: a review on the global status and future challenges. *Atmosphere.* 2020;11:449. doi:10.3390/atmos 11050449

25. European Centre for Disease Prevention and Control. Cholera worldwide overview. Accessed June 25, 2021. https://www.ecdc.europa.eu/en/all -topics-z/cholera/surveillance-and-disease-data /cholera-monthly

26. Asadgol Z, Badirzadeh A, Niazi S, et al. How climate change can affect cholera incidence and prevalence? A systematic review. *Environ Sci Pollut Res.* 2020;27(28):34906–34926. doi:10.1007 /s11356-020-09992-7

27. Odeyemi OA. Incidence and prevalence of *Vibrio parahaemolyticus* in seafood: a systematic review and meta-analysis. *SpringerPlus.* 2016;5(1):464. doi:10.1186/s40064-016-2115-7

28. Li L, Meng H, Gu D, et al. Molecular mechanisms of Vibrio parahaemolyticus pathogenesis. *Microbiol Res.* 2019;222:43–51. https://doi.org/10.1016/j .micres.2019.03.003

29. Baker-Austin C, Trinanes J, Gonzalez-Escalona N, et al. Non-cholera vibrios: the microbial barometer of climate change. *Trends Microbiol.* 2017;25(1): 76–84. https://doi.org/10.1016/j.tim.2016.09 .008

30. Centers for Disease Control and Prevention. Cholera and Other *Vibrio* Illness Surveillance (COVIS). March 5, 2019. Accessed December 13, 2020, https://www.cdc.gov/vibrio/surveillance .html

31. United States Global Change Research Program. *The Impacts of Climate Change on Human Health in the United States: A Scientific Assessment.* 2016. Accessed March 8, 2021. https://health2016 .globalchange.gov

32. Nguyen H, Smith M, Swoboda H. Shellfish toxicity. *StatPearls.* July 21, 2020. Accessed December 10,

2020. https://www.ncbi.nlm.nih.gov/books/NBK4
70225/

33. Maine Center for Disease Control and Prevention.
Shellfish poisoning. Published February 11, 2019.
Accessed December 10, 2020. https://www.maine
.gov/dhhs/mecdc/infectious-disease/epi/shm
/Shellfish-Poisoning-SHM.pdf

34. Gobler CJ. Climate change and harmful algal
blooms: insights and perspective. *Harmful Algae*.
2020;91:101731. https://doi.org/10.1016/j.hal.2019
.101731

35. Maine Department of Marine Resources. Maine
biotoxin closures: shellfish biotoxin area inventory.
Accessed December 8, 2020. https://www.maine
.gov/dmr/shellfish-sanitation-management
/closures/psp.html

36. Trainer VL, Moore SK, Hallegraeff G, et al. Pelagic
harmful algal blooms and climate change: lessons
from nature's experiments with extremes. *Harmful
Algae*. 2020;91:101591. https://doi.org/10.1016/j
.hal.2019.03.009

37. Ralston DK, Moore SK. Modeling harmful algal
blooms in a changing climate. *Harmful Algae*.
2020;91:101729. https://doi.org/10.1016/j.hal.2019
.101729

38. Environmental Protection Agency. Learn about
cyanobacteria and cyanotoxins. Accessed
December 19, 2020. https://www.epa.gov
/cyanohabs/learn-about-cyanobacteria-and
-cyanotoxins

39. California Department of Water Resources. Current
conditions for major reservoirs. June 24, 2021.
Accessed June 25, 2021. http://cdec.water.ca.gov
/resapp/RescondMain

40. Wahabu E, Patel P. Climate change and water
scarcity disrupting youth livelihoods in Ghana.
International Water Management Institute.
May 13, 2020. Accessed December 14, 2020. https://
www.iwmi.cgiar.org/2020/05/climate-change-and
-water-scarcity-disrupting-youth-livelihoods-in
-ghana/

41. Omer A, Elagib NA, Zhuguo M, et al. Water
scarcity in the Yellow River Basin under future
climate change and human activities. *Sci Total
Env*. 2020;749:141446. https://doi.org/10.1016/j
.scitotenv.2020.141446

42. World Health Organization. A global overview of
national regulations and standards for drinking-
water quality. June 17, 2018. Accessed June 25,
2021. https://www.who.int/publications/i/item/9789
241513760

43. Environmental Protection Agency. National
primary drinking water regulations. January 5,
2021. Accessed June 25, 2021. https://www.epa
.gov/ground-water-and-drinking-water/national
-primary-drinking-water-regulations

44. European Commission. Legislation: The revised
Drinking Water Directive. Accessed June 25, 2021.
https://ec.europa.eu/environment/water/water
-drink/legislation_en.html

CHAPTER 9

Food and Nutrition

KEY TERMS

Hunger
Malnutrition
Undernutrition
Stunting
Underweight
Wasting

Severe acute malnutrition
Global Hunger Index (GHI)
Protein-energy malnutrition
  (PEM)
Kwashiorkor
Marasmus

Ready-to-use therapeutic food
  (RUTF)
Micronutrient deficiencies
Overnutrition
Food insecurity
Aflatoxins

LEARNING OBJECTIVES

- Describe the global burden of hunger and recent trends.
- Define the various forms of undernutrition and recent trends.
- Define food insecurity and understand its many causes, including climate change.
- Understand how agriculture and food systems contribute to greenhouse gas emissions that fuel climate change and how they can be made more climate-resilient to protect human health and food security.

In 2020, the Nobel Peace Prize was awarded to the United Nations (UN) **World Food Programme (WFP)** for its continued efforts to fight hunger in countries around the world. WFP recognizes that climate change is a major driver of global hunger and food insecurity, and in 2020, the COVID-19 pandemic and associated economic downturn, along with war and conflict in many regions, added to the threat of hunger and famine for millions, especially children. "Humankind's greatest success was to consign famine to history. That we now face it again is heart-wrenching and obscene when we produce enough food to nourish every person on the planet," wrote UN humanitarian chief Mark Lowcock and WFP executive director David Beasley.[1]

In late 2020, WFP identified six countries on the brink of famine: Afghanistan, Burkina Faso, Democratic Republic of the Congo, northeastern Nigeria, South Sudan, and Yemen. Complex intersecting factors create famine conditions, and in these countries, armed conflict has played a major role. But climate change is also an important stressor and one that can worsen both conflict and food systems. Also in 2020, a La Niña weather cycle threatened Ethiopia, Somalia, and Kenya with devastating drought and hunger.[2] Aid experts noted that the timing could not have been worse—not only because of the COVID-19 pandemic, but also because of destructive outbreaks of crop-destroying locusts. They warned that some governments may not be able to react quickly enough or sufficiently to prevent severe food shortages.[2]

A famine in East Africa three years earlier, caused in large part by climate-change-fueled drought, left 25 million people in Somalia, Ethiopia, South Sudan, and Kenya, including 15 million children, in dire need of food aid and other humanitarian assistance.[3] Thirty percent of children suffered from acute undernutrition, and without food, water, or livestock, people scrambled for remaining resources. This created conflict and forced people to travel long distances in search of food and water, weakening children and adults alike. One mother in Somalia took her seven children on a journey in search of food, but one by one, they succumbed to starvation. "I was not in my senses anymore," she said. "I was lost in my grief. I don't even know how I got here."[3] A Kenyan family lost all 100 of its sheep and goats to the drought, and in another family, two boys were left orphaned after their mother died of starvation and their father was killed in a cattle raid.[3]

Climate change is stressing agriculture and food systems all over the world, with negative consequences for food production and human health. In addition to drought-fueled food insecurity, climate change threatens vital fishing industries in places like Kiribati, a low-lying island nation in the Pacific Ocean that is extremely vulnerable to ocean warming and reef damage.[4]

Coastal Nigeria faces sea level rise and resulting erosion and salinization of water and soil, harming agriculture and fishing in a region with high population pressure.[4] California produces much of the fruits and vegetables supply in the United States, and nearly 90% of harvested crops are irrigated. With extreme temperatures, historically low rainfall, prolonged drought, and diminished snowpack in the region, water availability in California may decrease by two-thirds, severely destabilizing food production and supply.[5]

Most climate models predict that human undernutrition and food insecurity will worsen in the future, particularly for populations living in regions prone to drought, erratic rainfall, flooding, and saltwater intrusion. Within these regions, how people actually experience these climate impacts relates significantly to health status, socioeconomic conditions, dominant food crops, food prices, household expenditures for food, land and agricultural policies, and globalized food markets that often disadvantage smallholder farmers.

Hunger

The UN **Food and Agriculture Organization (FAO)** defines **hunger** as "an uncomfortable or painful physical sensation caused by insufficient consumption of dietary energy."[6] Hunger is thus the physical manifestation of food deprivation. FAO measures hunger as **prevalence of undernourishment (PoU)**, the estimated percentage of the population whose habitual food consumption is insufficient to provide the dietary energy intake needed for a healthy life.[6] PoU is based on country-level data on food availability, food consumption, and energy needs. **Table 9.1** shows PoU by region of the world in 2019, with sub-Saharan Africa having the highest prevalence. Within these regions and within individual countries,

Table 9.1 Prevalence of Undernourishment (PoU) by Region in 2019.

FAO Region	PoU (%)
World	8.9
Sub-Saharan Africa	22.0
Southern Asia	13.4
South-eastern Asia	9.8
Northern Africa and Western Asia	9.0
Latin America and the Caribbean	7.4
Oceania	5.8
Central Asia	2.7
Eastern Asia	<2.5
Northern America and Europe	<2.5

Data from Food and Agriculture Organization, International Fund for Agricultural Development, UNICEF, World Food Programme, World Health Organization. *The State of Food Security and Nutrition in the World 2020: Transforming Food Systems for Affordable Healthy Diets.* Food and Agriculture Organization; 2020. doi:10.4060/ca9692en

hunger may vary widely among populations. Where the burden of hunger is severe, people may also be at high risk of the adverse health effects of climate change.

Over the past decade, little progress has been made on global PoU, and in fact, it is now on the rise. Nearly 690 million people (8.9% of the global population) were undernourished in 2019, 55% of whom live in Asian countries and 36% in African countries.[6] This number has increased by 60 million since 2014 and at current trends will be 841 million by 2030 (nearly 10% of the world's population) (**Figure 9.1**). In 2019, FAO estimated that fully one-quarter of all the world's people experienced hunger or lacked regular access to sufficient and nutritious food.[6]

FAO attributes this recent and ominous uptick in undernutrition primarily to conflict and climate-change-fueled extreme weather, drought, and floods that affect food production. Measuring how food-related diseases and conditions are being impacted by climate change is difficult, however, especially in populations with high baseline prevalence of malnutrition and food insecurity. Understanding the effects of climate change also requires careful consideration of the many complex and often interacting risk factors for hunger.

Malnutrition

Malnutrition is defined as conditions that result from deficiencies, excesses, or imbalances in consumption of energy, macronutrients (proteins), or micronutrients (vitamins and minerals). Malnutrition comes in two forms: undernutrition and overnutrition. **Undernutrition** is defined as inadequate intake of food quantity or quality that leads to unhealthy dietary deficiencies. Undernutrition may also result from infections or other illnesses that cause poor utilization of nutrients by the body. Undernutrition itself increases the risk of infectious diseases and is a contributing factor in 45% of all deaths in children under 5.[7] Adults who were undernourished as children tend to face higher rates of communicable and noncommunicable disease burdens, diminished cognitive abilities, and lower economic productivity.[8]

In children, undernutrition is measured as stunting, underweight, and wasting. **Stunting** is low height-for-age that is due to chronic or recurrent malnutrition. Stunted children are deficient in physical and cognitive growth, and stunting before age 2 predicts reduced educational outcomes in later childhood and significant loss of economic productivity in adulthood.[9] **Underweight** is low

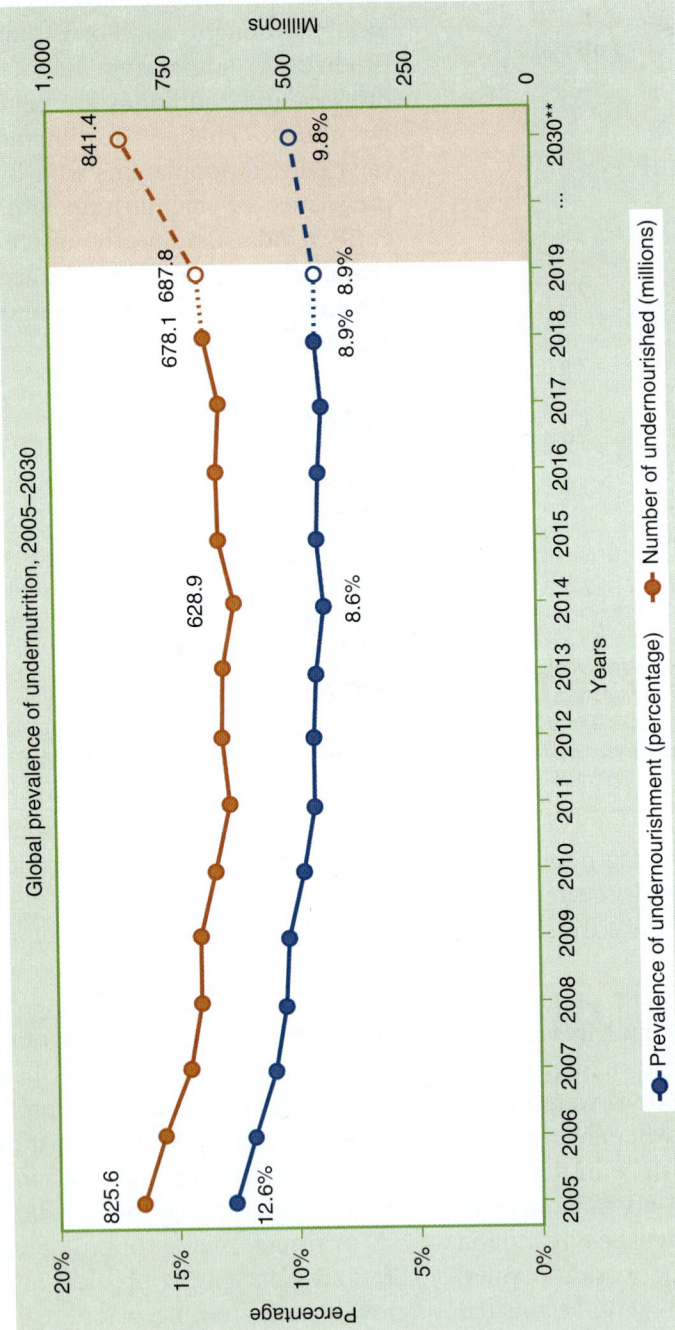

Figure 9.1 Global Undernourishment, 2005–2030. Prevalence of undernourishment (% of population) is shown in blue and number of undernourished people in millions is shown in red.

Notes: Projected values in the figure are illustrated by dotted lines and empty circles. The shaded area represents projections for the longer period from 2019 to the 2030 target year. **Projections to 2030 do not consider the impact of the COVID-19 pandemic.

Food and Agriculture Organization of the United Nations. Reproduced with permission.

weight-for-age that is often due to a lack of adequate food intake. Many children who are underweight had a low birth weight, defined as a weight below 2,500 g (5.5 lb) at birth. **Wasting** is low weight-for-height that is due to sudden or acute malnutrition, with immediate risk of death. Wasting is the most dangerous of these three conditions. Approximately 10% of all deaths in children under 5 in low- and middle-income countries are attributed to wasting.[10]

Each condition can be diagnosed using standardized child growth charts showing population distributions of height-for-age, weight-for-age, and weight-for-height measurements (**Figure 9.2**). On the relevant growth chart, stunting, underweight, or wasting is diagnosed if the measurement for an individual child is more than two standard deviations (z-scores) below the median value on the chart. This corresponds approximately to the lowest 2% of measurements in the population. Using the chart in Figure 9.2, an 18-month-old girl whose length is below 75 cm is considered stunted, and she is considered underweight if her weight is below 8 kg. Child growth and growth faltering occur on a continuum, however, and crossing a standardized threshold does not mean that a child shifts suddenly from well to ill. Good health or health deterioration may occur at a range of growth measurements. Without using growth charts, cases of undernutrition can also be diagnosed based on physical appearance and clinical symptoms.

Severe stunting, underweight, or wasting is defined as having measurements more than three z-scores below the median value, corresponding approximately to the lowest 0.1% of measurements. Severe wasting, also called **severe acute malnutrition**, is the most dire and visible form of undernutrition, affecting millions of children and causing perhaps one million child deaths each year.[11]

In 2019, 21% of the world's children were stunted, and 7% were wasted.[6] According to the Global Burden of Disease report, in 2019, child wasting, underweight, and stunting contributed to one-quarter of all deaths and disability-adjusted life years (DALYs) in children under 5.[12] Specifically, 875,000 deaths were attributed to child wasting, 241,000 deaths to child underweight, and 164,000 deaths to child stunting. The burden of child undernutrition has declined significantly since 1990, with DALYs rates down 81% for stunting, 82% for underweight, and 61% for wasting (**Figure 9.3**).[12] In 2020, however, an additional 6.7 million children, most in South Asian and sub-Saharan African countries, were at risk of wasting because of the economic crisis brought on by the COVID-19 pandemic.[10]

Another indicator of undernutrition is the **Global Hunger Index (GHI)**, a composite metric that incorporates prevalence of undernourishment and data on wasting, stunting, and mortality in children under 5 (**Figure 9.4**).[13] GHI is published annually by Concern Worldwide, an international humanitarian organization, and Welthungerhilfe, a private international aid organization based in Germany. In 2020, three countries had "alarming" hunger severity: Chad, Madagascar, and Timor-Leste. Eight others were "provisionally alarming": Burundi, Central African Republic, Comoros, Democratic Republic of the Congo, Somalia, South Sudan, Syria, and Yemen.

Undernutrition is also indicated by measuring the burden of **protein-energy malnutrition (PEM)**, a condition that results from inadequate consumption of protein-rich foods, total calories, or both. The two forms of PEM, kwashiorkor and marasmus, are both considered forms of severe acute malnutrition. **Kwashiorkor** is undernutrition resulting from deficient intake of protein. It is most common in children over the age of 1 and is

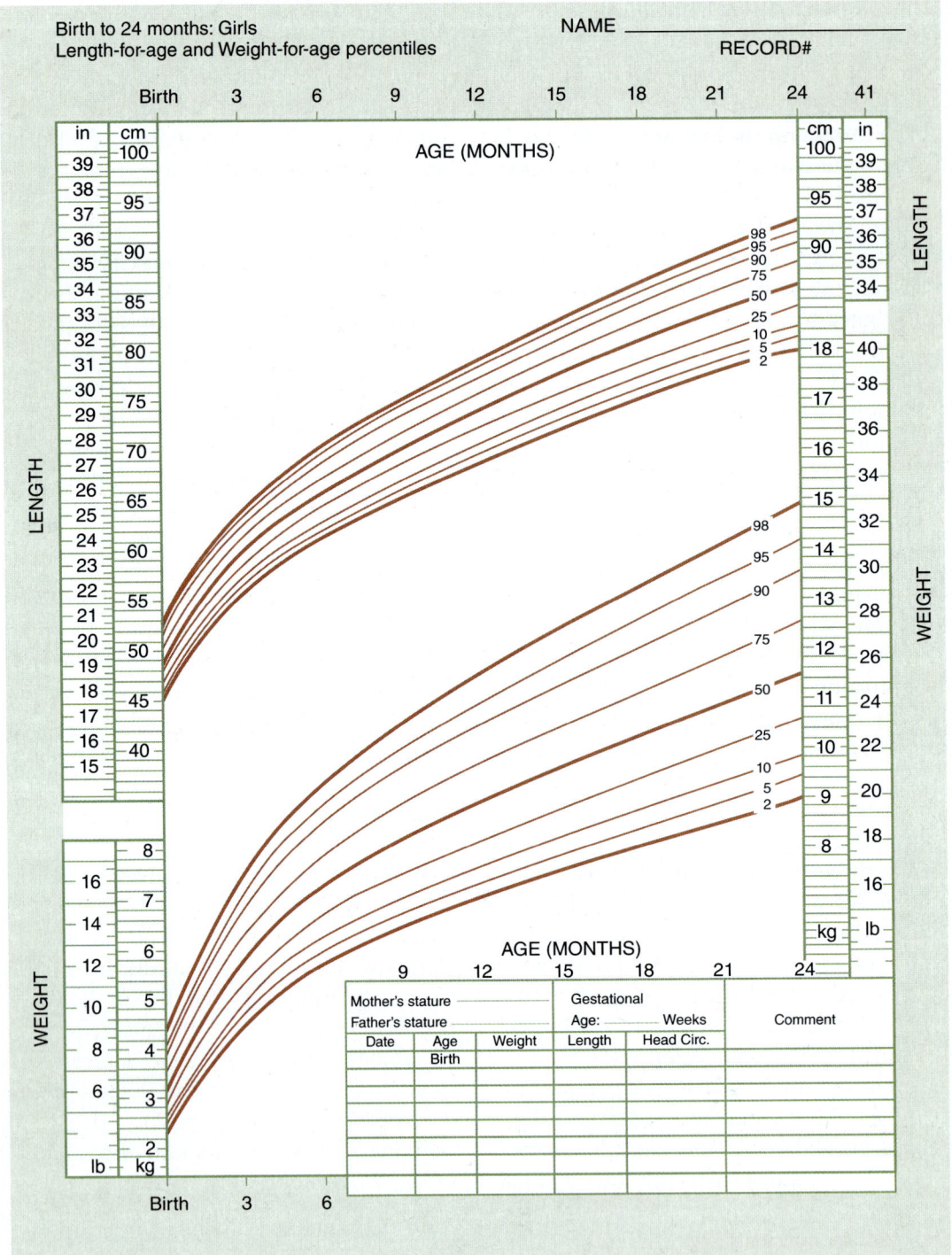

Figure 9.2 Girl Child Growth Chart from Birth to 24 Months Used to Measure Stunting (Length-for-Age) and Underweight (Weight-for-Age). Measurements two z-scores below the median are at or below the 2% line.

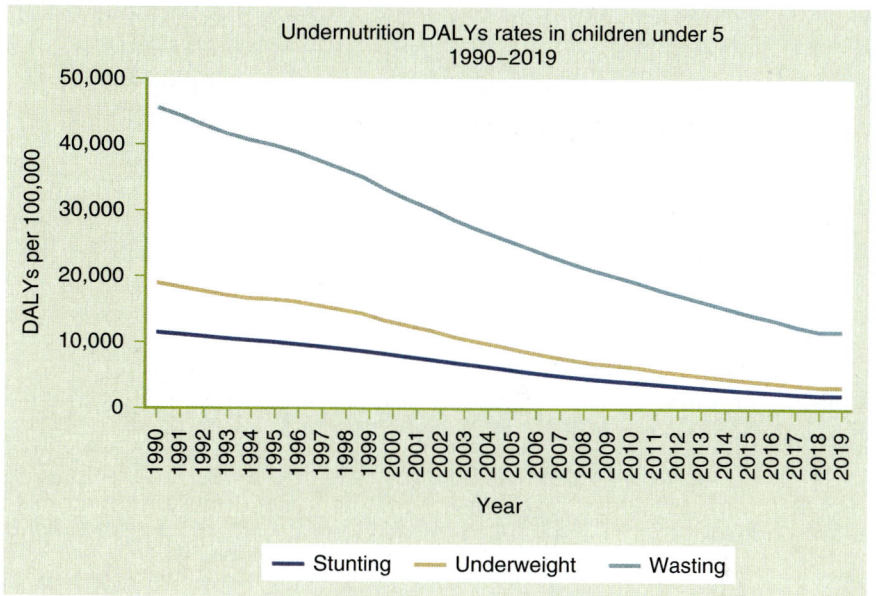

Figure 9.3 DALYs Rates Attributed to Stunting, Underweight, or Wasting in Children under 5, 1990–2019.

Data from Institute for Health Metrics and Evaluation. GBD Compare. Accessed October 15, 2020. https://vizhub.healthdata.org/gbd-compare

characterized by fluid retention and swelling, especially of the belly, due to an osmotic imbalance in the gut caused by low protein intake. Other symptoms are changes in skin pigment, decreased muscle mass, diarrhea, failure to thrive, a damaged immune system, and, in some cases, death in late stages. **Marasmus** is undernutrition resulting from deficient intake of total calories. It is most common in infants and is characterized by

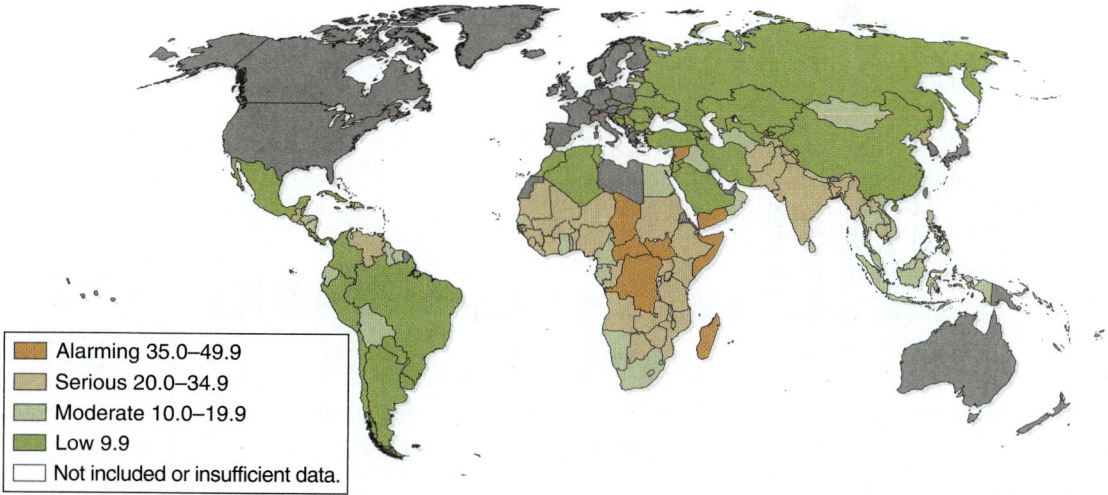

Figure 9.4 Global Hunger Index by Country, 2020. Countries shaded in orange face the highest hunger risk.

Reproduced from von Grebmer K, Bernstein J, Wiemers M, et al. 2020 global hunger index: one decade to zero hunger: linking health and sustainable food systems (Figure 1.4). Welthungerhilfe/Concern Worldwide; 2020. https://www.globalhungerindex.org/pdf/en/2020.pdf

severe low body weight, wasting, and loose, dry, peeling skin. Other symptoms are subnormal body temperature, decreased pulse and metabolic rate, constipation, diarrhea, and, in severe cases, starvation and death. Overall, PEM prevalence rate in children under 5 is **nearly 5 times higher** than in people of all ages (**Figure 9.5**).[12]

DALYs rates due to PEM have declined significantly since 1990 (79% in children under 5 and 75% in people of all ages). This reduction is due mostly to lower *mortality* in children under 5, which reduces the component of DALYs that accounts for years of life lost to premature deaths (YLLs). YLLs declined 82% in children under 5 from 1990 to 2019 (**Figure 9.6**). The YLDs component of DALYs, which accounts for years of healthy life lost due to suffering or disability from PEM, declined only 16% in children under 5 during this period.[12]

PEM *incidence* rates declined much less than YLLs during this period, down only 15% overall since 1990 in children under 5, and increasing slightly since 2017. This mirrors the recent uptick in hunger prevalence measured by FAO. The difference in incidence and mortality trends indicates that young children are still suffering significantly from PEM but are at a lower risk of dying. Trends in PEM incidence and YLLs rates in young children are similar to those for diarrheal disease burdens.

In 2019, the 10 countries with the highest *mortality* rates for PEM in children under 5 were all in sub-Saharan Africa, the region with the highest PoU (**Table 9.2**). Mali had the highest mortality rate, more than double that of the second-highest country (Somalia) and **nearly 22 times higher** than the global average for children under 5.[12] Mali's extreme burden of undernutrition among children (and adults) makes its population particularly susceptible to the devastating impacts of climate change on agriculture and food systems (see **Box 9.1**). PEM death rates in

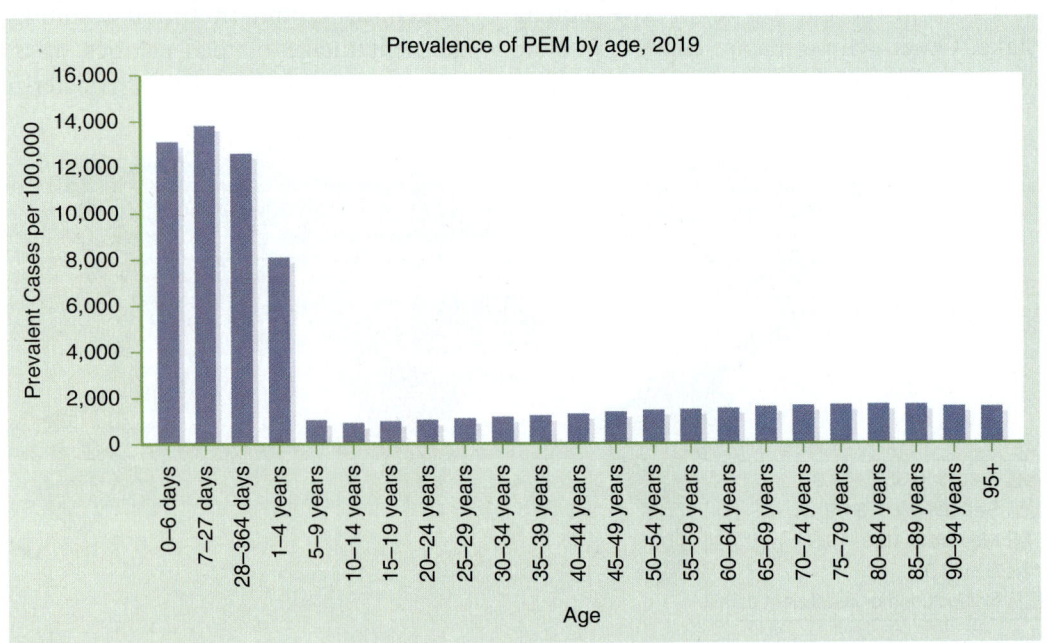

Figure 9.5 Age Distribution of Protein-Energy Malnutrition Prevalence Rate in 2019.

Data from Institute for Health Metrics and Evaluation. GBD Compare. Accessed October 15, 2020. https://vizhub.healthdata.org/gbd-compare

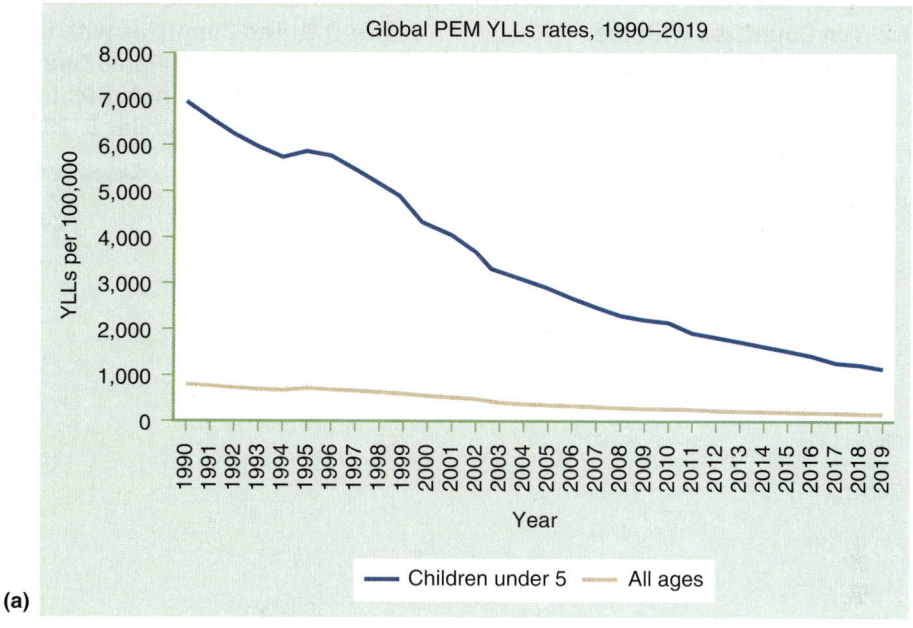

(a)

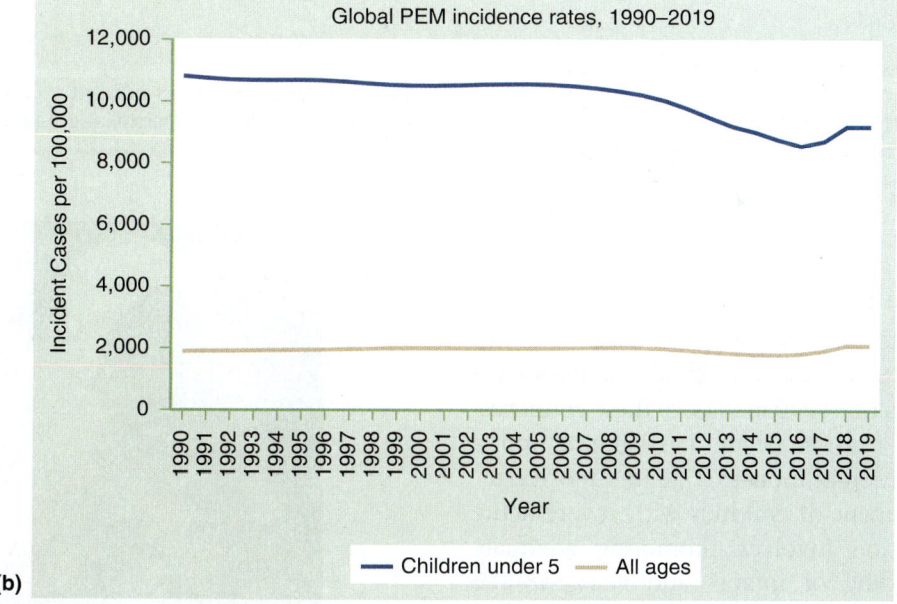

(b)

Figure 9.6 Global Burden of Protein-Energy Malnutrition from 1990 to 2019 Measured by **(a)** YLLs Rates and **(b)** Incidence Rates in Children under 5 and in People of All Ages (Age-Standardized Rates).

Data from Institute for Health Metrics and Evaluation. GBD Compare. Accessed October 15, 2020. https://vizhub.healthdata.org/gbd-compare

children under 5 in these countries declined since 1990, ranging from a 6% reduction in Zimbabwe to 83% in South Sudan. Globally, the mortality burden in young children is **more than 5 times higher** than in people of all ages.

Table 9.2 Ten Countries with the Highest Death Rates Due to Protein-Energy Malnutrition in Children under 5, 2019.

Rank	Country	Deaths per 100,000
1	Mali	302
2	Somalia	143
3	South Sudan	108
4	Central African Republic	96
5	Sierra Leone	77
6	Burkina Faso	77
7	Zimbabwe	76
8	Madagascar	69
9	Djibouti	67
10	Guinea	63
	World	14

Data from Institute for Health Metrics and Evaluation. GBD Compare. Accessed October 15, 2020. https://vizhub.healthdata.org/gbd-compare

Table 9.3 Ten Countries with the Highest Incidence Rates Due to Protein-Energy Malnutrition in Children under 5, 2019.

Rank	Country	Cases per 100,000
1	India	20,634
2	Sri Lanka	18,657
3	Maldives	17,238
4	Timor-Leste	14,429
5	Yemen	14,259
6	Sudan	14,054
7	Pakistan	13,774
8	Mali	12,664
9	Indonesia	11,503
10	Burkina Faso	11,418
	World	*9,206*

Data from Institute for Health Metrics and Evaluation. GBD Compare. Accessed October 15, 2020. https://vizhub.healthdata.org/gbd-compare

India, Sri Lanka, and Maldives had the highest PEM *incidence* rates in children under 5 in 2019, **double** the global rate (**Table 9.3**). Six of the top 10 countries were in South Asia and Southeast Asia. Incidence rates have declined since 1990 in most of these countries, although less than death rates, ranging from a 6% reduction in India to 37% in Timor-Leste.

Treatment of children with severe acute malnutrition involves promoting adequate breastfeeding of infants and **ready-to-use therapeutic food (RUTF)** in children over 6 months of age (**Figure 9.7**). RUTF is a high-energy, high-nutrient food paste that does not require processing or cooking and has a long shelf life.[11] RUTF contains lipid- and protein-rich foods, typically peanuts, oils, and milk powders (or locally available ingredients), along with sugars, vitamins, and minerals.

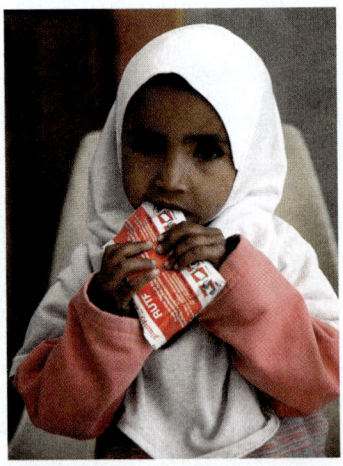

Figure 9.7 A Child in Yemen Consumes a Packet of Ready-to-Use Therapeutic Food (RUTF). Children in Yemen are suffering from very high rates of malnutrition as a result of years of conflict and limited access to humanitarian aid.
© MOHAMMED HUWAIS/Stringer/AFP/Getty Images.

Progress on severe acute malnutrition has also been made through control of infectious diseases in children that affect nutritional status, such as worm infections and malaria.

The UN Children's Fund (UNICEF) distributes 80% of the world's RUTF to children suffering from severe acute malnutrition.[11] One recent case was that of 3-year-old Pierre in Central African Republic, who weighed under 20 pounds and suffered from severe diarrhea. His mother had died, and his father had no regular employment in a country marred by conflict. As Pierre grew sicker, his father initially tried traditional medicine because he didn't think he could afford hospital care. But eventually he walked 7 miles with his child from their village to a hospital in the capital city, where Pierre received RUTF therapy and recovered.[11] Severe acute malnutrition is common in Central African Republic and other countries torn apart by conflict and war. Said one doctor working with UNICEF, "It's really sad because malnutrition is an illness that can be prevented but we see cases every day, dozens of cases every day. Sometimes children arrive here in a very serious condition. They go straight to the emergency ward and some die. It's really painful. It's sad."[11]

The global decline in PEM burden since 1990 was catalyzed in large part by international efforts to achieve Millennium Development Goal 1 (*Eradicate extreme poverty and hunger*), which set a target "**to halve, between 1990 and 2015, the proportion of people who suffer from hunger**." This target was nearly achieved, with an estimated reduction of 45% between 1990 and 2015.[14] Current efforts are focused on achieving Sustainable Development Goal 2 (*Zero hunger*), which has set targets to, "**by 2030, end hunger and ensure access by all people, in particular the poor and people in vulnerable situations, including infants, to safe, nutritious and sufficient food all year round**" and "**by 2030, end all**

forms of malnutrition, including achieving, by 2025, the internationally agreed targets on stunting and wasting in children under 5 years of age, and address the nutritional needs of adolescent girls, pregnant and lactating women and older persons."[15] These internationally agreed-upon targets (World Health Organization's **Global Nutrition Targets 2025**) set the following goals[16]:

- Reduce by 40% the number of children under 5 who are stunted (which would still leave 100 million children under 5 stunted)
- Reduce by 30% the number of children born at low birth weight (which would still leave 14 million low-birth-weight babies)
- Reduce and maintain childhood wasting to less than 5% of children under 5

Global progress to end hunger and malnutrition by 2030 is off pace, as indicated by FAO's prediction that undernutrition will increase in the next decade and possibly affect up to a billion people by 2030.[6]

Undernutrition also manifests in the form of **micronutrient deficiencies** that cause significant human illness and suffering. Micronutrient deficiencies are rarely the direct cause of mortality, but they cause millions of DALYs and exacerbate other diseases that are deadly. An estimated 340 million children had one or more micronutrient deficiencies in 2019.[6] These deficiencies may be exacerbated by climate change if production of key nutrient-rich foods worsens.

Iron deficiency is the most common nutritional disorder in the world, leading to impaired physical and cognitive development in children, reduced work productivity in adults, poor pregnancy outcomes, and maternal mortality. An estimated 2 billion people suffer from anemia, which is often caused by iron deficiency.[7] Treatment of iron deficiency requires increasing iron intake through consumption of iron-rich foods, iron fortification of foods, and iron supplementation. Iron deficiency can be exacerbated in people suffering from infectious diseases, including malaria and worm infections.[17]

Vitamin A deficiency results from inadequate consumption of foods rich in vitamin A, primarily fruits and vegetables. Vitamin A is a general term for a group of related compounds that includes a component of the visual pigment rhodopsin, and deficiency can lead to blindness if untreated. Vitamin A is also critical for immune function, and deficiencies lead to increased child mortality from infectious diseases, such as malaria, measles, diarrhea, and pneumonia. Nearly 200 million preschool-age children around the world may suffer from vitamin A deficiency, mostly in countries in Africa and Southeast Asia.[18] Treatment is with vitamin A supplementation, which is a part of routine pediatric care for many populations at high risk.

Iodine deficiency results from inadequate consumption of iodine-rich foods. It is a leading cause of cognitive deficiencies in children whose mothers were iodine-deficient during pregnancy. Adequate iodine is required for normal thyroid function in pregnant women, which is critical for proper fetal brain development. Serious iodine deficiency in pregnancy is also linked to stillbirth, spontaneous abortion, and congenital abnormalities. In adults, iodine deficiency leads to low thyroid hormone and metabolic problems. Iodine deficiency is treated at the population level with iodized salt.

Zinc deficiency results from inadequate consumption of zinc-rich foods. Zinc is an essential micronutrient for many cellular functions, gene expression, and immune system health. Symptoms include skin lesions, growth retardation, hair loss, diarrhea, infections, and behavioral changes. In children, zinc deficiency increases the risk of infectious diseases, such as diarrheal diseases, pneumonia, and malaria, and is linked to an overall increase in child mortality.[7] As with iron deficiency, treatment of zinc deficiency is with targeted supplementation, food fortification, and dietary strategies.

Another form of malnutrition is **overnutrition**, which is excessive intake of nutrients that leads to accumulation of body fat that may impair health. Globally, 6% of children and 40% of adults are overweight, and the numbers are rising.[6] Overweight and obesity are associated with numerous chronic illnesses and conditions, including diabetes, cardiovascular disease, and certain cancers. Overnutrition and climate change share certain contributing factors, including high human consumption of animal products, livestock farming, processed food production, and use of sedentary forms of transportation that are motorized and burn fossil fuels.

Food Insecurity

According to the UN, "food security exists when all people, at all times, have physical and economic access to sufficient safe and nutritious food that meets their dietary needs and food preferences for an active and healthy life."[6] FAO defines **food insecurity** as lacking "regular access to enough safe and nutritious food for normal growth and development and an active and healthy life."[6] All people experiencing hunger are food insecure, but not all food-insecure people are hungry because some may consume foods that provide adequate calories but lack important nutrients.[8]

The four key dimensions of food security are availability, access, utilization, and stability. All four must be fulfilled simultaneously to achieve food security.[8]

- **Availability**: Sufficient food supply for the population (determined by levels of food production, food stocks, and trade)
- **Access**: Sufficient household-level food access (determined by income, expenditures, markets, and prices)

- **Utilization**: Healthy biological uptake and metabolism of nutrients from foods consumed (influenced by individual health status, adequate nutritious food supply, good feeding practices and food preparation, dietary diversity, and intra-household distribution of food)
- **Stability** (of the other three dimensions simultaneously): Sufficient food availability, access, and utilization *at all times*

Food insecurity is experienced at varying levels of severity, which can be indicated at the household or individual level by FAO's Food Insecurity Experience Scale (FIES). FIES is based on direct yes or no answers to eight brief survey questions about access to adequate food[19]:

- During the last 12 months, was there a time when you were worried you would not have enough food to eat because of a lack of money or other resources?
- Still thinking about the last 12 months, was there a time when you were unable to eat healthy and nutritious food because of a lack of money or other resources?
- Was there a time when you ate only a few kinds of foods because of a lack of money or other resources?
- Was there a time when you had to skip a meal because there was not enough money or other resources to get food?
- Still thinking about the last 12 months, was there a time when you ate less than you thought you should because of a lack of money or other resources?
- Was there a time when your household ran out of food because of a lack of money or other resources?
- Was there a time when you were hungry but did not eat because there was not enough money or other resources for food?
- During the last 12 months, was there a time when you went without eating for a whole day because of a lack of money or other resources?

Food insecurity ranges from mild to severe, and FIES captures individual and household experiences on a continuum that includes worrying about running out of food, compromising on the quality and variety of foods in the diet, reducing quantities of food consumed and skipping meals, and experiencing hunger.[19] Food insecurity may be temporary, seasonal, or long-term.

Levels of food insecurity vary by region, with sub-Saharan Africa having the highest prevalence (**Table 9.4**). Food insecurity also

Table 9.4 Prevalence of Food Insecurity by Region in 2019.

Region	Prevalence of Moderate or Severe Food Insecurity (%)
World	25.9
Sub-Saharan Africa	56.8
Southern Asia	36.1
South-eastern Asia	18.6
Northern Africa and Western Asia	28.5
Latin America and the Caribbean	31.7
Oceania	13.9
Central Asia	13.2
Eastern Asia	7.4
Northern America and Europe	7.9

Data from Food and Agriculture Organization, International Fund for Agricultural Development, UNICEF, World Food Programme, World Health Organization. *The State of Food Security and Nutrition in the World 2020: Transforming Food Systems for Affordable Healthy Diets*. Food and Agriculture Organization; 2020. doi:10.4060/ca9692en

varies widely among people within countries and regions.

A significant portion of the U.S. population faces food insecurity. In 2019, 10.5% of U.S. households were food insecure at any point during the year, and 4.1% (5.3 million households) had "very low food security."[20] Food insecurity varies by U.S. state, with the highest prevalence in Mississippi (15.7% of households) and the lowest prevalence in New Hampshire (6.6% of households). (**Figure 9.8**). Cities and rural areas had similar levels of food insecurity.

Food insecurity also varies by household characteristics, with highest prevalence in households with income levels below the U.S. poverty level, followed by households led by single mothers, and then Black households (**Figure 9.9**). A higher proportion of households with children experienced food insecurity (13.6%), but in half of these households, only the adults were food insecure because they prioritized making sure the children had enough food to eat. The U.S. Department of Agriculture offers several food and nutrition assistance programs that in 2019 totaled over $92 billion in expenditures.[20] Not all food-insecure people qualify for federal assistance, however, or are aware of or choose to take advantage of these benefits. Other sources of food aid include local food pantries and hunger relief organizations. Food assistance fills a critical gap in food access, but is often insufficient and will be

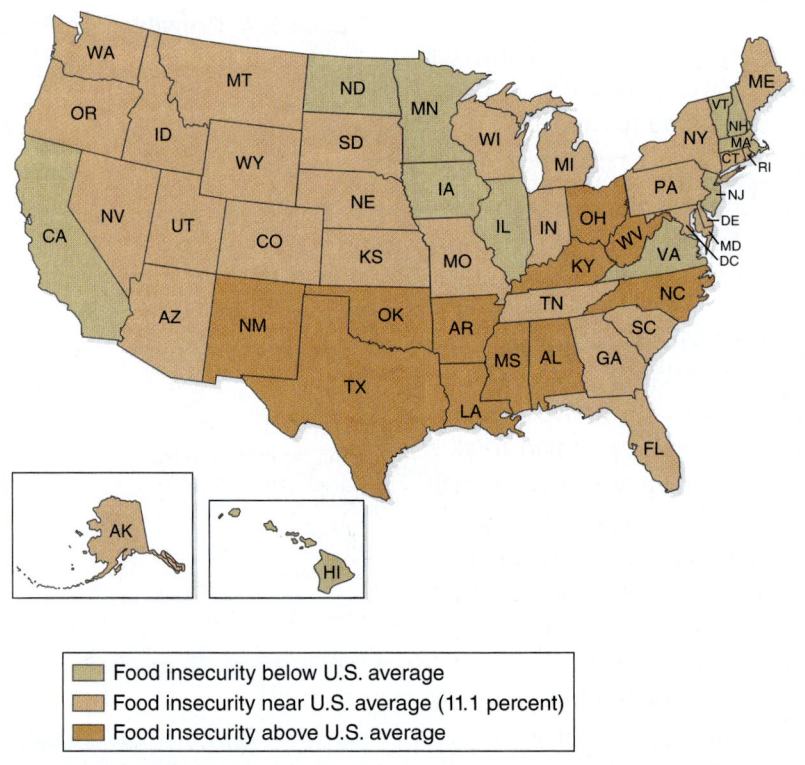

Legend:
- Food insecurity below U.S. average
- Food insecurity near U.S. average (11.1 percent)
- Food insecurity above U.S. average

Figure 9.8 Prevalence of Food Insecurity in U.S. States, 2017–2019 Average, Compared with the 2018 National Average of 11.1% of Households.

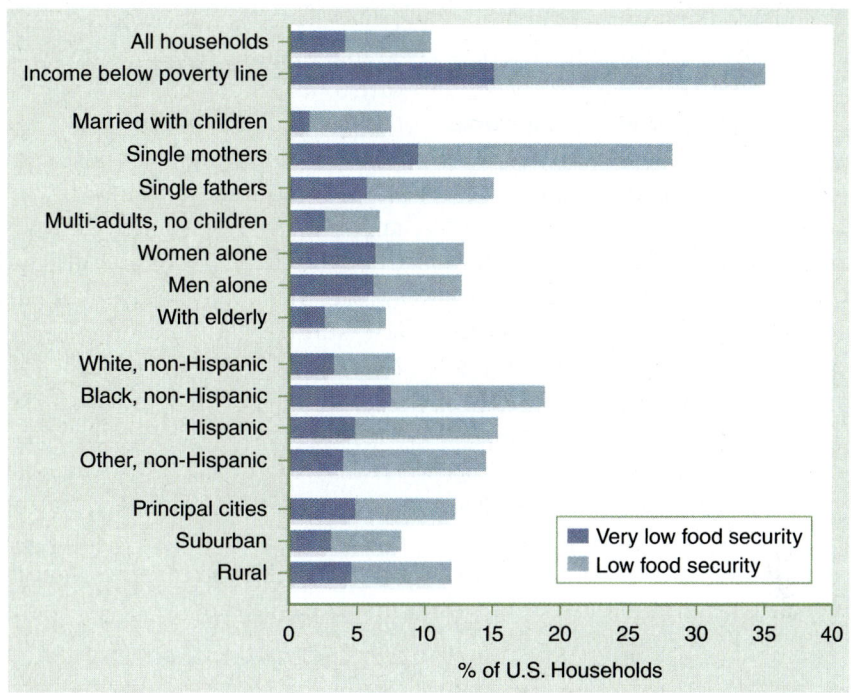

Note: Food-insecure households include those with low food security and very low food security.

Figure 9.9 Prevalence of Food Insecurity among U.S. Households with Selected Characteristics. Both low (light blue) and very low (dark blue) food insecurity levels are indicated.

Reproduced from U.S. Department of Agriculture Economic Research Service. Food security and nutrition assistance. Updated December 16, 2020. Accessed March 21, 2021. https://www.ers.usda.gov/data-products/ag-and-food-statistics-charting-the-essentials/food-security-and-nutrition-assistance

further strained as climate change impacts on food systems become more severe.

Many factors influence food security status, including adverse weather and climate conditions, environmental degradation, political instability, conflict and unrest, and economic factors (poverty, unemployment, and rising food prices).[21] Climate change increases the risks of droughts and dry spells, floods, extreme storms, precipitation, and sea level rise, all of which can affect agricultural output and food availability.

It is estimated that climate change and natural disasters created acute food insecurity for 29 million people in 2018, including 23 million people living in 20 African countries.[21] These food crises were caused or exacerbated by widespread drought,

prolonged dry spells, erratic rainfall, and cyclone-related flooding. Resulting impacts included 28% lower wheat production in Afghanistan, 34% lower maize production in Zambia, and displacement of 300,000 people in Kenya as a result of crop and livestock failures.[21] Maize is a staple crop in many countries, including providing 40–70% of the calorie requirements in Kenya, so a reduced supply significantly increases malnutrition risks.[22] In Zimbabwe, climate change is creating erratic precipitation patterns, a later start to the rainy season, and drought conditions. Most Zimbabweans are subsistence farmers and rely on rain-fed agriculture to feed their families. The government estimated that in 2018, more than 2.4 million people in rural areas faced food insecurity, and food shortages in Zimbabwe continue to be a persistent threat.[23]

Food insecurity in 2020 was made worse by the COVID-19 pandemic and associated economic crises, as well as devastating desert locust outbreaks in numerous countries in East Africa, South Asia, and the Middle East. Locusts are usually present in low numbers and dormant, but when conditions are suitable, they reproduce in large numbers and can travel long distances, creating a crop-destroying plague (**Figure 9.10**).[24] Locust swarms occur periodically, but 2020 saw the worst disaster in 70 years in Kenya. Because locusts feed on crops grown for human consumption, these swarms cause enormous crop losses and billions of dollars in damage.

Favorable conditions for locusts include sandy soil, persistent rain, and vegetation near arid areas. Climate change is linked to this particular locust disaster because two years earlier, two powerful cyclones hit the Arabian peninsula and added extreme moisture to its sandy soil, creating perfect conditions for locust growth. The early stages of locust swarm formation went undetected, notably in Yemen, which has been devastated by civil war since 2014. At the end of 2019, another cyclone hit, and its winds spread the locusts to eastern Africa. Rising sea surface temperatures in the western Indian Ocean are fueling cyclones in this region.[25]

Food insecurity fueled by climate change is also a major threat for small island developing states (SIDS) in the Caribbean.[26] For example, changes in precipitation patterns in Jamaica that create even mild water deficits threaten the yields of commonly cultivated crops such as sweet potato, tomato, and pigeon pea. Animal husbandry is impacted by rising temperatures because heat stress decreases livestock fertility, growth, and productivity and raises the risk

Figure 9.10 Locust Swarm.
© John Carnemolla/Corbis Documentary/Getty Images.

of animal diseases. Fisheries provide critical protein and nutrient sources for people living in Caribbean SIDS, and changes in ocean temperatures, intense storms, sea level rise, and coral reef damage are disrupting fishing operations and livelihoods (**Figure 9.11**). Additional factors for household food insecurity in Caribbean SIDS are low-income status and low educational achievement. In addition, an emphasis on export-oriented agriculture and cash economies weakens food self-provisioning by households.[26]

Climate change also threatens food security among Indigenous peoples who rely on traditional food sources. For example, Native Alaskan subsistence hunters have for millennia hunted seals, walruses, whales, and other ocean species, which requires predictable winter weather and ice conditions (Figure 9.11). Now, they are facing record high temperatures and thinning and disappearing sea ice.[27] The Gwich'in people in northern Alaska rely on the Porcupine caribou herd, which is threatened by climate change because increasingly common winter rains create ice layers on land that make it difficult for caribou to walk, and these ice layers also cover lichen, an essential food for caribou. Habitats for caribou and other land animals in this region are at risk from changing environmental conditions, wildfires, and the ever-present threat of oil and gas development in Alaska.[28]

On St. Lawrence Island, Alaska, the diet of Yup'ik people includes traditionally harvested fish and seal meat that is air-dried during the summer months for sufficient stored food to get through the lean winter months. As the climate changes, this practice is threatened by unpredictable storms and extreme rainfall. Some people store meat in the permafrost rather than in freezers, but this practice is also being undermined as climate change causes the permafrost to melt.[28] Consuming traditional meats—such as fish, seal, whale, and walrus, which are rich in protein, vitamins, and fats—confers health benefits on Native Alaskans, and traditional diets have significant cultural value as well. With failing hunts, people increasingly rely on a limited selection of store-bought groceries, which tend to be expensive, high in fats and sugars, and highly processed for long shelf life. Fresh, canned, or frozen fruits or vegetables are scarce. For many Indigenous people, shifting to a largely processed food diet has led to increases in obesity, diabetes, hypertension, and other diseases.[28]

Figure 9.11 Left: A Man Prepares Fish for Sale at a Street Market in the Caribbean. Right: A Seal Hunter Collects His Catch on a Floating Ice Sheet.

Another major driver of food insecurity is that many people cannot afford the costs of food. The world produces sufficient food for everyone, yet according to FAO, more than 1.5 billion people cannot afford foods that provide the required levels of essential nutrients.[6] The costs of a healthy diet exceed average household food expenditures in many low- and middle-income countries, including for nearly 60% of people in sub-Saharan African and South Asian countries.[6] Price shocks result from severe acute disasters, including those fueled by climate change, and affect food availability, access, and security.

Agricultural Systems

Agricultural productivity depends on a predictable and stable climate. In general, climate change is expected to shift global agro-ecological zones around the world, due mostly to increasing temperatures and changing water availability and usage conditions. These shifts will improve agricultural conditions in some areas and degrade conditions in other areas (**Figure 9.12**).[29] Crops have optimal temperature ranges for plant growth and yields, and livestock require a suitable range of conditions with adequate water and grazing land. Depending on which agricultural practices are used, crop yields and livestock grazing at middle and high latitudes may improve because of higher temperatures, longer growing periods, and extension of frost-free zones. In many low-income countries, crop yields, especially cereals, are predicted to decrease. As a result, food prices may spike and more food imports may be needed, putting adequate food out of reach for many.[30]

Eighteen crops—wheat, maize, soybeans, rice, barley, sugar beet, cassava, cotton, groundnuts, millet, oats, potatoes, pulses, rapeseed, rye, sorghum, sunflower, and sweet potatoes—are grown on 70% of global crop land area and account for 65% of global caloric intake.[31] Considering future temperature increases and accounting for use of agricultural inputs such as fertilizers and pesticides, countries with already high yields of these crops are projected to benefit from warming, whereas countries that struggle to produce high yields of these crops will suffer further.[31] Many sub-Saharan African countries have very low agricultural output by global standards, and climate change threatens to further reduce productivity, which puts people's health, nutritional status, and food security at risk.[6] Important factors that may modulate these climate impacts are soil conditions, crop varieties grown, land use changes, regional population pressure, and capacity for climate change adaptation.[30] Climate impacts now and in the future burden small-holder farmers the most because they are more likely to be poor, to farm on marginal lands, and to lack the capacity to recover from lower crop yields, a less nutritious food supply, and extreme weather impacts.

Climate-change-fueled extreme storms often destroy farms and food supplies. For example, in March 2019, Cyclone Idai ripped through central Mozambique and caused flooding on more than 700,000 hectares of agricultural land. Many crops were destroyed, particularly maize. Desperate farmers salvaged unripe maize cobs from the floodwaters and dried them, but when consumed, they made people sick.[32] Food prices spiked rapidly, putting food purchases out of reach of many low-income people and increasing reliance on food imports and humanitarian aid. After Hurricane Katrina hit the U.S. Gulf Coast in 2005, its storm surge of 25–28 feet disrupted coastal grain transport by rail and barge, destabilizing the food supply and leading to food price spikes.

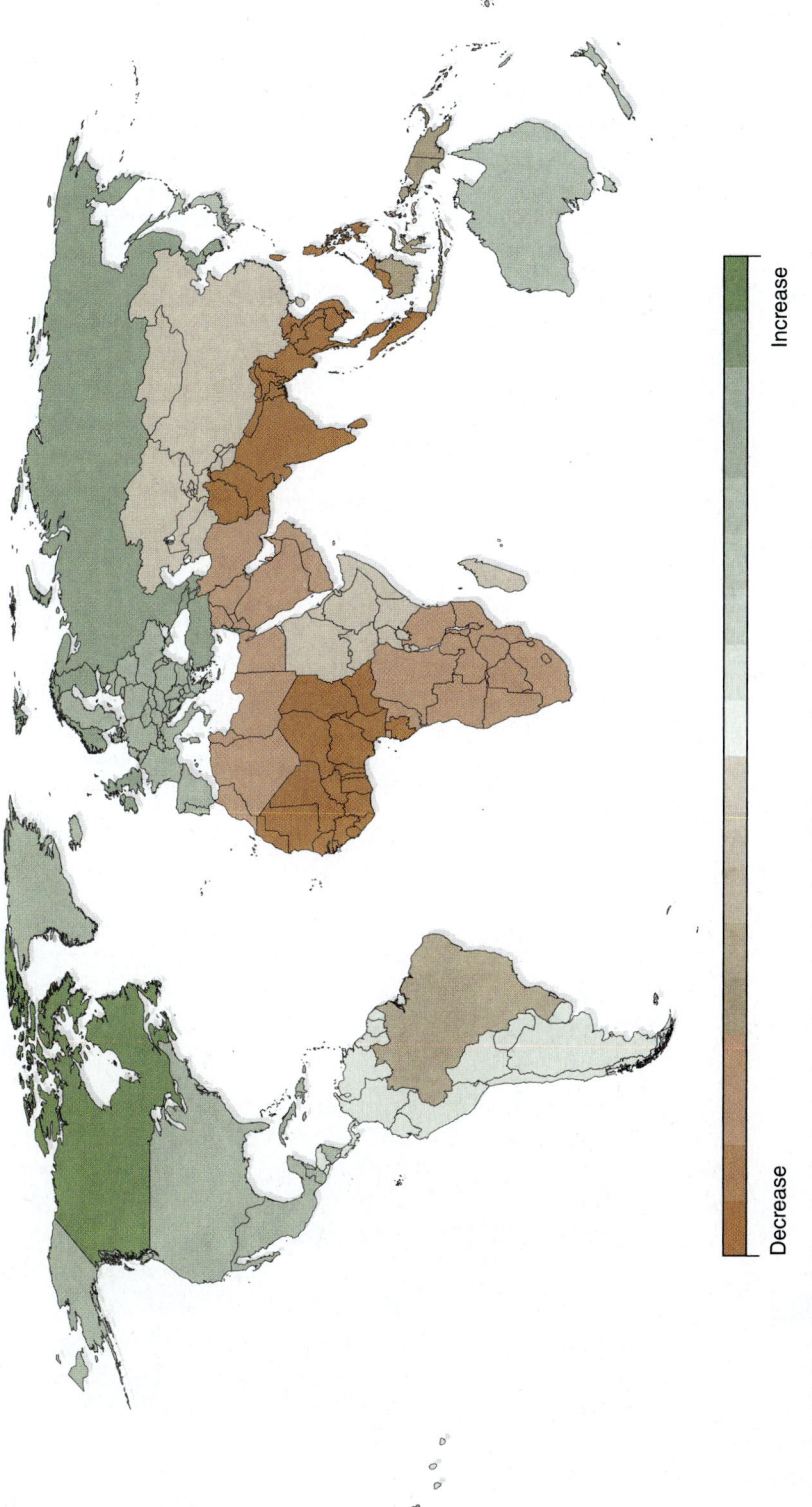

Figure 9.12 Predicted Changes in Agricultural Production in 2050 Due to Climate Change. Brown areas have decreased production and green areas have increased production.

Food and Agriculture Organization of the United Nations. Reproduced with permission.

Box 9.1 Food Insecurity in Mali

"When the Sahel is in the news, it is often because millions of people are at risk of going hungry. A humanitarian crisis usually unfolds on the back of a food crisis."[33]

One of the regions of the world most vulnerable to climate change impacts on agriculture and food security is the Sahel, especially the country of Mali, which faces the greatest malnutrition burden in the world.[12] More than half of Mali's population lives in poverty, two-thirds of Mali is arid and persistently under prolonged drought conditions, and the country has been destabilized by armed conflict for many years.[34] Poverty, food insecurity, and conflict create conditions for widespread malnutrition, particularly in children, and the country relies heavily on international food aid.

The northern region of Mali is part of the Sahara Desert. Much of the rest of the country is in the Sahel, a semiarid, barren, sandy, and rock-strewn transition zone between the Sahara to the north and fertile tropical regions to the south (**Figure 9.13**). The Sahel region faces a slow-onset climate disaster with strong climatic variations, irregular rainfall, persistent and severe drought, deterioriating soil and water resources, and high food insecurity. Deforestation and poor crop production and grazing practices have degraded the land significantly, population growth pressure is very high, and the region lacks effective environmental and sustainable development policies.[33] Farming is the main source of employment in the Sahel, and agricultural production contributes significantly to GDP.[33]

Mali experienced a military coup in 2012 that shattered 20 years of stable rule, and violent uprisings are common, particularly in Northern Mali, due to a lack of government presence. This conflict, coupled with prolonged drought, is driving human migration southward, with an estimated 335,000 people fleeing their

Figure 9.13 People Walk across the Sahel Region of Mali.
© Timothy Allen/The Image Bank Unreleased/Getty Images.

homes.[34] Water scarcity is widespread in Mali and is forcing people to relocate to areas with freshwater access for drinking, crop irrigation, and livestock watering. Migration of farmers has thus become a crucial survival strategy.

Climate resilience and improvements to agricultural productivity are needed for Mali to become food secure, including expanding the amount of arable land in cultivation and irrigation. FAO estimates that nearly

1 million people in Mali faced high acute food insecurity in 2021, more than 300,000 were displaced, and thousands were affected by COVID-19. As a result, FAO created a humanitarian response plan for Mali, with recommendations for projects ranging from diversifying agriculture-based livelihoods and improving irrigation systems to replenishing livestock. FAO also emphasized the need to train farmers on sustainable agricultural practices.[35]

This devastated households already at risk for food insecurity.[36] The world's dependence on global food markets means greater reliance on food grown far away from where it is consumed. This requires food processing, transport, distribution, and storage—all of which are fossil fuel intensive and vulnerable to climate disruptions.

Atmospheric carbon dioxide (CO_2), the principal greenhouse gas (GHG) driving climate change, is taken up by plants for photosynthesis, and thus, increasing atmospheric CO_2 levels impact plant growth, metabolism, and nutritional content. Plants tend to grow faster at higher CO_2 concentrations, with plant biomass and harvestable yields increasing. The macromolecular composition of plants also changes, including increases in nonstructural carbohydrates in leaves and decreases in protein in grains and tubers.[29] Micronutrients may also be reduced because higher CO_2 levels reduce plant uptake of key minerals such as nitrogen, phosphorus, magnesium, and calcium from the soil.[29] How these plant changes will impact human nutrition is uncertain, but one study estimated that under a high future GHG emissions scenario, an additional 1.4 billion children under 5 and women of childbearing age will be iron-deficient, 175 million additional people will be zinc-deficient, and an additional 122 million will be protein-deficient.[37]

Climate change alters the abundance and seasonality of agricultural pests, including weeds, insects, and molds, in much the same way that it affects vector-borne disease vectors and pathogens. **Aflatoxins** are naturally occuring proteins made by the molds *Aspergillus flavus* and *Aspergillus paraciticus* that frequently infect crops such as corn and groundnuts (peanuts) (**Figure 9.14**). These molds are common in tropical and subtropical areas where temperatures are suitably warm. If ingested, mold toxins may cause cancer, birth defects, and immunosuppression. As much as 5–28% of global liver cancer cases are attributed to aflatoxin exposure, particularly in China, Southeast Asia, and sub-Saharan Africa, where maize and groundnut consumption is high.[38] At very high exposure to aflatoxins, **aflatoxicosis** can result, leading to potentially fatal acute liver failure.

Groundnuts are a staple crop in many areas and an important source of protein and beneficial fats. The groundnut crop also fixes atmospheric nitrogen in the soil to improve soil fertility. Climate change appears to be the major environmental driver for crop infection with aflatoxin, mainly by causing increased temperatures and prolonged drought that, in the final weeks of groundnut growth, spark aflatoxin growth.[39] In the United States, aflatoxin contamination

Figure 9.14 Aflatoxin-Producing Molds Growing on Corn (Top) and Peanuts (Bottom).

Top: Reproduced from FtF Aflatoxin. U.S. Department of Agriculture; 2020. Accessed April 10, 2020. https://www.ars.usda.gov/office-of-international-research-engagement-and-cooperation/unpublished-old-oirp-pages/ftf-aflatoxin; Bottom: Reproduced from Mycotoxin Research. U.S. Department of Agriculture; 2020. Accessed August 8, 2020. https://www.ars.usda.gov/southeast-area/dawson-ga/national-peanut-research-laboratory/docs/mycotoxin-research/page-1

is predicted to spread across corn-growing states and in a worst case may cause losses of $1 billion annually.[40] Aflatoxin-contaminated corn is also predicted

to increase in range in Europe, especially across southern regions.[41]

Increased pest growth is expected to lead to increased pesticide usage, which carries its own health risks. Pesticide exposure is linked to a number of diseases in humans, including cancers, neurological disorders, and birth defects. In addition, at higher atmospheric CO_2 concentrations, some weeds become more herbicide-tolerant, requiring more herbicide application on fields for the same efficacy as at lower CO_2 levels.[42] Human exposure to agricultural chemicals can occur via ingestion of contaminated foods and drinking water and by inhalation of airborne residues. Climate-change-induced prolonged dry spells, in which increased irrigation is used, may increase water pollution via runoff of chemicals and other contaminants into water sources. Climate-change-induced flooding may disperse existing pollutants onto farm fields or into irrigation or drinking water.[43] With severe water supply limitations, people may be forced to use poorer quality water sources, increasing health risks. Under dry conditions and with intensive industrial agriculture, soil drying and erosion are increasing, resulting in blowing soil dust that may increase human exposure to pesticides and bacterial and fungal spores that may cause disease.[43]

Many **foodborne illnesses** spike in summer months because warmer temperatures favor pathogen growth and food spoilage (**Figure 9.15**). At the same time, more outdoor food-related activities take place, such as picnics and barbeques, during which refrigeration, washing facilities, and other safety controls may be lacking.[36] *Salmonella*, *Campylobacter*, and *E. coli* are some of the most common foodborne infections in the United States, and each year sees 2 million cases of serious gastrointestinal illnesses from these infections occur.[36] Higher temperatures are predicted to lead to greater disease incidence as pathogen growth on foods increases. For example, *Salmonella* growing on raw chicken doubles in number in 60 minutes at 70°F and 22 minutes at 90°F.[36] In addition, antibiotic resistance of foodborne bacteria has been shown to be increasing with climate change.[44]

Not only are agriculture and food systems impacted by climate change, but they also *contribute* a significant portion of the world's GHG emissions (approximately 12% in 2015),[45] which come from farming practices, energy use in agriculture, and food processing, distribution, consumption, and waste. GHG emissions in agriculture are primarily nitrous oxide (N_2O), methane (CH_4), and CO_2 emitted during a range of activities.[46]

- Soil management practices can increase the availability of soil nitrogen and lead to emissions of N_2O.
- Ruminant livestock (e.g., cattle) emit CH_4 from digestive pathways.
- Livestock manure is a source of N_2O and CH_4.
- Rice cultivation produces high levels of CH_4.
- Burning crop waste releases N_2O and CH_4.
- Fossil fuel use during agricultural practices and to produce fertilizer, pesticides, and other inputs emits CO_2 and other GHGs.

GHG emissions attributed to the production of specific foods vary widely, with highest emissions from beef and dairy production and lower emissions from plant-based foods.[47] Agricultural practices, dietary choices, and food consumption patterns significantly impact GHG emissions. Human diets that

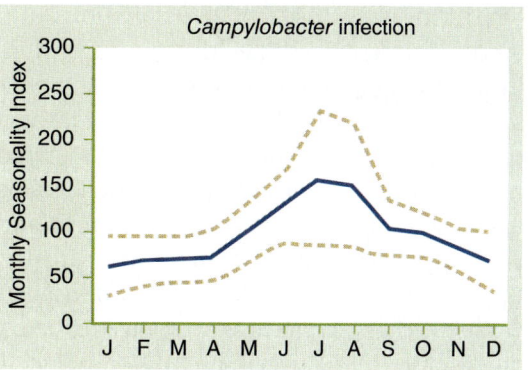

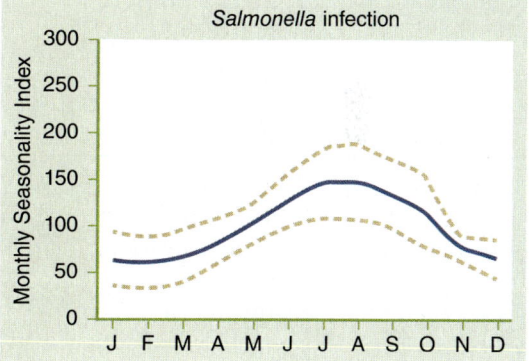

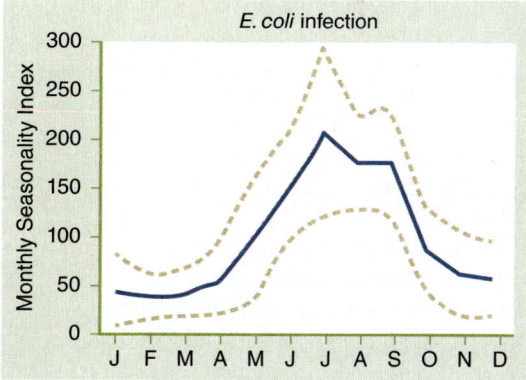

Figure 9.15 Seasonality of Foodborne Illnesses in the U.S. Human infections with *Campylobacter*, *Salmonella*, and *E. Coli* spike in the summer months.

Reproduced from Ziska L, Crimmins A. Food safety, nutrition, and distribution. In: Crimmins A, Balbus J, Gamble JL, et al., eds. *The Impacts of Climate Change on Human Health in the United States: A Scientific Assessment*. U.S. Global Change Research Program; 2016:Figure 3. doi:10.7930/J0R49NQX

contribute the most to GHG emissions also tend to be the least healthy, and are often linked to higher rates of overweight and obesity,[47] which are risk factors for many chronic diseases. Switching from a diet high in red meats, processed meats, and refined grains and low in fruits, vegetables, legumes, and whole grains to an alternative diet with the opposite composition leads to *health benefits* (reduced risk of heart disease, type 2 diabetes, and colorectal cancer), as well as *climate benefits* (reduced GHG emissions).[48] Global health costs linked to consuming a poor diet are estimated to exceed $1.3 trillion per year by 2030, so a switch to a climate-friendly diet will yield enormous health cost savings.[6]

Adaptations

Mitigating GHG emissions is key to reducing problems for local, regional, and global agriculture and food systems. Strengthening health systems to improve food security and nutrition is also critical. Strategies to create adaptive capacity are needed, particularly for people living in areas highly prone to drought or floods, crop failures, food insecurity, and famine (see **Box 9.2**).

Box 9.2 Climate Adaptation by Farmers in Vietnam

Hannah Richelieu

Vietnam is at high risk of climate change impacts, due primarily to its geographic location, long coastline, and resulting threat of human displacement on a large scale. Sea level rise is increasingly inundating land and causing saltwater intrusion and lower availability of fresh water.[49] Projected increases in temperatures, droughts, and floods are expected to degrade soil and make agriculture less productive.[50] About 70% of Vietnam's population live in rural areas and depend on agriculture for their livelihoods.[51] One-quarter of children suffer from stunting, an indicator of chronic undernutrition.[52]

Rice is the most important crop in Vietnam, grown primarily in the Red and Mekong River Deltas, and often by small-scale farmers (**Figure 9.16**). Vietnam is among the top three rice exporters in the world. Over half of the rice crop is grown in flood-prone areas and thus vulnerable to the effects of climate change, and freshwater salinization alone threatens to reduce rice production by about 50%.[50,51] These effects will likely be made worse by increasingly extreme weather events. In October 2020 alone, Vietnam was hit by four destructive tropical cyclones and torrential rain events, which led to flooded rice fields and thousands of livestock animals being swept away. As a result, food shortages occurred, and food price spikes threatened those who were food insecure.[50]

Hà Tĩnh province, a coastal region of Vietnam highly dependent on rice farming and with a high poverty rate, is experiencing increasing mean annual temperatures and extreme high temperatures during the hottest months of the year.[49] In 2007, prolonged floods destroyed most rice seedlings, and in 2008, abnormal temperature variations and flooding reduced the overall rice yield by 40%. These events increased food insecurity in Hà Tĩnh province, and at the time, only 20% of families reported having sufficient amounts of rice to last nine months after the extreme weather events.[49]

As a result, farmers in Hà Tĩnh province and across Vietnam are paying close attention to changing weather and climate, and they have implemented innovative strategies to increase crop resiliency, including shifting their growing season to avoid extreme weather months. This has been somewhat

Figure 9.16 Woman Working in a Rice Field in the Mekong Delta.
© 1905HKN/iStock Unreleased/Getty Images.

successful, but storms and floods still pose a threat. Some farmers in Hà Tĩnh province have begun practicing **agroforestry**, in which woody perennials like trees, shrubs, bamboo, and palm are intentionally planted on farms and in agricultural areas along with crops to provide diversified production and enhance food supplies, health status, and income. Several woody plants are climate-resilient, including bamboo, rattan, jack fruit, tea, acacia, eucalyptus, and banana, and have been planted near rice paddies to act as windbreakers, prevent soil erosion, and provide protective fencing.[53] Tea can be sold year-round to supplement income in case of failed rice crops, and fruit trees provide additional food for households. One study found that economic recovery after severe climatic events for families who practiced agroforestry in Hà Tĩnh province was up to three months faster than for families who did not practice agroforestry.[53] One major constraint to adoption of agroforestry practices, however, is that poor farmers often lack the resources to make long-term investments in trees, which may not mature for 7–14 years.[49]

Effective government policies could incentivize more agroforestry, as well as climate resilient landscape-level land management and crop protection, and social safety nets for subsistence farmers. These adaptive measures are necessary because climate change-attributable reductions in rice production will have significant effects on food security and human health in Vietnam and many other parts of the world.

Targets under SDG 2 (*Zero hunger*) outline a number of agricultural solutions.[15]

Target 2.3: "**By 2030, double the agricultural productivity and** incomes of small-scale food producers, in particular women, indigenous peoples, family farmers, pastoralists and fishers, including through secure and equal access to land, other productive resources and inputs, knowledge, financial services, markets and opportunities for value addition and non-farm employment."

Target 2.4: "**By 2030, ensure sustainable food production systems and implement resilient agricultural practices that increase productivity and production, that help maintain ecosystems, that strengthen capacity for adaptation to climate change, extreme weather, drought, flooding and other disasters and that progressively improve land and soil quality.**"

Target 2.a: "**Increase investment, including through enhanced international cooperation, in rural infrastructure, agricultural research and extension services, technology development and plant and livestock gene banks in order to enhance agricultural productive capacity in developing countries, in particular least developed countries.**"

Target 2.b: "**Correct and prevent trade restrictions and distortions in world agricultural markets. . . .**"

Target 2.c: "**Adopt measures to ensure the proper functioning of food commodity markets . . . to help limit extreme food price volatility.**"

According to the U.S. Environmental Protection Agency, GHG emissions from agricultural practices can be minimized by the following actions.[46]

- Switch to renewable energy sources in agriculture
- Avoid overapplication of nitrogen fertilizers

- Drain water from rice soils to reduce methane emissions
- Adjust livestock feed and improve pasture quality to reduce methane emissions from livestock
- Use better manure management practices

Developing climate-smart food systems also requires shifting to healthier and decarbonized diets, and minimizing food waste so that what is grown is eaten.[54] In addition, farming on existing agricultural lands rather than converting new lands to agriculture protects soils, forests, and ecosystems, preserving carbon sequestration capacity.[54] Many aspects of dominant human food systems drive up the cost of nutritious foods, and according to FAO, key solutions include supporting small-scale food producers to get nutritious foods to markets at low cost.[6]

The agricultural impacts of climate change may burden women more than men, and both food insecurity and gender inequalities are predicted to worsen with climate change.[55] Many of the world's subsistence and small-holder farmers are women, and even though they produce much of the world's food that is consumed by people, they often lack access to land, resources, and decision-making power, and thus have unique susceptibilities and lower adaptive capacity. Agriculture is the primary economic activity for more than three-quarters of women in the least developed countries, and household food provision falls mainly on women, which becomes harder when crops fail. Women's household and agricultural work is labor-intensive and reduces time available for education, livelihood diversification, and other activities.[55] High nutritional disease burdens increase women's workload and stress and reduce their own and their children's health status. Gender-sensitive climate adaptations in agriculture recognize and respond to socially differentiated roles and responsibilities for men and women.[55] If women had greater access to agricultural land and resources, it is projected that crop yields could increase 20–30%, and the number of hungry people could decrease by 150 million.[55]

Specific policies that protect and empower women include[56]:

- Protecting rural women's rights and increasing access to and control of lands and other natural resources
- Protecting communities from land and resource grabs by agribusiness, mining, or fossil fuel industries
- Scaling up food sovereignty (the right to healthy and culturally appropriate food produced through ecologically sound and sustainable methods, and the right to define one's own food and agriculture systems)
- As for all health impacts, strengthening national climate change plans

When planning for climate change mitigation, careful attention to hunger and food insecurity is needed to avoid inadvertently exacerbating these problems. For example, pricing carbon to mitigate GHGs is predicted to increase food prices and household food expenditures, decrease food availability, increase consumption of less expensive foods, and raise the risk of hunger.[57] Switching to biofuels as alternatives to fossil fuels increases the demand for growing biofuels crops on existing agricultural land instead of growing food, or conversion of forested land that otherwise would serve as a carbon sink into land for biofuels. Large-scale and industrial biofuels production contributes to food instability and unequal land access for the world's poor and small-holder farmers.

Effective climate adaptation strategies must also create food safety-net programs to minimize food insecurity and malnutrition and help pull people up out of poverty. Careful attention to equity is needed so that decarbonized climate-smart and resource-efficient food systems and farming practices are accessible to all.

Discussion Questions

1. Discuss how we measure the human health impacts of food insecurity. What are the latest trends?
2. How are child stunting, underweight, and wasting diagnosed?
3. What are the treatments for severe acute malnutrition and micronutrient deficiencies?
4. What are the undernutrition targets in the Sustainable Development Goals and the World Health Organization's Global Nutrition Targets 2025 and how likely is it that these targets will be achieved?
5. How is climate change exacerbating the problems of undernutrition and food insecurity?
6. How are certain foodborne illness risks made worse by climate change?

7. Discuss examples of specific populations at particular risk of food insecurity. How are their situations affected by a changing climate?
8. Discuss the ways that our agricultural practices and food systems not only are vulnerable to climate change but also actually contribute to climate change.
9. Discuss climate adaptation strategies for agriculture and food systems that will benefit women, poor and small-holder farmers, and people most at risk of food insecurity and malnutrition.

References

1. Lowcock M, Beasley D. We've averted famine in the past–we must do it again. *The Times* [London]. November 17, 2020. Accessed January 20, 2021. https://www.thetimes.co.uk/article/weve-averted-famine-in-the-past-we-must-do-it-again-jb7fz8s8g
2. Funk C. Ethiopia, Somalia and Kenya face devastating drought. *Nature*. 2020;586(7831):645. doi:10.1038/d41586-020-02698-3
3. Constanza K. Facing famine: battling hunger with hope in East Africa. July 12, 2017. Accessed January 20, 2021. https://www.worldvision.org/hunger-news-stories/facing-famine-hunger-hope-east-africa
4. Law T. The climate crisis is global, but these 6 places face the most severe consequences. *Time*. September 30, 2019. Accessed January 20, 2021. https://time.com/5687470/cities-countries-most-affected-by-climate-change/
5. Pathak T, Maskey M, Dahlberg J, et al. Climate change trends and impacts on California agriculture: a detailed review. *Agronomy*. 2018;8:25. doi:10.3390/agronomy8030025
6. Food and Agriculture Organization, International Fund for Agricultural Development, UNICEF, World Food Programme, World Health Organization. *The State of Food Security and Nutrition in the World 2020. Transforming food systems for affordable healthy diets*. Food and Agriculture Organization; 2020. doi:10.4060/ca9692en
7. World Health Organization. Micronutrients. Accessed January 21, 2021. https://www.who.int/health-topics/micronutrients
8. Food and Agriculture Organization. An introduction to the basic concepts of food security. 2008. Accessed January 21, 2021. http://www.fao.org/3/al936e/al936e.pdf
9. World Health Organization. Comprehensive implementation plan on maternal, infant and young child nutrition. May 19, 2014. Accessed January 22, 2021. https://www.who.int/publications/i/item/WHO-NMH-NHD-14.1
10. Headey D, Heidkamp R, Osendarp S, et al. Impacts of COVID-19 on childhood malnutrition and nutrition-related mortality. *The Lancet*. 2020; 396(10250):519–521. doi:10.1016/S0140-6736(20)31647-0
11. Hart M, Ferguson S. What is ready-to-use therapeutic food? March 6, 2019. Accessed January 22, 2021. https://www.unicefusa.org/stories/what-ready-use-therapeutic-food/32481
12. Institute for Health Metrics and Evaluation. GBD Compare. Accessed October 15, 2020. https://vizhub.healthdata.org/gbd-compare
13. Concern Worldwide, Welthungerhilfe. 2020 Global Hunger Index. 2021. Accessed January 22, 2021. https://www.globalhungerindex.org/results.html
14. United Nations. News on Millennium Development Goals. Accessed August 11, 2020. https://www.un.org/millenniumgoals/
15. Sustainable Development Goals. The 17 Goals, United Nations. ©(2021) United Nations. Reprinted with the permission of the United Nations. https://www.undp.org/content/undp/en/home/sustainable-development-goals/goal-2-zero-hunger/targets.html
16. World Health Organization. Global Targets 2025: To improve maternal, infant and young child nutrition. Accessed March 21, 2021. https://www

.who.int/teams/nutrition-and-food-safety/global-targets-2025

17. Hailu T, Yimer M, Mulu W, et al. Synergetic effects of *Plasmodium*, hookworm, and *Schistosoma mansoni* infections on hemoglobin level among febrile school age children in Jawe Worda, northwest Ethiopia. *J Parasitol Res.* 2018;2018:9573413-9573413. doi:10.1155/2018/9573413

18. World Health Organization. Guideline: vitamin A supplementation in infants and children 6–59 months of age. 2011. Accessed January 23, 2021. https://www.who.int/publications/i/item/9789241501767

19. Food and Agriculture Organization. *The Food Insecurity Experience Scale: Measuring Food Insecurity through People's Experiences.* September 2017. Accessed January 23, 2021. http://www.fao.org/3/i7835e/i7835e.pdf

20. U.S. Department of Agriculture Economic Research Service. Food security and nutrition assistance. Updated December 16, 2020. Accessed March 21, 2021. https://www.ers.usda.gov/data-products/ag-and-food-statistics-charting-the-essentials/food-security-and-nutrition-assistance

21. Food Security Information Network. 2019 Global Report on Food Crises. April 2, 2019. Accessed January 23, 2021. https://www.wfp.org/publications/2019-global-report-food-crises

22. Grace K, Brown M, McNally A. Examining the link between food prices and food insecurity: a multi-level analysis of maize price and birthweight in Kenya. *Food Policy.* 2014;46:56–65. doi:10.1016/j.foodpol.2014.01.010

23. Thornton J. Fueled by climate change, Zimbabwe's erratic harvests cause farmers with HIV to struggle. *CNN.* February 13, 2019. Accessed January 24, 2021. https://www.cnn.com/2019/02/13/health/climate-change-zimbabwe-farmers-hiv-intl/index.html

24. Meynard CN, Lecoq M, Chapuis M-P, et al. On the relative role of climate change and management in the current desert locust outbreak in East Africa. *Global Change Biol.* 2020;26(7):3753–3755. doi:10.1111/gcb.15137

25. Salih AAM, Baraibar M, Mwangi KK, et al. Climate change and locust outbreak in East Africa. *Nat Clim Change.* 2020;10(7):584–585. doi:10.1038/s41558-020-0835-8

26. Lenderking H, Robinson S-A, Carlson G. Climate change and food security in Caribbean small island developing states: challenges and strategies. *Int J Sustainable Dev World Ecol.* 2020. doi:10.1080/13504509.2020.1804477

27. Struzik E. Food insecurity: Arctic heat is threatening indigenous life. *Yale Environment 360.* March 17, 2016. Accessed January 24, 2021. https://e360.yale.edu/features/arctic_heat_threatens_indigenous_life_climate_change

28. Grossman E. Alaska's uncertain food future. *High Country News.* August 18, 2014. Accessed January 24, 2021. https://www.hcn.org/issues/46.14/alaskas-uncertain-food-future

29. Food and Agriculture Organization. *The State of Agricultural Commodity Markets 2018: Agricultural Trade, Climate Change and Food Security.* 2020. Accessed January 24, 2021. http://www.fao.org/3/I9542EN/i9542en.pdf

30. Das HP. *Climate Change and Agriculture: Implications for Global Food Security.* CRC Press; 2016:612.

31. Agnolucci P, Rapti C, Alexander P, et al. Impacts of rising temperatures and farm management practices on global yields of 18 crops. *Nature Food.* 2020;1(9):562–571. doi:10.1038/s43016-020-00148-x

32. Eisenhammer S. Hunger stalks Mozambique after deadly cyclone destroys farmland. *Reuters.* April 1, 2019. Accessed January 25, 2021. https://www.reuters.com/article/us-africa-cyclone-farmers/hunger-stalks-mozambique-after-deadly-cyclone-destroys-farmland-idUSKCN1RD24Q

33. Essoungou A-M. The Sahel: one region, many crises. *Africa Renewal.* December 2013. Accessed January 24, 2021. https://www.un.org/africarenewal/magazine/december-2013/sahel-one-region-many-crises

34. Norwegian Refugee Council. Sahel: the world's most neglected and conflict-ridden region. Accessed January 24, 2021. https://www.nrc.no/shorthand/fr/sahel---the-worlds-most-neglected-and-conflict-ridden-region/index.html

35. Food and Agriculture Organization. Mali: Humanitarian Response Plan 2021. March 11, 2021. Accessed March 21, 2021. http://www.fao.org/emergencies/appeals/detail/en/c/1372786/

36. United States Global Change Research Program. The impacts of climate change on human health in the United States: a scientific assessment. 2016. Accessed January 22, 2021. https://health2016.globalchange.gov

37. Myers S, Smith M. Impact of anthropogenic CO_2 emissions on global human nutrition. *Nat Clim Change.* 2018;8:834–839. doi:10.1038/s41558-018-0253-3

38. Tillett T. Carcinogenic crops: Analyzing the effect of aflatoxin on global liver cancer rates. *Environ*

Health Perspect. 2010;118(6):A258. doi:10.1289/ehp.118-a258a

39. Haerani H, Apan A, Basnet B. The climate-induced alteration of future geographic distribution of aflatoxin in peanut crops and its adaptation options. *Mitig Adapt Strateg Global Change.* 2020;25(6):1149–1175. doi:10.1007/s11027-020-09927-0

40. Mitchell NJ, Bowers E, Hurburgh C, et al. Potential economic losses to the U.S. corn industry from aflatoxin contamination. *Food Addit Contam A.* 2016;33(3):540–550. doi:10.1080/19440049.2016.1138545

41. Battilani P, Toscano P, Van der Fels-Klerx HJ, et al. Aflatoxin B1 contamination in maize in Europe increases due to climate change. *Sci Rep.* 2016;6(1):24328. doi:10.1038/srep24328

42. Waryszak P, Lenz TI, Leishman MR, et al. Herbicide effectiveness in controlling invasive plants under elevated CO_2: sufficient evidence to rethink weeds management. *J Environ Manage.* 2018;226:400–407. doi:10.1016/j.jenvman.2018.08.050

43. Boxall ABA, Hardy A, Beulke S, et al. Impacts of climate change on indirect human exposure to pathogens and chemicals from agriculture. *Environ Health Perspect.* 2009;117(4):508–514. doi:10.1289/ehp.0800084

44. MacFadden DR, McGough SF, Fisman D, et al. Antibiotic resistance increases with local temperature. *Nat Clim Change.* 2018;8(6):510–514. doi:10.1038/s41558-018-0161-6

45. Environmental Protection Agency. Climate Change Indicators: global greenhouse gas emissions. April 14, 2021. Accessed June 26, 2021. https://www.epa.gov/climate-indicators/climate-change-indicators-global-greenhouse-gas-emissions

46. Environmental Protection Agency. Sources of greenhouse gas emissions. April 14, 2021. Accessed June 26, 2021. https://www.epa.gov/ghgemissions/sources-greenhouse-gas-emissions

47. Hjorth T, Huseinovic E, Hallström E, et al. Changes in dietary carbon footprint over ten years relative to individual characteristics and food intake in the Västerbotten Intervention Programme. *Sci Rep.* 2020;10(1):20–20. doi:10.1038/s41598-019-56924-8

48. Hallström E, Gee Q, Scarborough P, et al. A healthier U.S. diet could reduce greenhouse gas emissions from both the food and health care systems. *Clim Change.* 2017;142(1):199–212. doi:10.1007/s10584-017-1912-5

49. Nguyen Q, Hoang MH, Öborn I, et al. Multipurpose agroforestry as a climate change resiliency option for farmers: an example of local adaptation in Vietnam. *Clim Change.* 2013;117(1):241–257. doi:10.1007/s10584-012-0550-1

50. Schneider P, Asch F. Rice production and food security in Asian Mega deltas—a review on characteristics, vulnerabilities and agricultural adaptation options to cope with climate change. *J Agron Crop Sci.* 2020;206(4):491–503. doi:10.1111/jac.12415

51. Rutten M, van Dijk M, van Rooij W, et al. Land use dynamics, climate change, and food security in Vietnam: a global-to-local modeling approach. *World Dev.* 2014;59:29–46. doi:10.1016/j.worlddev.2014.01.020

52. Beal T, Le DT, Trinh TH, et al. Child stunting is associated with child, maternal, and environmental factors in Vietnam. *Matern Child Nutr.* 2019;15(4):e12826. doi:10.1111/mcn.12826

53. Simelton E, Dam BV, Catacutan D. Trees and agroforestry for coping with extreme weather events: experiences from northern and central Viet Nam. *Agrofor Syst.* 2015;89(6):1065–1082. doi:10.1007/s10457-015-9835-5

54. Project Drawdown. Food, agriculture, and land use. 2020. Accessed February 1, 2021. https://drawdown.org/sectors/food-agriculture-land-use

55. Nyasimi M, Huyer S. Closing the gender gap in agriculture under climate change. *Agric Dev.* 2017;30:37–40.

56. Wijeratna A. Public policies that advance or hinder rural women's and young people's livelihoods and climate justice for all. ActionAid. October 9, 2019. Accessed February 1, 2021. https://actionaid.org/publications/2019/public-policies-advance-or-hinder-rural-womens-and-young-peoples-livelihoods-and

57. Hasegawa T, Fujimori S, Havlík P, et al. Risk of increased food insecurity under stringent global climate change mitigation policy. *Nat Clim Change.* 2018;8:699–703. doi:10.1038/s41558-018-0230-x

CHAPTER 10

Mental Health

KEY TERMS

Mental health
Solastalgia
Eco-anxiety

Major depressive disorder
(MDD)
Anxiety disorder

Posttraumatic stress disorder
(PTSD)
Posttraumatic growth

LEARNING OBJECTIVES

- Understand the range of emotional and psychological responses to climate change, both negative and positive.
- Understand the range and burdens of diagnosable mental disorders that may be exacerbated by climate change.
- Describe specific climate-related exposures linked to exacerbation of mental disorders.
- Describe which populations are at particular risk of mental health impacts and why.
- Understand the range of adaptations to protect and strengthen mental well-being in the face of climate impacts.

The strongest storm of the 2019 Atlantic hurricane season was Category 5 Hurricane Dorian, which battered the Bahamas for 48 hours. The storm reached a strength at which the U.S. National Hurricane Center acknowledged "catastrophic damage will occur," and "most of the area will be uninhabitable for weeks or months."[1,2] Much of the Abaco Islands in the Bahamas was destroyed and more than 200 people died. Emergency health workers reported that victims suffered severe head injuries, puncture wounds and other types of physical trauma, as well as worsened chronic diseases such as diabetes and high blood pressure, for which treatment was disrupted.[3] In addition, acute psychological distress was common.

Residents of the Bahamas were exposed to both a life-threatening storm and major hardships in its aftermath as people lost their homes, power, safe water or sanitation, food supply, and incomes (**Figure 10.1**). People experienced acute grief due to dead, injured, or missing loved ones, seeing dead bodies floating in the storm water, or fear that they themselves were going to die. Many had

Figure 10.1 Left: Satellite View of Hurricane Dorian. Right: Destruction Caused by Hurricane Dorian in the Marsh Harbour Settlement Known as "The Mudd" in the Bahamas.

Left: © Roberto Machado Noa/Moment/Getty Images; Right: © Branden Hoolehan/Moment/Getty Images.

debilitating and persistent clinical symptoms of major depression, generalized anxiety, and post-traumatic stress disorder (PTSD).[3] At the same time, some felt gratitude for having survived. A very heavy burden was put on clinical health workers and emergency medical personnel to treat everyone affected, and there was a lack of sufficiently trained mental health professionals to deliver the psychosocial services needed to help this population deal with such a catastrophic event.[3] Further hampering efforts was stigma, common in the Bahamas, that prevented people from talking about their mental health and seeking professional help. Two years after the storm, the Abaco Islands still struggled to recover and rebuild.

The devastating physical and mental health impacts of Hurricane Dorian are common for people who directly experience extreme storms and other climate change-fueled stressors, and climate change can be thought of generally as a threat to mental well-being. "It's pretty hard now to live on planet Earth and not have any clue that something's happening that's not good," said California-based psychotherapist Linda Buzzell.[4] It is natural and rational to *feel* something, to react emotionally and

psychologically to the dangers and uncertainties that climate change is bringing to people's lives. Not everyone reacts the same way in stressful situations, and mental health status can fluctuate over time. Some responses may be beneficial to mental health status, whereas others may cause harm. Some conditions require support services, therapies, and treatments, and others do not. Reactions to climate change may compound preexisting vulnerabilities to mental disorders, tip people from states of mental wellness to illness, or create opportunities for emotional and psychological growth.[5,6]

The World Health Organization (WHO) defines **mental health** as "a state of well-being in which an individual realizes his or her own abilities, can cope with the normal stresses of life, can work productively and is able to make a contribution to his or her community."[7] Others define mental well-being as "a subjective and dynamic state of feeling healthy and happy that ties into life satisfaction and influences a person's (or a collective's) psychological and social function."[8]

Many mental health impacts arise from direct exposure to climate change hazards. Others result from people's reactions to current and future *threats* of climate change, awareness of climate disasters impacting other communities, or a perceived lack of climate action, especially with the realization that climate change and its attributed disasters are human-caused and thus can be prevented or mitigated.[9] Mental health status may also *improve* through increased compassion, optimism, altruism, and personal growth as people and communities come together to support each other and take action.[10]

As was seen in Hurricane Dorian, climate change is already affecting mental health in certain populations, and research suggests this is quite common.[4] Populations of particular concern include those at high risk of exposure to climate-related devastation, who have little power in decision-making about climate change or emergency preparedness, or who live in communities with high baseline mental health burdens, less mental health awareness and care, or inadequate support systems. Specific groups at risk include people who experience disasters, people with preexisting mental illnesses, emergency workers and first responders, people experiencing homelessness, people with limited mobility, people who rely on natural resources for their survival or livelihoods, Indigenous peoples, displaced persons, and women and children.

Climate change interacts with a range of individual and collective risk factors in complex causal pathways that may lead to adverse emotional and psychological responses (**Figure 10.2**).[11] In general,

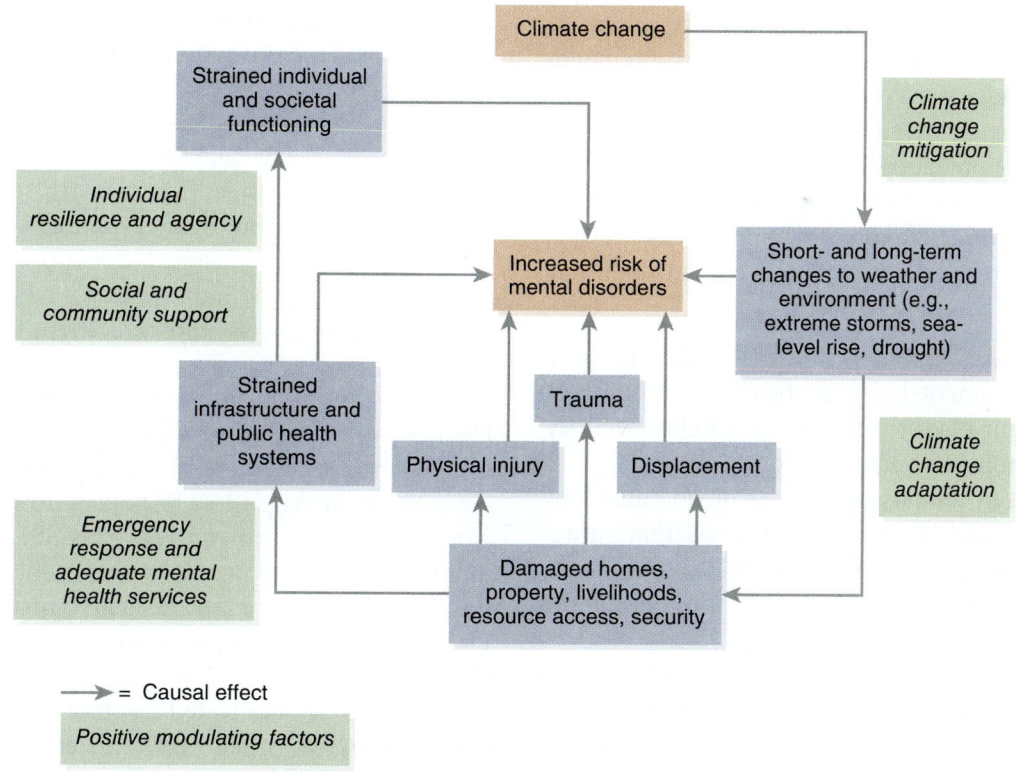

Figure 10.2 Causal Pathways through Which Climate Change Increases the Risk of Mental Disorders.

Modified from Berry HL, Waite TD, Dear KBG, et al. The case for systems thinking about climate change and mental health. *Nat Clim Change.* 2018;8:282–290. doi:10.1038/s41558-018-0102-4

determinants of mental wellness or illness include the capability to manage emotions, behaviors, relationships, and social support networks. In addition, living and working conditions, environmental threats, access to health care and support services, and experiences of impoverishment, discrimination, and violence also play a role.[6] Climate-related stressors include heat waves, sudden-onset extreme weather events such as storms and floods, longer term events such as heat waves and droughts, and slow-onset changes such as prolonged drought, desertification, and sea level rise. Mental health impacts may be *direct* (due to acute loss, stress, or trauma) or *indirect* (due to disruption of infrastructure, healthcare or community services, or forced displacement).[10] Positive modulating factors include individual resilience, family and community stability and support, access to mental health services, emergency preparedness, effective communications about climate change disaster risks, and ultimately, mitigation of greenhouse gas (GHG) emissions.

Climate change may induce a range of **emotional responses** specifically related to perceived or experienced *environmental* destruction or change.[12-14]

- **Solastalgia**—sense of loss due to environmental change, usually to an important place that provides stability, security, and personal identity
- **Eco-anxiety**—stress or dread caused by phobic fear of extreme weather, unrelenting day-to-day despair about climate change impacts, or worrying about the future
- **Eco-paralysis**—inability to take action because of a perception that the environmental threat is intractable and thus difficult to solve

- **Eco-nostalgia**—perception that a place or geographic location was better before the environmental change occurred
- **Eco-anger**—anger or frustration about climate change or other environmental crisis
- **Ecological grief**—grief associated with ecological losses, loss of environmental knowledge, or anticipated future losses

Precise diagnostic criteria for these conditions are not well-defined, and they may overlap with each other or co-occur with diagnosable mental disorders. Evidence suggests that they are prevalent, particularly in young people. In a 2019 survey of teens in the United States, 57% reported that climate change made them feel afraid,[12] and another 2019 study found that two-thirds of American adults said they worried about climate change.[4] In a 2020 survey in Great Britain, 70% of 18- to 24-year-olds reported being more worried about climate change than the year before.[15] In Greenland, more than 75% of people surveyed said that climate change has affected them personally, and 38% reported experiencing "very" or "moderately" strong fear responses when they think of climate change.[14] Eco-anxiety in particular may be common in people who are environmentally aware and care about environmental issues.

Many of these emotional reactions are healthy and rational, and people can learn effective behavioral, cognitive, and emotional strategies to manage these responses. In some cases, worry and fear can be forms of adaptation that motivate people to change behaviors or take climate action. The American Psychological Association reported in 2019 that 87% of people reporting eco-anxiety said they were motivated to reduce their contribution to climate change, compared with 40% of those who did not report eco-anxiety.[12] Taking climate action may promote the critical mental wellness qualities of perceived efficacy and competence, and may reduce stress, anxiety and depression.[13]

In contrast, some psychologists argue that eco-anxiety may lead people to disengage if they find climate change too upsetting or if they feel personally helpless. "For many of us, we'd literally rather not know because otherwise it creates an acutely distressing experience for us as humans," said Dr. Renee Lertzman, founding member of the Climate Psychology Alliance, a network of psychologists helping individuals and organizations respond to climate stress.[16] This reaction is not a form of denial of climate change but rather a blocked ability to self-actualize and take action. A major contributing factor to the epidemic of eco-anxiety is that most people have few opportunities to publicly discuss how climate change is distressing, uncertain, and fear-provoking or acknowledge that resulting emotional responses can be healthy and vital. Said Caroline Hickman, another member of the Climate Psychology Alliance, "A measure of mental health is having the capacity to accurately emotionally respond to the reality in our world. So it's not delusional to feel anxious or depressed. It's mentally healthy."[16]

In addition to emotional reactions, climate change may induce or exacerbate certain **diagnosable mental disorders**.[17] These include:

- Major depressive disorder
- Anxiety disorders
- Posttraumatic stress disorder
- Substance use disorders

Major depressive disorder (MDD) is a serious mood disorder that causes persistent sadness, loss of interest, and interference with daily functioning. To be diagnosed with MDD, an individual must experience five or more of the following symptoms nearly every day during the same two-week period, and at least one of the symptoms must be *depressed mood* or *loss of interest or pleasure*[17]:

- Depressed mood most of the day
- Markedly diminished interest or pleasure in all or most activities most of the day
- Significant weight loss (when not dieting), weight gain, or change in appetite
- A slowing down of thought and a reduction of physical movement (observable by others)
- Fatigue or loss of energy
- Feelings of worthlessness or excessive or inappropriate guilt
- Diminished ability to think or concentrate, or indecisiveness
- Recurrent thoughts of death, recurrent suicidal ideation, a specific plan for committing suicide, or a suicide attempt

An **anxiety disorder** may take many forms, including generalized anxiety (persistent and excessive worry that is difficult to control), social anxiety, separation anxiety, panic disorders, and phobias. Anxiety is diagnosed if at least three of the following disturbances are present and are not attributed to another mental disorder, condition, or substance abuse[17]:

- Restlessness or feeling keyed up or on edge
- Being easily fatigued
- Difficulty concentrating or mind going blank
- Irritability
- Muscle tension
- Sleep disturbance (difficulty falling or staying asleep, or having restless, unsatisfying sleep)

Posttraumatic stress disorder (PTSD) develops in a subset of people who have experienced a shocking, dangerous, or life-threatening event. It is natural to have stress reactions to this type of event, but if symptoms last more than a month and are severe enough to interfere with relationships, work, or the ability to carry out everyday tasks and routines, a PTSD diagnosis may be made. Most people recover within months, but a small number may develop chronic PTSD and be at higher risk of other mental health disorders, substance abuse, and self-harm.[17] Adults diagnosed with PTSD must have all the following symptoms for at least one month[17]:

- At least one *reexperiencing* symptom
  - Flashbacks—repeatedly reliving the trauma (may include physical symptoms such as sweating or a racing heartbeat)
  - Bad dreams
  - Frightening thoughts
- At least one *avoidance* symptom
  - Staying away from places, events, or objects that are reminders of the traumatic experience
  - Avoiding thoughts or feelings related to the traumatic experience
- At least two *arousal and reactivity* symptoms
  - Being easily startled
  - Feeling tense or on edge
  - Having difficulty sleeping
  - Having angry outbursts
- At least two *cognition and mood* symptoms
  - Trouble remembering the traumatic experience
  - Negative thoughts about oneself or the world
  - Distorted feelings (e.g., guilt or blame)

- Loss of interest in previously enjoyable activities

A **substance-use disorder** (SUD) is "a mental disorder that affects a person's brain and behavior, leading to a person's inability to control their use of substances such as legal or illegal drugs, alcohol or medications."[18] SUDs result from recurrent use of alcohol and/or drugs that leads to significant impairment, health problems, disability, and inability to carry out responsibilities at home, work, or school. People suffering from an SUD often have another co-occurring mental disorder, such as anxiety, depression, or personality disorder, and those with a mental disorder are at higher risk of substance abuse.

Onset of these mental disorders is affected by a wide range of health determinants, and in many cases, climate change may not be the primary driver but rather a threat magnifier. As with other adverse health impacts, understanding the burdens of diagnosable mental health disorders aids in identifying specific populations at risk and which disorders to keep in focus as climate change worsens. Global burdens of selected mental disorders are listed in **Table 10.1**. These disorders accounted for 135 million disability-adjusted life years (DALYs) in 2019, more than 5% of all global DALYs for all causes.[19] Mental health burdens differ widely among countries and among populations within countries. Because awareness and monitoring vary significantly, there is likely a high degree of uncertainty associated with these data, and they likely underestimate the true burdens in many places.

These DALYs rates are high and reflect a significant burden of mental disorders. The burdens of depression and anxiety are approximately 60% more prevalent in females, and substance-use disorders are much more prevalent in males. Since 1990, the global DALYs

Table 10.1 Global Burden of Selected Mental Health Disorders in 2019, Measured by Total DALYs and Age-Standardized DALYs Rates for Females, Males, and Both Sexes, and Change since 1990.

Disorder	Sex	Total DALYs (millions)	DALYs Rate (per 100,000)	% Change since 1990
Major depressive disorder	Both	37.2	460	−2
	Female	23.0	564	−5
	Male	14.2	354	+0.4
Anxiety disorders	Both	28.7	360	---
	Female	17.7	445	−1
	Male	11.0	275	+1
Alcohol-use disorders	Both	17.0	207	−19
	Female	3.3	80	−22
	Male	13.7	336	−18
Drug-use disorders	Both	18.1	225	+20
	Female	7.2	178	+14
	Male	11.0	272	+24

Data from Institute for Health Metrics and Evaluation. GBD Compare. Accessed October 15, 2020. https://vizhub.healthdata.org/gbd-compare

rates for major depression and anxiety have not changed appreciably, and alcohol-use disorders have declined about 20%. In contrast, drug-use disorders increased 14% in females and 25% in males.

Burdens of mental disorders vary widely among countries, and data gaps prevent accurate comparisons. Australia, Palestine, and Greenland had the highest recorded DALYs rates in 2019.[19] The United States has by far the highest burden of drug-use disorders, mostly driven by skyrocketing use of opioids, a class of drugs that includes illegal substances (e.g., heroin) and synthetic pain relievers available legally by prescription (e.g., oxycodone and morphine) and also produced and/or used illicitly (e.g., fentanyl). Drug-use disorders led to more than 65,000 U.S. deaths in 2019, 72% due to opioid use and most in males.[19] Since 1990, prevalence of drug-use disorders in the United States has risen nearly 50% (compared with 2% globally), and the death rate has increased nearly 900% (compared with 58% globally).[19]

Depending on severity, mental disorders can lead to significantly increased risk of death. For example, persons with major depression and schizophrenia have a 40–60% higher chance of dying prematurely than the general population, from suicide as well as physical illnesses such as cancers, heart disease, diabetes, and HIV infection, especially when health care is not sought. Mental disorders put people at risk of substance-use disorders and experiences of poverty, homelessness, and incarceration, and all these effects take a toll on individuals, families, and communities.[20]

Globally, health systems tend to inadequately respond to people suffering from mental disorders. According to WHO, 76–85% of people with a severe mental disorder receive no treatment for their disorder in low-income and middle-income countries, and 35–50% are not treated in high-income countries.[16] Even when care is available, quality and access vary widely. Nearly half the world's population live in countries where there is just one psychiatrist for every 200,000 people.[20] In many populations, mental health awareness is low, and sufferers may be stigmatized as weak or dangerous. Self-stigma is also common, in which people may feel internalized shame and have deficient self-esteem and self-efficacy.

Mental disorders may be exacerbated by climate change and may also put people at higher risk of certain physical health impacts of climate change, including heat-related illnesses. Large gaps in awareness and surveillance create barriers to the development of effective mental health interventions as climate adaptation strategies. Given that mental health impacts may be more common overall than physical health impacts for many people experiencing climate change, this is a big problem.

Climate Change Impacts on Mental Health

Heat

Exposure to extreme heat is a climate stressor that significantly affects mental health (described here and in Chapter 4). Heat waves increase hospitalizations for mental disorders,[21] and having a preexisting psychiatric illness increases the odds of heat wave mortality by more than 3-fold, a greater effect than for preexisting heart or lung diseases.[22] In the summer of 2012, for example, the U.S. state of Wisconsin recorded 27 heat-related deaths, more than half in people with a preexisting mental health condition.[23] Many victims had been taking psychotropic medications, which can impair the body's temperature regulation and thus increase sensitivity to ambient heat. The risk of heat mortality during this particular event was lower if people had a working home air conditioner, visited a cooling center, or increased their social contact. Seeking heat-protective measures, particularly social interaction, may be constrained, however, if one's mental health is poor.[23]

Extreme heat in the summer of 2021 smashed high temperature records across North America. On June 28, Portland, Oregon, set a record at 113°F—40° above the average high temperature for June.[24] Hundreds of people across the Pacific Northwest region went to hospital emergency departments for heat-related illnesses, and many excess deaths were reported. Though less visible, the mental health of many exposed to this unprecedented heat reportedly suffered as well. That same day, many cities across the United States issued heat alerts that warned people at high risk of health impacts— including those with "mental health conditions" and those who "use medications that impede body temperature regulation"—to "pay special attention to the weather."[25] They also emphasized the importance of regularly checking on those with psychiatric disorders.

Extreme Storms

One of the most common climate-change-related exposures that lead to adverse mental health impacts is extreme storms (discussed here and in Chapter 5). Much of the world's population (40%) lives within 100 kilometers (60 miles) of a coast, and

10% live in coastal areas less than 10 meters (33 feet) above sea level, making them vulnerable to tropical cyclones and other coastal storms, which can be particularly destructive.[26] In general, experiencing extreme weather events is linked to anxiety, depression, PTSD, grief, guilt, trauma, fatigue, substance abuse, and suicidal ideation.[10] After experiencing a cyclone, residents of coastal Bangladesh, most of whom were farmers, exhibited stress, sleep disorders, anxiety, depressions, PTSD, and suicidal ideation.[27] Many of these impacts of extreme storms come from experiencing flooding (see next section).

Accurately assessing these mental health impacts is made difficult because storms are often dangerous and chaotic events that disperse and displace people, making affected populations difficult to monitor. In addition, impacts vary within affected populations because people have differing abilities to evacuate or protect themselves, their loved ones, and their property, and a range of emotional and psychological reactions. Symptoms may present rapidly or be delayed by days, weeks, or even months after the event, making attribution to the extreme weather event difficult. In many cases, the number of people who experience psychological trauma after a natural or human-caused disaster may be much higher than the number of people with physical injuries.[28] Some impacts may even appear in the time period *before* the disaster hits as people in harm's way experience heightened fear, anxiety, hopelessness, or fatalism.[10]

Because extreme storms damage and disrupt property and infrastructure, recovery often diverts scarce resources away from public health systems and social services. Also, a lack of awareness or acceptance of psychological impacts hampers efforts to detect and respond to mental disorders and acknowledge the role of climate change in their onset. If untreated, psychological effects in the wake of disasters may develop into epidemics of mental illness in climate-vulnerable regions.[29]

In 1998, **Hurricane Mitch** hit parts of Nicaragua, killing 10,000 people. The storm's survivors, especially young people in the most severely impacted cities, reported high prevalence of depression and PTSD symptoms, and nearly all reported being scared that they or a loved one were going to die.[30] In 2005, **Hurricane Katrina** devastated the coasts of Louisiana and Mississippi, killing thousands and displacing millions—at the time, 10% of all the world's internally displaced persons. In the wake of the storm, PTSD, suicide ideation, and other serious mental conditions increased and were still prevalent a year later.[31] As many as one-third of survivors experienced a mental health problem, and nearly half of low-income Black women in New Orleans, one of the hardest hit areas, had symptoms of PTSD.[10] Among displaced people who received shelter after the storm in trailers provided by the government, 72% reported feeling "down, depressed or hopeless," 57% reported having major depressive disorder, nearly 25% reported contemplating suicide, and nearly 5% attempted suicide.[32] Among women living in these trailers, suicide attempts were 79 times higher than the regional average, and the suicide completion rate was 15 times higher.[33] Three-quarters of these residents reported receiving no counseling or mental health services.

In 2017, **Hurricane Harvey** devastated the city of Houston, Texas, dumping more than 50 inches of rain, displacing 30,000 residents, and causing $70–$170 billion in property damage (**Figure 10.3**).[34] Among Houston residents, the odds of having PTSD symptoms in the weeks after the storm increased 50% if the person's property was damaged by the hurricane and by 40% if they directly

Figure 10.3 Left: Residents of Houston, Texas, Affected by Flooding Caused by Hurricane Harvey, August 2017. Right: Destruction Caused by Hurricane Eta in Puertos Cabezas, Nicaragua, November 7, 2020.

Left: © Joe Raedle/Staff/Getty Images News/Getty Images; Right: © Jeiner Huete_P/Shutterstock.

experienced the hurricane in any way. Similar results were seen for anxiety symptoms.

Also in 2017, **Hurricane Maria** devastated Puerto Rico, killing nearly 3,000 people and leaving 1.5 million without power for months. Measurable surges in the incidence of mental health problems, substance abuse, domestic violence, and homicides were reported.[35] The number of schoolchildren who met clinical standards for PTSD doubled, and in the 9 months after the hurricane, the suicide rate increased 18%. Six months after the storm, 44% of adults on the island and 66% of Puerto Ricans who relocated to Florida because of the storm reported PTSD symptoms.[8] Among mental health workers, "compassion fatigue" was common, with high levels of exhaustion, burnout, and anxiety after treating hurricane victims for such a long time. In addition, extensive hurricane damage meant that mental health infrastructure and services were weakened and difficult to access.[35] In storm-prone places like Puerto Rico, not knowing when the next devastating storm will hit or whether adequate preparations have been made often causes serious adverse emotional reactions.

One of the strongest tropical cyclones in history was **Typhoon Haiyan**, which struck the Philippines in 2013 with sustained winds of 190–195 miles per hour. It is one of only four storms on record at that strength.[36] More than 6,000 people died, and destruction occurred on a massive scale. Among survivors, 39% lost a loved one, 57% were forced to relocate, and 42% reported psychological problems related to the storm. Health professionals deployed during this disaster also experienced significant adverse mental health conditions.[36]

Floods

Direct experience of a flood event resulting from extreme storms, heavy precipitation, or sea level rise can be a major mental health shock, one that is becoming more prevalent with climate change. In the wake of flooding disasters, immediate impacts such as physical injury, risk to property, and displacement tend to be the primary focus, whereas mental health impacts may not be recognized right away or at all and may have delayed onset. Floods are not experienced uniformly within affected populations, but some psychological impacts are reported generally across time, regions, and flooding events.

For example, in the winter of 2013–2014, southern England experienced severe flooding caused by persistent heavy rainfall, strong winds, high tides, and extreme storm surges. More than 6,500 homes ended up flooded. Prevalence of probable anxiety, depression, and PTSD one year after the floods was highest among people who directly experienced flooding of their homes compared with people who were unaffected or whose lives were disrupted by the floods but whose homes were not impacted (**Figure 10.4**).[37] These results were dose-dependent in that people in the *Flooded* group who had more than 100 cm of floodwater in their homes had nearly 15-times-higher odds of PTSD than the *Unaffected* group, whereas those with less than 30 cm of floodwater

in their homes had 4.5-times-higher odds. *Flooded* people who had to move out of their homes experienced higher odds of PTSD than those who were flooded but stayed in their homes.[37]

In 2010, Pakistan experienced one of the worst flooding events in recent history after heavy monsoon rains inundated more than half of the country, causing 2,000 deaths and destroying more than 1 million homes (**Figure 10.5**). As many as 20 million people were affected, mostly via displacement.[38] Flooding destroyed infrastructure and disrupted health care, communications, educational services, and transportation. People reported anxiety, panic, fear, and trauma as their family members, homes, possessions, crops, and livelihoods were lost, and social relationships suffered.[39] In the years afterward among Pakistanis exposed to

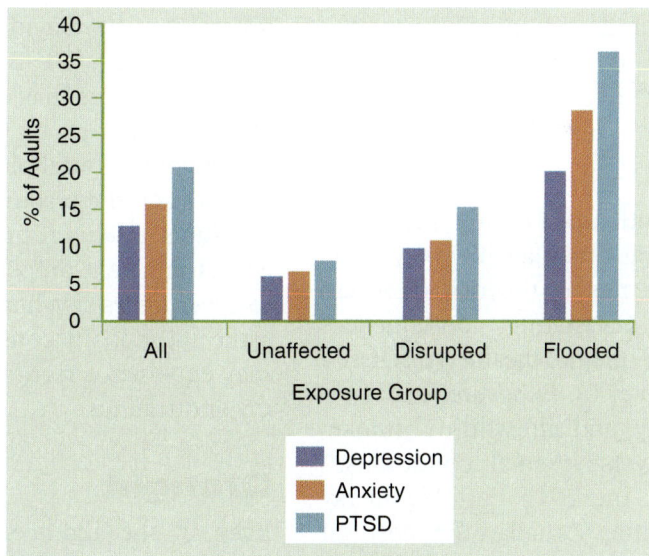

Figure 10.4 Prevalence of Mental Disorders in People Who Experienced Severe Floods in England, 2013–2014. Three exposure groups were studied: *Unaffected* people, who did not experience flooding (n = 285), *Disrupted* people, who were impacted by flooding but whose homes were not flooded (n = 1099), and *Flooded* people, who experienced at least one livable room in their homes with floodwater (n = 622). Probable depression, anxiety, and PTSD were diagnosed using validated psychology instruments.

Data from Waite TD, Chaintarli K, Beck CR, et al. The English national cohort study of flooding and health: cross-sectional analysis of mental health outcomes at year one. *BMC Public Health*. 2017;17:129. doi:10.1186/s12889-016-4000-2

Figure 10.5 Millions of Pakistanis Were Displaced in 2010 by Severe Floods Similar to This One in Karachi in 2020.
© Asianet-Pakistan/Shutterstock.

Figure 10.6 Property Destroyed in the 2016 Fort McMurray Wildfire in Alberta, Canada.
© Bloomberg Creative Photos/Getty Images.

these floods, positive mental health impacts were also seen, particularly **posttraumatic growth**—a phenomenon in which people become more resilient to future traumatic events because of improved recognition of personal strength, relationships, appreciation for life, and new possibilities that arise because of the experience of living through trauma.[38]

Wildfires

Climate-change-induced warming, prolonged drought, and erratic rainfall in many parts of the world are raising the risk of severe wildfires, with devastating consequences for physical and mental health (discussed here and in Chapter 6). Exposure to air pollutants, generally and in wildfire smoke, is linked to adverse mental conditions.[40] Six months after the 2016 Fort McMurray wildfire in Alberta, Canada, the costliest natural disaster in that country's history (**Figure 10.6**), an eightfold spike in prevalence of generalized anxiety disorder and a fourfold spike in major depressive disorder were measured in the affected population compared with the general population.[41] Two important factors in postdisaster anxiety

were witnessing burning homes and having a preexisting anxiety disorder, and 60% of anxiety patients reported increased alcohol and drug use after the fires. In California, where wildfires are becoming more frequent and more extreme, 10–30% of wildfire survivors may be diagnosed with mental disorders, notably PTSD and depression, and half of survivors may suffer from subclinical effects.[42] Wildfires hit poor communities and low-income households particularly hard if people lack the resources to take preventive measures, evacuate, protect their property, or access sufficient property insurance coverage to cover losses.[42] When wildfires threaten the same location in consecutive years, people may experience recurring or persistent anxiety and trauma.

Drought

Even in the absence of an acute disaster, the slow-onset impacts of climate change, including prolonged drought and sea level rise, may have chronic psychological consequences.[6] In addition, human displacement due to slow-onset disasters may itself challenge mental health (see Chapter 11). Experiences of persistent drought challenge mental well-being, particularly among farmers in drought-prone regions. For example, Australians exposed to chronic drought and weather disasters that disrupt agriculture report psychological distress and hopelessness, and severe drought over the past decade has been linked to chronic distress, generalized anxiety, depression, and increased suicide incidence.[9] Older farmers in Australia are most severely impacted and report a higher sense of loss than younger farmers.[9] Higher susceptibility among farmers is attributed largely to dependence on suitable climate and weather conditions for agricultural productivity and income, relative socioeconomic disadvantages in farming communities, and limited access to mental health care in rural areas. In low- and middle-income countries, a large fraction of the population participates in farming and mental health services are often lacking. Drought-fueled crop and livestock failures may lead affected people to migrate to other areas, which is associated with adverse mental health impacts.[43]

Climate-change-fueled agricultural losses often exacerbate the burden of food insecurity in areas prone to persistent drought, extreme storms, flooding, and other disasters that impact food production. Food insecurity and poor mental health status are often reciprocally associated. Among adults, failure to provide sufficient food for the family may lead to negative emotional and psychological reactions, including vulnerability, isolation, and despair.[44] Poor diets low in fresh fruits and vegetables and rich in processed calorie-dense, low-cost foods may also increase the likelihood of anxiety and depression.[44] In turn, these mental disorders can lower the motivation for food-insecure people to seek food from local food banks and other emergency food services. These associations are particularly serious in communities already burdened by high rates of food insecurity and a lack of access to healthy food, and those facing disproportionate burdens of environmental and social inequities.

At-Risk Populations

Women

As global data show, women face higher risks of depression and anxiety generally, and growing evidence points to significant associations between climate change and adverse mental health status in women. For example, women were found to be at higher risk for emotional distress, depression, and PTSD following exposure to tropical cyclones, other extreme storms, and severe flooding in countries in Asia (China, Myanmar) and in the United States, England, and Australia.[45] Risk factors for women include experiencing food insecurity after storms, destruction of agriculture and natural resources, social and financial dependence, displacement after disasters, and gender-based violence.[45] Women are often less likely to bounce back from disasters, in part because they are the primary caregivers in most households and must continue to provide food and collect water and fuel for their families in the wake of destruction caused by disasters, which can be extremely difficult. Women in Bangladesh face a particularly high burden in part because they are often restricted to the home and cannot or choose not to evacuate when facing destructive storms, flooding, and seawater inundation.[46] Women, children, and the elderly are often left behind when men migrate to urban areas in Bangladesh in search of work or in response to environmental disasters.[46] These stressors have caused Bangladeshi women to report feeling despondent about the future.[47] The combination of economic and social inequality and psychological stress in mothers also places significant mental and physical health risks on their children, made worse without community-level support.[48]

Children

Most of the world's children live in low- and middle-income countries highly susceptible to climate change and lacking adequate mental health infrastructure, services, and support. In many populations, mental distress in children may be as prevalent as in adults.[48,49] Children tend to have strong responses to experiencing extreme weather, including excessive worry, sleep disorders, and PTSD. In fact, children often suffer more from prolonged PTSD than adults in the years after a flood, and they are prone to becoming terrified when subsequent extreme weather or climate-related events occur.[49] It is estimated that PTSD prevalence in children who experience any type of disaster is 15–30%.[49] Disruption of daily activities, including by an extreme weather event, makes children vulnerable to distress, fear, and irritability, and they may have altered capacity to regulate emotions, impaired emotional and cognitive development and function, and behavioral problems.

In general, young people are at risk because their emotional and psychological resilience is still developing. Their experiences are also affected by household dynamics and social support networks. Children have greater dependence on adult family members for mental and physical health support and for actions to avoid disaster impacts. They may also rely on adults in their social safety nets, which are often disrupted in disasters. High stress levels in adults can impact their children, in some cases contributing to neglect, conflict, and abuse. In children, early experiences of trauma and other accumulated stresses may have long-lasting impacts, including on behavior and cognition throughout life.[12] Young people are also more likely to express concern and fear about the threats of climate change and report feeling anger because of persistent climate inaction. In some cases, these reactions may lead young people to take action themselves, and many are leaders in the climate movement, having benefited from the resulting feelings of empowerment, self-reliance, and positivity.[49]

Migrants

People who migrate as a result of climate disasters are at risk of adverse mental health conditions when they end up in unfamiliar environments far from home, family members, or support systems (see Chapter 11). Loss of home, community, culture, identity, or sense of belonging may cause harm, along with a lack of acceptance, xenophobia, or racism from others in the new location. Migrants may face pressure to procure adequate resources for survival or means of employment, often in unfamiliar places, which may be stressful or traumatic.[43] Relocating to a camp or settlement without adequate housing, water, sanitation, food, and health care, and where social networks are disrupted and exposure to violence increases, particularly for women, greatly destabilizes both mental and physical well-being. In contrast, migration may be a *voluntary* choice, which could confer significant mental and physical health benefits if relocation offers adequate or improved resources and protection from the health impacts of disasters to which migrants were previously exposed.

Indigenous Peoples

The world's Indigenous peoples face unique climate change impacts and are distinctly at risk of mental and physical harm. Even though there is significant diversity among Indigenous groups, they tend to share a cultural connection to the land for sustenance, spirituality, and livelihoods, including through hunting, fishing, harvesting, trapping, and herding. They also have in common complex legacies of trauma from European colonization, land and resource expropriation, loss of sovereignty, and discrimination, all of which increase their health burden and contribute to the erosion of social, familial, cultural, and linguistic systems.[50] Indigenous peoples tend to live in regions undergoing rapid environmental and socioeconomic change, and they inhabit some of the world's most biodiverse and fossil-fuel-rich lands that remain under intense development and extraction pressures despite rising awareness of the importance of conservation, renewable energy, and climate action. Traditional knowledge and practices of Indigenous peoples, which serve as important examples of sustainable climate adaptation, are often under threat.

The world's estimated 370 million Indigenous peoples reside in 90 countries and are often isolated geographically. They are among the poorest and most marginalized peoples, and the poverty gap between Indigenous and non-Indigenous groups is increasing around the world.[51] Integration into global economies and dominant cultures has increased the risk of substance abuse, addiction, and suicide, and many of these groups have disproportionately high mental health burdens.[51] For Indigenous peoples in particular, mental health status is associated in part with seasons, meteorological factors, and exposure to weather events.[8] Particularly in remote regions, they may not be able to access mental health services, and services that are available may be underfunded and not culturally sensitive.[50]

Non-Indigenous mental health frameworks often clash with Indigenous peoples' concepts of mental well-being, which may be communal and incorporate relationships with ancestors, land, and nature. In 1999, WHO issued a **Declaration on the Health and Survival of Indigenous Peoples**, in which it defined Indigenous health as "both a collective and an individual inter-generational continuum encompassing a holistic perspective . . . [of the] spiritual, intellectual, physical, and emotional . . . [and] where the past, present, and future co-exist simultaneously."[52] The United Nations **Declaration on the Rights of Indigenous Peoples** refers specifically to Indigenous peoples' right to improve the conditions for health, to determine their own health programs, and to maintain their traditional health practices.[51]

The Circumpolar North, home to numerous Indigenous groups, is warming faster than the rest of the world, with an average surface air temperature 4°C (7.2°F) warmer than in 1960 (**Figure 10.7**). At increased temperatures, sea ice is being lost rapidly (**Figure 10.8**). Sea ice is a critical feature of the Arctic environment, forming extensions of the land that allow for movement of people and access to important food resources and culturally valued places.[53] Ice loss also puts coastal communities at extremely high risk of destruction from ocean storms, from which sea ice normally provides a protective buffer. Climate change and loss of ice and land are disrupting traditional practices, lifestyles, livelihoods, access to food and

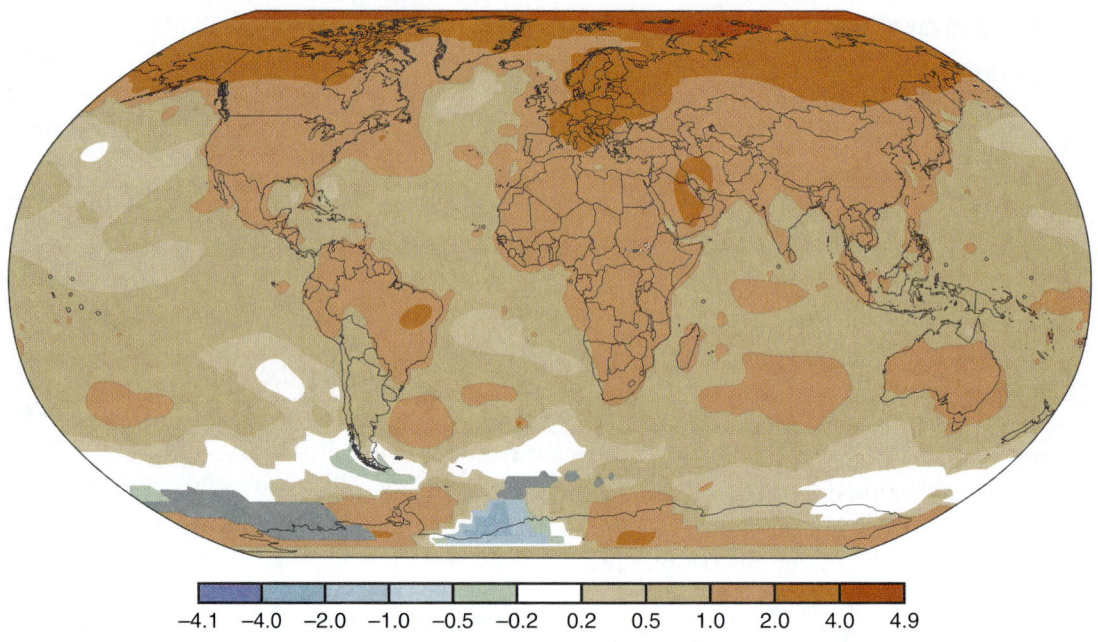

Figure 10.7 Geographic Variation in Mean Surface Air Temperature Showing the Arctic Region Has Warmed Nearly 4°C (7.2°F) Over the Period 1960–2019.

Reproduced from National Snow and Ice Data Center. All about arctic climatology and meteorology. Accessed May 4, 2020. https://nsidc.org/cryosphere/arctic-meteorology/climate_change.html

traditional plants for medicine, and physical and mental health among the many Indigenous groups across the Circumpolar North in Canada, the United States, Greenland, Scandinavian countries, and Russia.[54]

Global climate negotiations and scientific assessments have tended to overlook Indigenous knowledge, stories, traditions, and ecological observations. Indigenous communities have significant adaptive capacity underpinned by their traditions and experiences living in close connection to land, water, and ecosystems, and incorporating the past, present, and future. Many Indigenous peoples have called for explicit recognition of this accumulated knowledge in climate negotiations and adaptation plans. Strategies prioritized by Indigenous peoples emphasize restoring land rights and sovereignty, confronting past and present colonial history and trauma, and addressing inequalities in health and environmental hazards (see **Box 10.1**).[51]

Residents of Small Island Developing States

Small island developing states (SIDS) are among the most at-risk countries for the impacts of climate change because of their high vulnerability to storms and sea level rise that may entirely inundate some islands. An estimated 30–40% of people on SIDS have experienced adverse mental effects in response to climate-fueled disasters.[8] SIDS also have a long history of having endured major cultural and environmental threats with serious consequences, but in many cases long-term resilience has preserved local and cultural knowledge and place-specific identities.

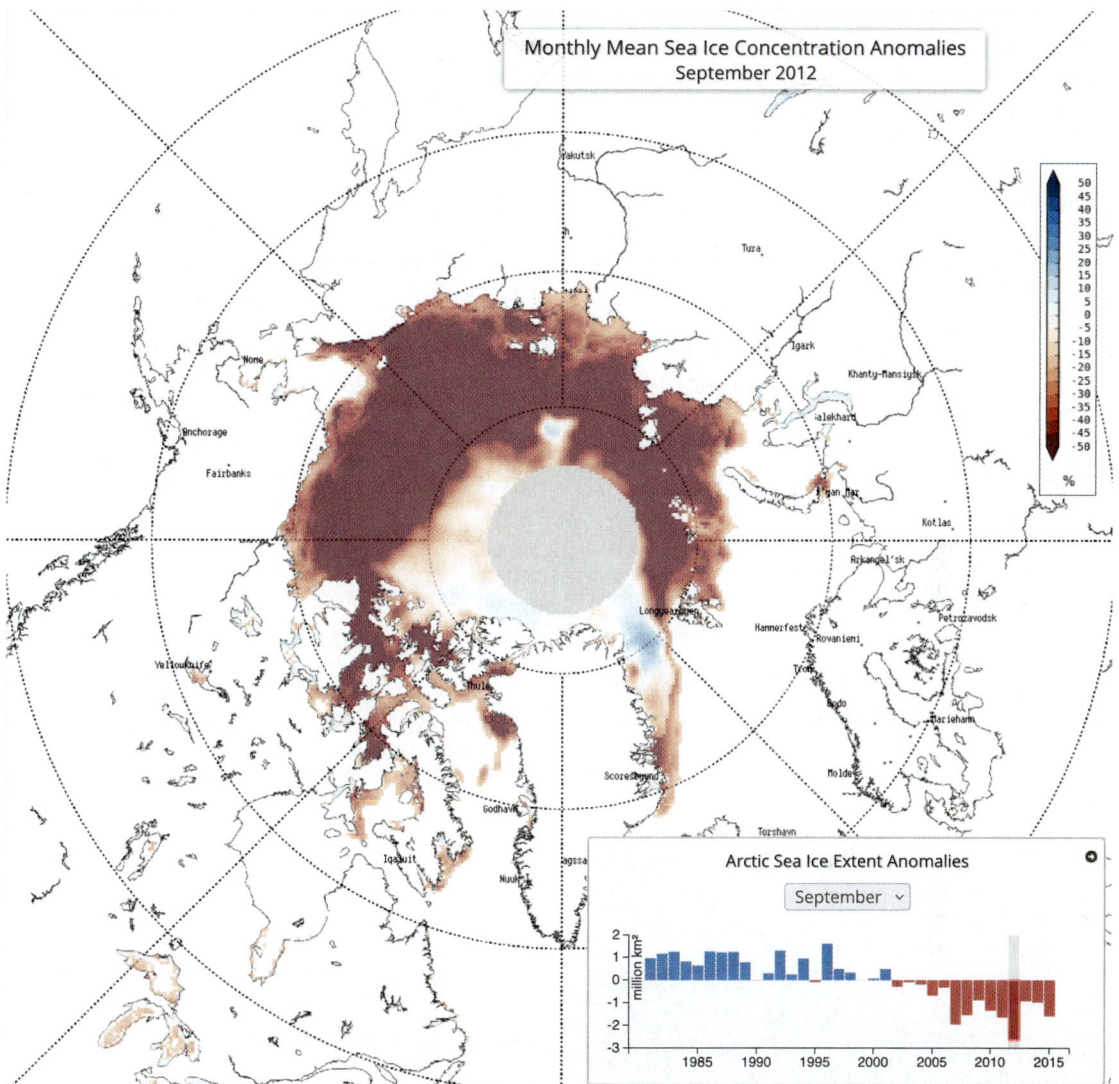

Figure 10.8 Sea Ice Extent Anomalies in the Circumpolar Regions of the World in September 2012 (the Most Sea Ice Loss on Record). The darkest red colors indicate up to 50% sea ice loss.

Reproduced from National Snow and Ice Data Center. All about arctic climatology and meteorology. Accessed May 4, 2020. https://nsidc.org/cryosphere/arctic-meteorology/climate_change.html

In Pacific SIDS, tropical cyclones cause feelings of loss, grief, sadness, anger, and stress, and observable increases in anxiety, depression, and PTSD.[8] In coastal parts of the Solomon Islands, for example, people have described sea level rise as a major stressor, causing feelings of loss, uncertainty, and powerlessness, as well as fear and worry for their family's future and for their society, culture, and entire country.[8] Loss of attachment to land, place, ancestors, and cultural traditions because of climate disasters can erode family and community connections, along with local knowledge and

Box 10.1 Inuit and Inupiat Mental Health

Indigenous residents of the Labrador Inuit Settlement Area of Nunatsiavut in Inuit Nunangat, the Inuit homeland in northern Canada, face significant climate change harm (**Figure 10.9** and **Figure 10.10**).[54] This group identifies a strong connection between traveling on sea ice and their mental, emotional, physical, spiritual, cultural, and social health and well-being. They acknowledge that predicting weather and ice conditions is becoming more difficult, and they experience emotional and psychological challenges when travel- and land-based activities such as hunting, fishing, trapping, and food gathering are disrupted.[54] People who cannot go on the ice feel that they would get sick, be sad and lost, go crazy, and "have no health."[53] Going on the ice is perceived as calming, stress-relieving, restorative, and "good for your spirit."[53] Some view ice and land as places relatively uninfluenced by the legacies of colonization and globalization—

thus, people report feeling more "free" there.[53]

The Inupiat peoples native to what is now the U.S. state of Alaska are also highly impacted by climate change. Loss of sea ice is making coastal Alaska Native villages much more vulnerable to destructive storms, which has triggered stress, anxiety, fear, and more clinical visits for behavioral health problems during storm seasons.[55] One community counselor in Kivalina reported, "Every time the waves get high, people get anxious. Some people walk all night. My ten-year-old son was worried the house would blow away."[55] Changing sea and ice conditions cause significant anxiety as well as increase the risk of injury, hypothermia, and death. "Falling through the ice" is now an epidemiological outcome that appears to be on the rise.[56]

At least 31 Alaska Native villages face imminent destruction due primarily to

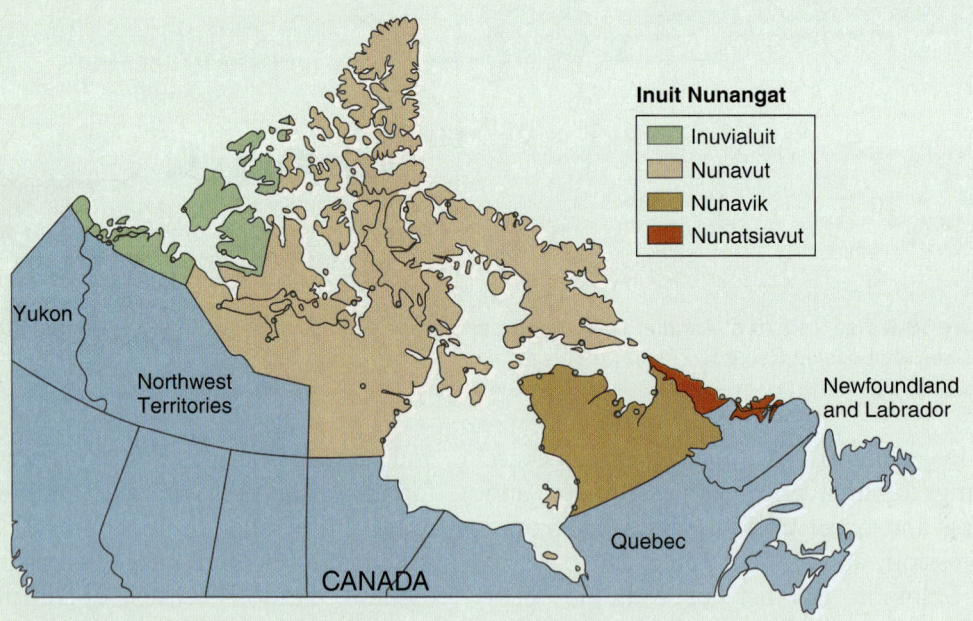

Figure 10.9 Map of Inuit Regions of Canada, with Nunatsiavut (Dark Brown) the Easternmost Region.

Reproduced from Maps of Inuit Nunangat (Inuit Regions of Canada). Inuit Tapiriit Kanatami; 2021. Accessed April 19, 2021. https://www.itk.ca/maps-of-inuit-nunangat

Figure 10.10 Rocky Coastline of Hopedale on the Nunatsiavut Coast.
© Posnov/Moment/Getty Images.

storms, flooding, and erosion that will displace residents permanently, although little support for relocation has been provided, and most villages are not adequately prepared to respond.[57] This eventual loss of homes, villages, and traditional lands is predicted to lead to widespread solastalgia, the profound sense of loss of place and home.[57] In addition, anxiety and depression among Alaska Natives is being brought on by changes to subsistence practices, transportation, culture, traditional knowledge systems, and threats to the health of future generations.[57]

Suicide rate is a proxy measure of severe psychological distress within a population, and suicide rates in some Arctic regions are among the highest in the world.[19] Inuit communities in Canada have suicide rates 11 times higher than the national average, and this disparity is particularly high for teenagers and young adults.[9] Substance abuse, depression, and other mental health disorders are also prevalent.[50] Although attributing these mental health impacts to climate change is difficult and uncertain, among Canadian Inuit people, climate change has been shown to be associated with increased suicide ideation and substance abuse.[12]

Indigenous people are observing changes in their climate and environment and how these changes are affecting their lives. Inuit youth in Nunatsiavut report worry, isolation, restlessness, fear, sadness, anger, frustration, stress, and anxiety, and in response, they have identified five criteria important to protecting mental well-being in the climate crisis.[58]

1. **Being on the land** is part of youth identity, has nurturing and restorative powers, and provides opportunities for youth to develop and practice traditional and transferrable skills and connect to ancestors. In this sense, it becomes a "pathway to culture."[58]
2. **Connection to Inuit culture** shapes youth identity and relationships.
3. **Connection to others** in strong communities offers support networks, mentoring, safety and security, and a sense of belonging, allowing youth to feel a part of something collective.
4. **Strong relationships with family and friends** are an important protective factor in youth well-being and provide trusted confidants and mentors.
5. **Staying active and busy** distracts youth from self-reported boredom, stress, and negativity, and creates a sense of purpose.

Inuit youth have identified key adaptation measures to preserve and strengthen these criteria.

(continues)

Box 10.1 Inuit and Inupiat Mental Health **(continued)**

- Finding new places for them to go safely on land and ice
- Changing the timing of activities that require ice travel
- Hunting, fishing, and trapping at times when the ice is safer

- Developing educational outreach and support for youth to get involved on the land and in community spaces
- Creating new activities for youth to stave off boredom, notably when they are not able to go out on land and ice

belief systems. In addition, some SIDS residents, including youth in Kiribati, had more anxiety about their own future as a result of witnessing extreme weather affecting *other* SIDS.[8] Migration in response to disasters is predicted to have destabilizing mental health impacts, but it may also be beneficial in some cases. For example, women from Tonga who migrated to New Zealand reported improved well-being attributed to increased income, more opportunities for a social life, and access to better public services such as education and health care.[8] SIDS have also been effective in promoting equitable adaptation strategies in international climate negotiations.

Health and Fossil Energy Workers

Specific occupations place workers at high risk for adverse mental health impacts, and some relate to climate change. Mental health professionals and emergency responders are at particular risk because they provide health care, interventions, and emergency first response in highly climate-change-impacted communities and may face pressure to deliver a greater level of care as climate change accelerates.[9] This burden has been reported during many destructive tropical cyclone and wildfire events. A much larger mental health workforce is needed, one that

is well trained to understand, recognize, and respond in culturally sensitive ways. In addition, the mental health of these workers will need to be protected.

Workers in the fossil fuel industry face job insecurity, anxiety, and suicide rates that are higher than for the general population. A study of suicide rates by occupation in England and Wales showed that among coal miners, the suicide rate more than tripled from 1979 to 2005, a period during which the coal industry declined in that region.[59] Among the 30 occupations with the highest suicide rates in the UK, "coal miner" went from #29 in 1979–1983 to #1 in 2001–2005. In the United States, among 20 occupational industries, "mining, quarrying, and oil and gas extraction" had the highest suicide rate for men in 2016.[60] Workers in extractive fossil fuel industries are likely to have high mental health burdens for many reasons, including the inherent dangers of the jobs, economic uncertainty brought on by the boom-and-bust nature of the industry, time spent away from home, unfamiliar or nonexistent social support networks, and high rates of substance abuse among workers.

In the Appalachian region of the United States, residents of coal mining communities report significantly fewer mental wellness days than people in communities without coal mining. In general, people in central Appalachia, where coal mining has been concentrated, have higher risks of major depression and severe psychological distress compared with people in other areas of Appalachia or the entire United States.[61] General health disparities persist in Appalachia and are correlated with living in coal-producing areas, even when confounding variables such as poverty, education, smoking, race, physician supply, and health insurance access are accounted for.[61] In Wyoming County, West Virginia, a coal mining area of Appalachia, there are no psychiatrists. One of the few family doctors there, Dr. Joanna Bailey, reported in 2017, "As a family doctor, I'm doing way more psychiatry than I am comfortable with." Bailey identified a lack of economic growth, high unemployment, and migration of people away from the area as being "traumatic to families on every level."[62]

Appalachian residents exposed to mountaintop removal (MTR) activity, a highly environmentally destructive form of coal mining practiced in the United States, had 37% higher odds of being diagnosed with depression and 41% higher odds of substance-abuse disorder during hospital emergency department visits, controlling for socioeconomic and demographic factors.[63] Many people living in these regions report distress that natural spaces, homes, clean water, and entire ecosystems, which contribute to their identity, stability, and security, are being lost. Said one resident, "Our place defines us. We're a distinct mountain culture, and our culture means something. The blasting literally makes you feel like you're in a war zone. . . . You feel like you're being attacked. It does something to your psyche."[63] MTR is so destructive that many predict widespread solastalgia in MTR areas as a result of perceived loss of familiar places. States in Appalachian coal country— West Virginia, Kentucky, and Ohio—have the highest rates of opioid use disorders in the United States, with West Virginia at more than twice the national average.[19] Drug use may be high among current and former coal workers for many reasons, including that mining is dangerous and often results in workplace injuries for which opioid painkillers may be prescribed. There is also a strong social stigma against seeking help for drug use disorders.

As the world moves away from fossil fuel use in order to mitigate climate change, net benefits in mental and physical health are likely to result in communities that have been negatively impacted by the fossil fuel industry. However, losing fossil energy jobs without replacing them with other opportunities will likely result in some people losing income, health insurance, and retirement benefits, and thus access to mental and physical health services. Improving public health in these areas and creating job training programs and new job opportunities will benefit mental and physical well-being.

Adaptations

Slow- and rapid-onset climate-change-related stressors to mental well-being are affecting people around the world. Rapid mitigation of GHG emissions is urgently needed to tackle all health impacts, including mental health, as is understanding how the burdens of mental disorders are exacerbated by climate change. Building support for mental well-being among individuals and in communities is a form of climate adaptation. This includes boosting people's coping and social networking capacity, belief in their own resilience, and connections to people, places, and culture, as well as increasing access to high-quality and responsive health services.[3] More mental health professionals need to be trained, and access to adequate mental health services strengthened, including in remote locations and for particularly at-risk populations. Other actions include investing in renewable energy, public transportation, and green spaces, and expanding opportunities for physical activities such as walking and biking. These activities improve physical and mental well-being and reduce air pollution, exposure to which is linked to poor mental and physical health outcomes.[64] Experiences with nature are shown to be psychologically therapeutic and beneficial.

One specific example for how to build mental health resilience came in Pakistan in the wake of the 2010 floods, after which affected people reported many forms of emotional and psychological distress.[29] Psychiatrists developed *psychotrauma response plans* to "regenerate hope, renew trust, promote social cohesion and lessen hopelessness and desperation."[17] The country implemented mental health awareness campaigns, new psychiatric outpatient services, and training of volunteers to do basic counseling.[17] This may in part account for the observed onset of posttraumatic growth in affected people after the floods.

A call for global action on mental health was outlined in WHO's **Comprehensive Mental Health Action Plan 2013–2020**. This plan calls for expanding services, research, and governance for mental health, as well as changing attitudes that perpetuate stigma and discrimination about mental health.[16] The Sustainable Development Goals (SDGs) make cursory mention of mental health in SDG 3 (*Ensure healthy lives and promote well-being for all at all ages*), which sets the target to, "**by 2030, reduce by one third premature mortality from noncommunicable diseases through prevention and treatment and *promote mental health and well-being*.**"[65]

The **American Academy of Nursing** recently proposed a set of actions to build resilience and respond effectively to the ways that climate change impacts mental health[66]:

- Promote mental health initiatives at the community level that target vulnerable populations, including children, the elderly, and low-income people.
- Increase access to health insurance for mental health care.
- Educate patients and families about the health risks of climate change and how to prepare for it and protect themselves.
- Increase community preparedness for climate change to minimize mental health impacts across the life span, including a focus on schools and nursing homes.
- Improve mental health awareness and care in hospital emergency departments.
- Improve treatment services for patients with psychiatric disorders associated with climate-change-related trauma.
- Increase the number of healthcare practitioners, counseling services, clinics, and other health-related facilities in high-impact areas.
- Increase funding at all levels to improve mental health care.

Practical steps for the public to take to protect mental well-being during climate disasters were recently developed as part of a collaboration among the American Public Health Association, the American Psychological Association, ecoAmerica, and Climate for Health.[67]

- Create emergency preparedness kits with food, water, first aid supplies, and items to reduce stress.
- Refill medications before the disaster, if possible.
- Evacuate early in advance of the disaster.
- Check in on vulnerable neighbors before, during, and after disasters.

- Familiarize oneself with available mental health services such as counseling and therapy, and seek treatment as needed for mental health conditions.
- As needed, practice self-care and seek out spiritual and community-based services.
- Engage with the community after a disaster.
- Monitor children for changes in psychology or behavior.

Focusing on mental health, as is true for all human health impacts of climate change, provides challenges and opportunities. It is of critical importance to understand the complex causal factors that interact with climate change in distinct ways to improve or worsen mental health in individuals, populations, and communities. The climate and public health benefits of strengthening mental well-being and care are immense.

Discussion Questions

1. What are the global burdens of diagnosable mental disorders? What are the trends over time? How do these burdens differ in males versus females? In what ways does the United States have unique burdens of mental disorders?
2. Discuss how mental health is impacted by extreme storms, floods, wildfires, heat, and drought.
3. How might we measure emotional responses such as solastalgia and eco-anxiety? Do you think it is important for public health professionals to monitor the incidence and impacts of these conditions? Why or why not?

4. Discuss why certain populations are at particular risk, including women, children, Indigenous peoples, and residents of SIDS.
5. What is needed to implement the solutions to the mental health impacts of climate change identified by Inuit youth?
6. In what ways might experiences of climate change promote rather than worsen mental well-being?
7. Which mental health interventions are necessary in the climate crisis, and how can they be built into climate action plans?

References

1. National Oceanic and Atmospheric Administration. 2019 Atlantic hurricane season summary table. Accessed June 27, 2021. https://www.nhc.noaa.gov/data/tcr/2019_Atlantic_Hurricane_Season_Summary_Table.pdf
2. National Hurricane Center. Saffir-Simpson Hurricane Wind Scale. Accessed June 27, 2021. https://www.nhc.noaa.gov/aboutsshws.php
3. Shultz JM, Sands DE, Holder-Hamilton N, et al. Scrambling for safety in the eye of Dorian: mental health consequences of exposure to a climate-driven hurricane. *Health Aff*. 2020;39(12):2120–2127. doi:10.1377/hlthaff.2020.01203
4. Harrington S. How climate change affects mental health. *Yale Climate Connections*. February 4, 2020. Accessed March 5, 2021. https://yaleclimateconnections.org/2020/02/how-climate-change-affects-mental-health/
5. Majeed H, Lee J. The impact of climate change on youth depression and mental health. *Lancet Planet*

Health. 2017;1(3):e94–e95. doi:10.1016/s2542 -5196(17)30045-1

6. Clayton S, Manning CM, Krygsman K, Speiser M. *Mental Health and Our Changing Climate: Impacts, Implications, and Guidance.* 2017. American Psychological Association and ecoAmerica Accessed March 5, 2021. https://www.apa.org/news/press /releases/2017/03/mental-health-climate.pdf

7. World Health Organization. Mental health: strengthening our response. March 30, 2018. Accessed June 29, 2021. https://www.who.int /news-room/fact-sheets/detail/mental-health -strengthening-our-response

8. Kelman I, Ayeb-Karlsson S, Rose-Clarke K, et al. A review of mental health and wellbeing under climate change in small island developing states (SIDS). *Env Res Lett.* 2021;16(3):033007. doi:10 .1088/1748-9326/abe57d

9. Bourque F, Willox AC. Climate change: the next challenge for public mental health? *Int Rev Psychiatry.* 2014;26(4):415–422. doi:10.3109/095 40261.2014.925851

10. Hayes K, Blashki G, Wiseman J, et al. Climate change and mental health: risks, impacts and priority actions. *Int J Ment Health Syst.* 2018;12(1):28. doi:10.1186/s13033-018-0210-6

11. Berry HL, Waite TD, Dear KBG, et al. The case for systems thinking about climate change and mental health. *Nat Clim Change.* 2018;8(4):282–290. doi:10.1038/s41558-018-0102-4

12. Clayton S. Climate anxiety: psychological responses to climate change. *J Anxiety Disord.* 2020;74:102263. doi:10.1016/j.janxdis.2020.102263

13. Stanley SK, Hogg TL, Leviston Z, et al. From anger to action: differential impacts of eco-anxiety, eco-depression, and eco-anger on climate action and wellbeing. *J Clim Change Health.* 2021;1:100003. doi:10.1016/j.joclim.2021.100003

14. Cunsolo A, Harper S, Minor K, et al. Ecological grief and anxiety: the start of a healthy response to climate change? *Lancet Planet Health.* 2020;4: e261–e263. doi:10.1016/S2542-5196(20)30144-3

15. Friends of the Earth. Over two-thirds of young people experience eco-anxiety as Friends of the Earth launch campaign to turn anxiety into action. January 21, 2020. Accessed March 23, 2021. https:// friendsoftheearth.uk/climate/over-twothirds-young -people-experience-ecoanxiety-friends-earth-launch -campaign-turn

16. Ambrose J. "Hijacked by anxiety": how climate dread is hindering climate action. *The Guardian.*

October 8, 2020. Accessed June 28, 2021. https:// www.theguardian.com/environment/2020/oct/08 /anxiety-climate-crisis-trauma-paralysing-effect -psychologists

17. National Institute of Mental Health. Health topics. Accessed March 5, 2021. https://www.nimh.nih .gov/health/topics/index.shtml

18. Substance Abuse and Mental Health Services Administration. Substance use and co-occurring mental disorders. March 2021. Accessed June 29, 2021. https://www.nimh.nih.gov/health/topics /substance-use-and-mental-health/

19. Institute for Health Metrics and Evaluation. GBD Compare. Accessed October 15, 2020. https:// vizhub.healthdata.org/gbd-compare

20. World Health Organization. Mental health action plan 2013-2020. January 6, 2013. Accessed March 6, 2021. https://www.who.int/publications/i/item /9789241506021

21. Semenza JC, Rubin CH, Falter KH, et al. Heat-related deaths during the July 1995 heat wave in Chicago. *N Engl J Med.* 1996;335(2):84–90. doi:10.1056/nejm199607113350203

22. Bouchama A, Dehbi M, Mohamed G, et al. Prognostic factors in heat wave–related deaths: a meta-analysis. *Arch Intern Med.* 2007;167(20):2170–2176. doi:10 .1001/archinte.167.20.ira70009

23. Christenson ML, Geiger SD, Anderson HA. Heat-related fatalities in Wisconsin during the summer of 2012. *Wis Med J.* 2013;112(5):219–223.

24. Templeton A, Cureton E, Haas R, et al. Oregon heat wave linked to hospitalizations and drownings. *OPB.* June 28, 2021. Accessed June 28, 2021. https://www.opb.org/article/2021/06/28/oregon -heat-wave-record-temperatures-health-risk/

25. City of Greenwich. Heat alert news release. June 28, 2021. Accessed June 28, 2021. https://www .greenwichct.gov/DocumentCenter/View/23336 /Heat—Press-Release-June-28_2021

26. United Nations. Factsheet: people and oceans. May 2017. Accessed March 21, 2021. https:// www.un.org/sustainabledevelopment/wp-content /uploads/2017/05/Ocean-fact-sheet-package.pdf

27. Hasan MT, Adhikary G, Mahmood S, et al. Exploring mental health needs and services among affected population in a cyclone affected area in coastal Bangladesh: a qualitative case study. *Int J Ment Health Syst.* 2020;14(1):1–9. doi:10.1186 /s13033-020-00351-0

28. Links J. Predicting community resilience and recovery after a disaster. August 2017. Accessed

March 21, 2021. https://blogs.cdc.gov/publichealth matters/2017/08/predicting-community-resilience -and-recovery-after-a-disaster/

29. Bhamani A, Sobani ZA, Baqir M, Bham NS, Beg MA, Fistein E. Mental health in the wake of flooding in Pakistan: an ongoing humanitarian crisis. *J Coll Physicians Surg Pak*. 2012;22(1):66-8.

30. Goenjian AK, Molina L, Steinberg AM, et al. Posttraumatic stress and depressive reactions among Nicaraguan adolescents after hurricane Mitch. *Am J Psychiatry*. 2001;158(5):788–794. doi:10.1176/appi.ajp.158.5.788

31. Kessler RC, Galea S, Gruber MJ, et al. Trends in mental illness and suicidality after Hurricane Katrina. *Mol Psychiatry*. 2008;13(4):374–384. doi:10 .1038/sj.mp.4002119

32. Shehab N, Anastario MP, Lawry L. Access to care among displaced Mississippi residents in FEMA travel trailer parks two years after Katrina. *Health Aff*. 2008. doi:10.1377/hlthaff.27.5.w416

33. International Medical Corps. Suicide, violence, and depression widespread in FEMA travel trailer parks. March 26, 2007. Accessed March 21, 2021. https:// internationalmedicalcorps.org/press-release/suicide -violence-and-depression-widespread-in-fema -travel-trailer-parks/

34. Schwartz RM, Tuminello S, Kerath SM, et al. Preliminary assessment of Hurricane Harvey exposures and mental health impact. *Int J Environ Res Public Health*. 2018;15(5). doi:10.3390/ijerph 15050974

35. Abrams Z. Puerto Rico, two years after Maria. *Monit Psychol*. 2019;50(8):28.

36. Hugelius K, Gifford M, Örtenwall P, et al. Health among disaster survivors and health professionals after the Haiyan Typhoon: a self-selected Internet- based web survey. *Int J Emerg Med*. 2017;10(1):13. doi:10.1186/s12245-017-0139-6

37. Waite TD, Chaintarli K, Beck CR, et al. The English national cohort study of flooding and health: cross-sectional analysis of mental health outcomes at year one. *BMC Public Health*. 2017;17(1):129. doi:10.1186/s12889-016 -4000-2

38. Aslam N, Kamal A. Light at the end of the tunnel: post-traumatic growth among individuals exposed to Flood 2010 in Pakistan. *J Pak Psychiatr Soc*. 2013;10(1):34–37.

39. Fatima N, Rana S. Repercussion of flood of 2010 on the mental health of Pakistani victims. *Pak J Soc Clin Psychol*. 2017;15(1):42–52.

40. Pun VC, Manjourides J, Suh H. Association of ambient air pollution with depressive and anxiety symptoms in older adults: results from the NSHAP Study. *Environ Health Perspect*. 2017;125:342–348. doi:10.1289/EHP494

41. Hrabok M, Delorme A, Agyapong VIO. Threats to mental health and well-being associated with climate change. *J Anxiety Disord*. 2020;76:102295. doi:10.1016/j.janxdis.2020.102295

42. Stern J. A mental-health crisis is burning across the American West. *The Atlantic*. July 20, 2020. Accessed March 21, 2021. https://www.theatlantic .com/health/archive/2020/07/mental-health-after math-california-wildfires/608656/

43. Bose I, Faleiro J, Singh H. *Climate migrants pushed to the brink*. May 2020. Accessed March 21, 2021. https://actionaid.org/publications/2020/climate -migrants-pushed-brink

44. Nagata JM, Ganson KT, Whittle HJ, et al. Food insufficiency and mental health in the U.S. during the COVID-19 pandemic. *Am J Prev Med*. 2021;60(4):453–461. doi:10.1016/j.amepre.2020 .12.004

45. Dunne D. Mapped: How climate change dis- proportionately affects women's health. *Carbon Brief*. October 29, 2020. Accessed March 2 2021, https://www.carbonbrief.org/mapped-how-climate -change-disproportionately-affects-womens -health?s=03

46. Ayeb-Karlsson S. When the disaster strikes: gendered (im)mobility in Bangladesh. *Clim Risk Manage*. 2020;29:100237. doi:10.1016/j.crm.2020 .100237

47. Bagri NT. Bangladesh's water crisis: a story of gender. *Al Jazeera*. April 25, 2017. Accessed October 9, 2018. https://www.aljazeera.com/features/2017/4/25 /bangladeshs-water-crisis-a-story-of-gender

48. Bartlett S. Climate change and urban children: impacts and implications for adaptation in low- and middle-income countries. *Environ Urban*. 2008;20(2):501–519. doi:10.1177/0956247 808096125

49. Burke SEL, Sanson AV, Van Hoorn J. The psychological effects of climate change on children. *Curr Psychiat Rep*. 2018;20(5):35. doi:10.1007 /s11920-018-0896-9

50. Cunsolo Willox A, Stephenson E, Allen J, et al. Examining relationships between climate change and mental health in the Circumpolar North. *Reg Environ Change*. 2015;15(1):169–182. doi:10.1007 /s10113-014-0630-z

51. United Nations. *State of the World's Indigenous Peoples: Indigenous Peoples' Access to Health Services.* February 2016. doi:10.18356/7914b045-en

52. Durie MH. The health of indigenous peoples. *Br Med J.* 2003;326(7388):510–511. doi:10.1136/bmj .326.7388.510

53. Durkalec A, Furgal C, Skinner MW, et al. Climate change influences on environment as a determinant of Indigenous health: relationships to place, sea ice, and health in an Inuit community. *Soc Sci Med.* 2015;136–137:17–26. doi:10.1016/j .socscimed.2015.04.026

54. Ford JD. Indigenous health and climate change. *Am J Public Health.* 2012;102(7):1260–1266. doi:10.2105/AJPH.2012.300752

55. Brubaker M, Berner J, Chavan R, et al. Climate change and health effects in Northwest Alaska. *Global Health Action.* 2011;4:10. doi:10.3402/gha .v4i0.8445

56. Fleischer NL, Melstrom P, Yard E, et al. The epidemiology of falling-through-the-ice in Alaska, 1990-2010. *J Public Health (Oxf).* 2014;36(2): 235–242. doi:10.1093/pubmed/fdt081

57. Yoder S. Assessment of the potential health impacts of climate change in Alaska. *State of Alaska Epidemiology Bulletin.* 2018;20(1).

58. Petrasek MacDonald J, Cunsolo Willox A, Ford JD, et al. Protective factors for mental health and well-being in a changing climate: perspectives from Inuit youth in Nunatsiavut, Labrador. *Soc Sci Med.* 2015;141:133–141. doi:10.1016/j.socscimed .2015.07.017

59. Roberts SE, Jaremin B, Lloyd K. High-risk occupations for suicide. *Psychol Med.* 2013;43(6): 1231–1240. doi:10.1017/S0033291712002024

60. Peterson C, Sussell A, Li J, et al. Suicide rates by industry and occupation—National Violent Death Reporting System, 32 states, 2016. *MMWR Morb Mortal Wkly Rep.* 2020;69:57–62. doi:10.15585 /mmwr.mm6903a1

61. Zullig KJ, Hendryx M. A comparative analysis of health-related quality of life for residents of U.S. counties with and without coal mining. *Public Health Rep.* 2010;125(4):548–555. doi:10 .1177/003335491012500410

62. Connor V. On back roads of Appalachia's coal country, mental health services are as rare as jobs. *KHN.* October 17, 2017. Accessed March 22, 2021. https://khn.org/news/on-back-roads-of -appalachias-coal-country-mental-health-services -are-as-rare-as-jobs/

63. Canu WH, Jameson JP, Steele EH, et al. Mountaintop removal coal mining and emergent cases of psychological disorder in Kentucky. *Commun Ment Health J.* 2017;53(7):802–810. doi:10.1007 /s10597-017-0122-y

64. Sass V, Kravitz-Wirtz N, Karceski SM, et al. The effects of air pollution on individual psychological distress. *Health Place.* 2017;48:72–79. doi:10.1016 /j.healthplace.2017.09.006

65. United Nations. The 17 Goals. Accessed March 2, 2021. https://sdgs.un.org/goals

66. Liu J, Potter T, Zahner S. Policy brief on climate change and mental health/well-being. *Nurs Outlook.* 2020;68(4):517–522. doi:10.1016/j.outlook.2020 .06.003

67. American Public Health Association, ecoAmerica. Climate changes mental health. Accessed March 21, 2021. https://www.apha.org/~/media/files/pdf/topics /climate/climate_changes_mental_health.ashx

CHAPTER 11

Human Displacement

KEY TERMS

Migration
Migrants
Refugees
Internally displaced persons
   (IDPs)

Displacement
Planned relocation
Immobility
Least developed countries
   (LDCs)

Small island developing states
   (SIDS)

LEARNING OBJECTIVES

- Understand the terms used to refer to human migration and persons on the move.
- Describe the climate change stressors most likely to cause human displacement.
- Understand why SIDS are among the highest risk countries for climate-change-fueled displacement.
- Describe the major health impacts of displacement.
- Describe examples of country-level actions to address the threat of displacement.

In 2020, climate-change-fueled disasters that were unprecedented in number and intensity led to significant human displacement. Hurricanes Eta and Iota displaced half a million people across Nicaragua, Honduras, and Guatemala.[1] Intense flooding in parts of Asia, where the monsoon season brought extreme rainfall for the second year in a row, displaced 4 million people in India and Nepal and nearly 750,000 in China.[2,3] Australia and the U.S. states of California and Colorado saw the largest wildfires in their history, which forced tens of thousands of people to evacuate in each location.[4,5] Heavy rains, flooding, and mudslides across parts of Africa caused half a million people in Somalia, 350,000 people in Democratic Republic of the Congo, and 300,000 in Ethiopia to become displaced.[6]

Each year, millions of people move in *anticipation of* or in *response to* environmental hazards. Without adequate early warning systems and advance planning, *rapid-onset* climate disasters such as extreme storms and floods may force affected populations to flee suddenly. *Slow-onset* disasters, such as sea level rise and prolonged drought, may gradually erode livelihoods, safety, and health, and eventually create tipping points beyond which people can no longer sustain themselves in a vulnerable location. The World Bank estimates that by midcentury, 143 million people in sub-Saharan Africa, South Asia, and Latin America may be forced to migrate because of climate-change-fueled crop failures, water shortages, and sea level rise.[7] Many people in other regions are also at risk.

Human mobility patterns are diverse and context-specific, and people move for many reasons, both positive and negative. People sometimes move voluntarily and sometimes forcibly, and relocation may be temporary or permanent. "People move all the time to manage risks and pursue their aspirations to live well, stably and peacefully," according to Koko Warner of the United Nations Framework Convention on Climate Change (UNFCCC) secretariat.[8] The links between migration and climate change are complex, and geographic, social, cultural, economic, and political factors interact to put certain populations at increased risk.[9,10] Most human displacement occurs in developing countries that lack adaptive capacity for disaster risk reduction and response. No country is immune, however, to the threat of displacement. Of critical importance is understanding how climate change is affecting the threat of migration and the health and well-being of people on the move.

Migration occurs on a continuum ranging from entirely by choice to entirely forced, with most situations somewhere in between.[10] Depending on conditions, moving may represent a conscious decision and can be seen as a form of climate adaptation, but it may also result from a failure to adapt to the climate hazard due primarily to limited economic, social, and political capacity. People who move may do so as individuals, households, or entire communities, and

decisions to move are based on personal knowledge and experiences, family and social networks, available infrastructure, and government policies and support.

Many terms are used to refer to the phenomenon of human migration and to people on the move, and they differ in meaning and legal recognition.

Migrants: People who move away from their place of habitual residence, temporarily or permanently, either within their country or across international borders, for a variety of reasons.

Refugees: People who are forced to flee their country because of war, violence, or persecution due to race, religion, nationality, political opinion, or membership in a particular social group. In most cases, refugees cannot return home or fear doing so. The rights of refugees and legal obligations of countries are outlined in the 1951 UN Convention Relating to the Status of Refugees.

Internally displaced persons (IDPs): People who move away from their place of habitual residence but do not cross international borders. IDPs are not protected by international law because they are under the legal protection of their own country.

Displacement: The forced or voluntary movement of people away from their places of habitual residence, in particular to avoid the effects of natural or human-made disasters, armed conflict, violence, and/or human rights violations.

Planned relocation: A coordinated process in which people or communities are moved preemptively with the assistance of governments to protect them from the risks and impacts of environmental disasters, including climate change.

Immobility: The inability of people exposed to threats to move because of limited social and economic capital, creating "trapped populations," or the voluntary choice of people not to move but rather to stay in place and adapt to threats in situ.

Circular mobility: A form of migration in which workers, families, or entire communities move between places of habitual residence and host areas, usually for employment opportunities.

People who move as a result of climate change are often referred to as "climate refugees," especially in the media and by decision-makers, but this term is inaccurate and misleading for several reasons.[11]

- Most people displaced by disasters are IDPs, not refugees (which requires crossing international borders).
- Stigma may be attached to the term "refugee," especially in reference to a person dependent on a host nation obliged to take them in.
- The term "refugee" invokes ideas of sudden flight from danger, but much of the human movement caused by climate change will be due to slow-onset disasters such as sea level rise and prolonged drought.
- Climate change is rarely the sole cause of human migration but rather is a tipping point on top of persistent poverty, conflict, human rights abuses, and poor governance. Attributing human movement to a single factor (by using the term "climate refugee") oversimplifies the issue and prevents the development of effective responses that target all relevant factors.
- Climate change as a cause of migration tends to be framed as a future threat, which may be perceived as not requiring immediate action even though it is actually happening today.
- Refugees from climate change and other environmental disasters are not protected under the 1951 UN Convention on Refugees. Some have called for amending this treaty or creating a new agreement, but

others minimize the protective power of an international treaty and highlight instead the need for countries to implement legal protections for people on the move due to climate change.

To avoid the limitations of the term "climate refugee," other descriptions are often used, including "environmentally displaced persons," "climate change displaced persons," or "disaster displaced persons."[8] The **UN High Commissioner for Refugees (UNHCR)** does not use the term "climate refugee," but instead refers to "persons displaced in the context of disasters and climate change." Even these terms may be inexact, however, and focus too much on climate-specific migration as a phenomenon distinct from other forms of migration.

Migration as an effective climate adaptation strategy must result in disaster risk reduction and economic opportunities for those moving.[12] Migration may not actually reduce risk, however, if people are exposed to the same or new threats in new locations, or if experiences of poverty, ill health, or insecurity are exacerbated. If people must move, voluntary migration is advantageous because people are more likely to be able to safeguard their health, protect their assets, and have control over decisions about who moves, along with when and where. In addition, situations may emerge in which migrant host countries, many of which are major greenhouse gas (GHG) emitters and thus have an ethical obligation to take climate action and help those displaced by climate change, instead respond by tightening immigration policies and border security measures to keep people out.[8]

In its 2014 *Fifth Assessment Report*, the **Intergovernmental Panel on Climate Change** recognized that human movement was already occurring because of drought, land degradation, floods, and sea level rise linked to climate change. The report high-lighted numerous examples.[13]

- In the United States, sea level rise has led to residents leaving offshore islands in Maryland.
- In Alaska, sea level rise and erosion are damaging coastal villages to such an extent that relocation is the only viable response.
- In Mexico, reduced crop yields have led to international migration to the United States.
- In the Sahel region of sub-Saharan Africa, prolonged drought has been a factor in increasing migration pressure since the 1970s.
- In Burkina Faso, people living in dry regions have been more likely to migrate than people in regions with more precipitation.
- In Kenya, agricultural soil degradation leads to people migrating for temporary employment opportunities.
- In Vietnam, seasonal flooding in the Mekong Delta has increased migration out of the region.
- In Bangladesh, many people impacted by tidal-surge-related floods and riverbank erosion have migrated to cities.

Most human mobility in response to disaster or conflict takes the form of **internal displacement** within the home country. The **Internal Displacement Monitoring Centre (IDMC)**, based in Geneva, Switzerland, maintains the Global Internal Displacement Database to monitor the number of IDPs around the world. IDMC's mission is to provide data that inform policy- and decision-making to "reduce the risk of future displacement and improve the lives of internally displaced people worldwide."[5]

In 2019, 25 million people were internally displaced by weather-related and geophysical disasters, 65% because of extreme storms and floods in just four countries—India, Bangladesh, China, and the Philippines (**Table 11.1**).[5] Most disasters were tropical cyclones, which caused three-quarters of the displacements in these four countries. In the United States in 2019, nearly 1 million people were internally displaced, mostly because of hurricanes, wildfires, and floods.[5] These data cover only people displaced within their country's borders, but many millions more are forced to relocate to another country because of disasters and/or conflict. In 2020, IDMC reported nearly 31 million new internal displacements due to disasters.

The top five countries for total number of IDPs as of the end of 2019, including those displaced in prior years, are Syria (6.5 million), Colombia (5.6 million), Democratic Republic of the Congo (5.5 million), Yemen (3.6 million), and Afghanistan (3 million).[5] Most of these displacements are attributed to armed conflict and other forms of violence. Most conflict-related displacements are not directly associated with climate change, although these two determinants may interact in some locations, particularly when competition for scarce resources leads to violence, and because residents of war-torn areas may already suffer from health problems that could be made worse by climate change.

In 2019, the UN **International Organization for Migration (IOM)** and the UN **Office of the High Representative for the Least Developed Countries, Landlocked Developing Countries and Small Island Developing States** released a report in which they argued that **least developed countries (LDCs)**, landlocked developing countries (LLDCs), and **small island developing states (SIDS)** face significant displacement risks and deserve particular attention in climate policies.[14] LDCs, LLDCs, and

Table 11.1 Countries with the Highest Number of New Internal Displacements in 2019 Due to Extreme Storm and Flood Events That Displaced at Least 100,000 People.

Country	Weather-Related Disaster in 2019	Number of IDPs	Total IDPs
India	Southwest monsoon	2,623,000	**4,919,000**
	Cyclone Fani	1,821,000	
	Cyclone Vayu	289,000	
	Cyclone Bulbul	186,000	
Bangladesh	Cyclone Bulbul	2,107,000	**4,080,000**
	Cyclone Fani	1,666,000	
	Monsoon	307,000	
China	Typhoon Lekima	2,097,000	**3,673,000**
	Flood season	1,576,000	
Philippines	Typhoon Kammuri	1,424,000	**3,408,000**
	Flooding/landslides	580,000	
	Typhoon Phanfone	567,000	
	Tropical depression Usman	552,000	
	Northeast monsoon	179,000	
	Flash flood	106,000	
	TOTAL		*16,080,000*

Data from Internal Displacement Monitoring Centre. Global internal displacement database: IDMC query tool – disaster. Accessed February 14, 2021. https://www.internal-displacement.org/database/displacement-data

SIDS include 91 countries with a cumulative human population of more than 1 billion. Climate change impacts are particularly worrisome in these countries because they threaten progress in economic and human development.

As of June 2021, the UN recognized 46 LDCs, considered the poorest countries in the world. Approximately one-third of people in LDCs live on less than $1.90 per day, and more than two-thirds live in poor rural areas.[14] According to the UN,

LDCs are "low-income countries confronting severe structural impediments to sustainable development. They are highly vulnerable to economic and environmental shocks and have low levels of human assets."[15] Environmental vulnerability in LDCs is often measured by the proportion of people living in low-lying coastal zones or drylands, the instability of agricultural productivity, and the threat of disasters. These impediments may leave affected people without sufficient employment opportunities and with significant burdens of illness and mortality.[14] LLDCs have unique features that make them vulnerable, including a lack of access to the sea, which forces them to rely heavily on neighboring countries. They typically have poor infrastructure and trade, which tends to impede economic and social development. Many LLDCs are also LDCs.[14]

SIDS are recognized as being at high risk for climate change impacts and the threat of displacement, particularly because of sea level rise, extreme storms, and other coastal impacts.[14] Vulnerability of SIDS is attributed largely to their small size and long coastlines, fragile ecosystems susceptible to natural disasters, limited resource bases and economies of scale, geographic remoteness, lack of access to global markets, and, in some cases, weak governance.[16] Because of recurring costs of disaster recovery, SIDS tend to lack the financial resources to invest in infrastructure projects and other adaptations that could build climate resilience.[14] They also face high debt burdens as developing countries reliant on international aid, including loans.

No single measure exists to assess a country's risk of human displacement, but several indicate disaster risk, which is associated with displacement. For example, the **WorldRiskIndex** is a statistical model that provides a global assessment of the risk of disasters arising directly from extreme natural events such as earthquakes, storms, floods, droughts, and sea level rise.[17] It assumes that disaster risk is likely to be highest where these extreme events hit vulnerable regions. The WorldRiskIndex incorporates country-level metrics of hazard *vulnerability* (susceptibility and a lack of coping and long-term adaptive capacities for disasters) and actual hazard *exposure*.[17] **Figure 11.1** shows a map of disaster risk by country as indicated by the WorldRiskIndex.

Countries with the highest WorldRiskIndex in 2020 are listed in **Table 11.2**. The top 6 countries and 9 of the top 15 are SIDS, and 3 are LDCs.[17] SIDS are high on the list because of their high *exposure* to these hazards. The United States is ranked 134 out of 181 countries for WorldRiskIndex. Europe is the continent with the lowest risk.[17]

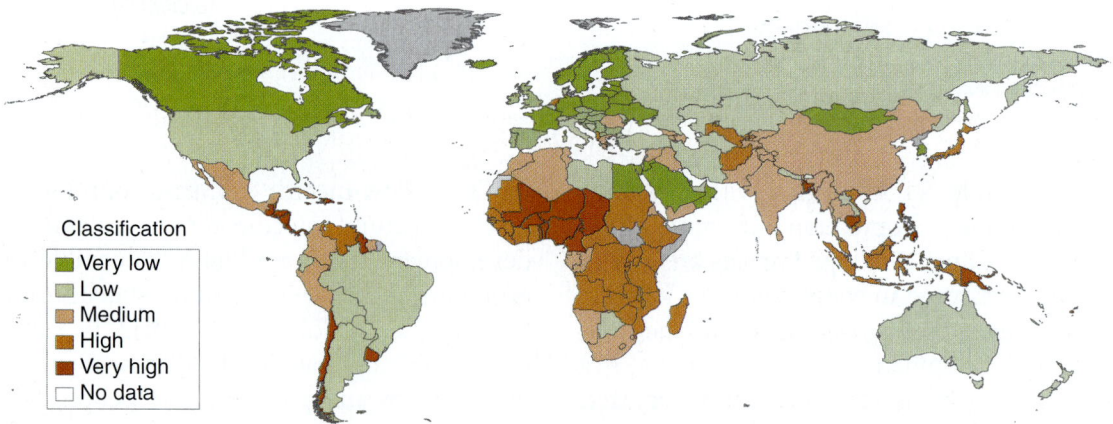

Classification
- Very low
- Low
- Medium
- High
- Very high
- No data

Figure 11.1 Classification of Countries Based on WorldRiskIndex 2020: Risk of Disasters That Arise from Extreme Natural Events (Earthquakes, Storms, Floods, Droughts, Sea Level Rise). Brown colors indicate high risk, and green indicates low risk.

Table 11.2 Top 15 Countries with the Highest WorldRiskIndex 2020, Indicating Overall Disaster Risk.

Rank	Country	Category
1	Vanuatu	SIDS
2	Tonga	SIDS
3	Dominica	SIDS
4	Antigua and Barbuda	SIDS
5	Solomon Islands	LDC, SIDS
6	Guyana	SIDS
7	Brunei Darussalam	
8	Papua New Guinea	SIDS
9	Philippines	
10	Guatemala	
11	Cabo Verde	SIDS
12	Costa Rica	
13	Bangladesh	LDC
14	Djibouti	LDC
15	Fiji	SIDS

Note. LDCs = least developed countries; LLDCs = landlocked developing countries; SIDS = small island developing states.

Data from Behlert B, Diekjobst R, Felgentreff C, et al. *WorldRiskReport 2020: Focus: Forced Displacement and Migration.* Bündnis Entwicklung Hilft; 2020. https://reliefweb.int/sites/reliefweb.int/files/resources/WorldRiskReport-2020.pdf

Table 11.3 Top 15 Countries Most Vulnerable to Climate-Related Hazards Based on WorldRiskIndex 2020.

Rank	Country	Category
1	Central African Republic	LDC, LLDC
2	Chad	LDC, LLDC
3	Democratic Republic of the Congo	LDC
4	Niger	LDC, LLDC
5	Guinea-Bissau	LDC, SIDS
6	Eritrea	LDC
7	Burundi	LDC, LLDC
8	Yemen	LDC
9	Liberia	LDC
10	Madagascar	LDC
11	Sierra Leone	LDC
12	Mozambique	LDC
13	Mali	LDC, LLDC
14	Papua New Guinea	SIDS
15	Haiti	LDC, SIDS

Note. LDCs = least developed countries; LLDCs = landlocked developing countries; SIDS = small island developing states.

Data from Behlert B, Diekjobst R, Felgentreff C, et al. *WorldRiskReport 2020: Focus: Forced Displacement and Migration.* Bündnis Entwicklung Hilft; 2020. https://reliefweb.int/sites/reliefweb.int/files/resources/WorldRiskReport-2020.pdf

When only *vulnerability* to disasters is considered, without *exposure*, the top 15 countries in 2020 were all LDCs except Papua New Guinea (**Table 11.3**).[17] Five were also LLDCs, and three were SIDS. Notably, all produce very low levels of GHG emissions. In contrast, the United States and European countries are major GHG emitters yet face relatively low disaster risks to their populations. This is perhaps one of the starkest indications of climate injustice, because the countries that have contributed very little to the problem of climate change are most at risk of its consequences. High risk status does not mean that climate impacts are inevitable, however. At-risk countries can, and in some cases already have taken action, unilaterally or aided by the international community.

Health Impacts of Displacement

Human migration has the potential to both positively and negatively affect human health. Epidemiological studies are lacking, however, in part because of the often unorganized and unrecorded nature of displacement. Study populations can be hard to identify and monitor, and some people who are identified have the right to choose not to participate. Most populations that have been studied are refugees in resettlement facilities and those who have been displaced by war, genocide, or famine. Nevertheless, people displaced as a result of climate change are likely to have similar health outcomes as those displaced for other reasons, and in certain regions, climate change may interact with conflict and other factors to cause migration and its associated health effects.

Which population is on the move and whether migration is forced or voluntary, internal or across borders, and temporary or permanent will determine the health impacts in places of origin, in transit, and in relocation sites.[18] Although health conditions may improve in some cases, the most predictable impacts are negative, with increased morbidity and mortality from preexisting and new-onset physical and mental health conditions. With migration comes changes in food and water supply, shelter, infectious disease patterns, air quality, healthcare access, social networks, and household incomes, any of which may influence health.[18]

Many health impacts have been shown to emerge in displaced populations, particularly communicable diseases.[18]

- Infectious diseases such as dysentery, typhoid, cholera, other diarrheal diseases, hepatitis A, and hepatitis E that are due to lack of access to safe water, sanitation, and hygiene (WASH)
- Acute malnutrition and nutrient deficiencies, particularly in children, that are due to lack of access to an adequate and safe food supply
- Respiratory infections such as pneumonia and tuberculosis, along with airborne viral diseases such as measles and meningitis that are due to overcrowding, low-quality housing, and poor ventilation
- Vector-borne diseases such as malaria and dengue that are due to moving into disease-endemic areas and lack of preventive measures or treatments
- Sexually transmitted infections, unplanned pregnancies, and mental disorders that are due to violence against women, sexual abuse, and interpersonal conflict
- Mental health disorders that are due to loss of home or livelihoods, disruption of family and social structures, experiences of violence, or loss of agency
- Generally, a lack of healthcare services that creates gaps in prevention, diagnosis, and treatment of many diseases and, importantly, childhood vaccinations and routine preventive care

As an example, after the 2004 Asian tsunami, 1 million people in Sri Lanka were displaced, and outbreaks of cholera, typhoid fever, hepatitis A and E, malaria, dengue, measles, respiratory infections, and meningitis followed.[18] The World Health Organization estimated that half of Sri Lankans also experienced mental health problems, including posttraumatic stress disorder (PTSD) in 40% of children.[19]

Evidence suggests that, generally, displaced persons risk developing mental disorders such as depression, anxiety, and PTSD at higher rates than nondisplaced populations, with most of this research focused on populations displaced by war and conflict.[20] Numerous factors associated with displacement, particularly if prolonged, can trigger mental problems, including inadequate food, water, and healthcare services, co-occurring physical or psychological disorders, experiences of violence, disrupted social networks, stigma at being displaced, and loss of sense of hope for the future. It is also possible that, in some cases, long-term displacement may lower the risk of mental disorders if safety improves, new opportunities for livelihoods arise, and feelings of hope for the future and a sense of belonging and wellness emerge.[20]

Migration may also be beneficial to health if opportunities arise in new locations for *improved* access to safe and adequate food and water supplies, healthcare facilities, and economic opportunities. Current health impacts of displacement are occurring at a time when major global efforts to reduce disease burdens have been ongoing for over two decades. Global progress has been particularly significant for many communicable diseases, including those prevalent among displaced people. Future impacts of displacement on human health will be modulated in part by the continued success or failure of these global efforts, and increased displacement may erode progress on public health in certain populations.

It is common for migrants to move from rural to urban areas. In doing so, they often settle in unmanaged areas with poor, crowded housing conditions that lack improved WASH services, employment opportunities, or health services.[18] Changes in diet, drug and alcohol use and physical activity, as well as isolation and stress may lead to increased incidence of chronic conditions such as high blood pressure, heart disease, and diabetes. Labor migrants, who move seeking employment, often face health threats such as extreme heat exposure during outdoor work and infectious diseases (e.g., malaria, tuberculosis, and HIV).[21] Other negative consequences of urban migration include social marginalization, cultural shifts, and fragmentation of families and social networks.[12] Health benefits may also arise if urban migrants are able to access education and job training and earn income. Funds sent back home to family members or communities may improve health if they are used to invest in food and water systems, improved housing, clinics, and schools.[18] Immobility (not moving when faced with environmental hazards) may increase psychological stressors, food insecurity, and existing disease burdens and prevent access to healthcare and social services that may be lacking or damaged by environmental hazards.[12]

At-Risk Countries

The population of an entire country may be at risk of displacement due to climate change, or certain sub-populations within countries may face the greatest risk. Selected examples of specific populations experiencing the health impacts of climate-change-fueled human migration are described in the next sections.

Small Island Developing States (SIDS)

SIDS include 58 developing countries in three geographic regions: the Caribbean, Pacific, and AIMS (Atlantic, Indian Ocean, Mediterranean, and South China Sea). As a group, SIDS are considered at very high risk of climate change impacts and the threat of human displacement, although each country is unique and differs in its precise vulnerabilities and adaptive capacities.[16] In particular, SIDS are the most at-risk countries for submersion because of sea level rise. For example, most of the population of the Marshall Islands (93%) lives where the land will be submerged with 4°C (7°F) of warming, along with 88% of the population in the Cayman Islands, 81% in Tuvalu, 77% in Kiribati, 76% in The Bahamas, and 73% in Maldives.[22] Many of these places are actually at risk of becoming uninhabitable well in advance of submersion, particularly low-lying atoll islands in the Pacific. SIDS also face hazards from extreme storms, coastal erosion, and saltwater intrusion into freshwater and soils, threatening settlements, agriculture, food security, and ecosystem services. SIDS residents tend to depend heavily on limited industries for their livelihoods, including fisheries, agriculture, and tourism, all of which are vulnerable to climate shocks.[14]

Residents of Pacific SIDS face a collective memory of prior experiences with forced relocation as a result of colonialism, as well as extractive activities (e.g., mining in Kiribati and nuclear weapons testing in the Marshall Islands).[23] Geophysical disasters such as volcanic eruptions and earthquakes have also led to both temporary and permanent migrations. A long history of experience with migration informs Pacific Islanders' responses to current and future climate migration threats.[23] Residents of Caribbean SIDS also face complicated and traumatic legacies of colonialism and resource extraction, along with slavery and plantation agriculture.

Numerous SIDS have experienced displacement in recent years because of tropical cyclone impacts and sea level rise. In 2015, Cyclone Pam, a Category 5 storm, hit **Vanuatu**, an archipelago of 83 primarily volcanic islands in the Pacific Ocean. The storm damaged 90% of homes, displaced thousands of people and caused nearly half a billion dollars in damage (**Figure 11.2**). More than three-quarters of residents rely on subsistence agriculture and fishing for their livelihoods, which were highly disrupted by the storm. Residents of several communities on the island of Mataso were temporarily

Figure 11.2 Aftermath of Cyclone Pam on the Islands of Vanuatu, March 14–15, 2015.
© ausnewsde/Shutterstock.

relocated internally with funding and oversight from IOM. However, they expressed dissatisfaction with the experience because they were not able to use existing social and kinship networks to relocate to places of their choice, such as moving in with extended family members on other islands, and they reported difficulty maintaining livelihoods and cultural traditions while relocated.[24]

Residents of Vanuatu are grappling with the likelihood of future planned relocation in response to climate change, although many have expressed resistance except as a last resort, with a preference for remaining in place and adapting in situ.[24] Also, strong connections to the land are culturally and spiritually important to residents of SIDS, which may make decisions to relocate difficult.[23] Several other Pacific SIDS are pursuing planned relocation, including Fiji and Papua New Guinea. Key factors influencing the success of planned relocation include full consultation and consent of affected people, support from host communities, creation of livelihoods for relocated people, and secure access to land.[12,23] A major limitation of planned relocation is its high cost.

Fiji, an archipelago of more than 300 islands in the South Pacific, has been hit with numerous destructive cyclones over the past decade. The village of Denimanu experienced Cyclone Evan in 2012, which prompted temporary evacuation and destroyed houses closest to the shoreline. These houses were rebuilt on a hillside 500 meters from their original location, and residents were provided with solar power, rainwater tanks, and other improvements that were not available in the original village.[23] The village of Vunidogoloa experienced decades of coastal flooding, saltwater intrusion, and shoreline erosion, and in 2014, leaders asked the Fijian government to relocate the village to higher ground (**Figure 11.3**). As a result, the entire village moved two kilometers inland, and residents received pineapple plants, livestock, and fish ponds, which helped reestablish their livelihoods.[23]

In 2016, Cyclone Winston, a strong Category 5 storm, hit the main island of Fiji. As a result, plans were made for more than 60 villages to be relocated to reduce exposure to future storm risks.[23] At the end of 2020, Fiji was hit by Category 5 Super Cyclone Yasa, which destroyed or damaged homes, healthcare facilities, schools, and farm fields, and displaced thousands. Recognizing the continued threat of storms like these and other climate impacts, Fiji has developed systematic guidelines for planned relocation, though as a last resort, recommending close consultation with those moving, those in host communities, and those who choose not to move.[10]

Atoll communities in **Papua New Guinea** are strongly affected by slow-onset climatic changes. For example, residents of the Carteret Islands face sea level rise, inundation, soil salinization, and loss of land, mangroves, and coral reefs. In 2007, the Council of Elders of the Carteret Islands organized a community-led voluntary relocation of its residents to Bougainville Island, 100 kilometers to the northeast. The location of the resettlement site was chosen to ensure enough land for relocated people to be economically self-sufficient, and food security was also prioritized by ensuring access to traditional fishing grounds. The project also provided upgraded healthcare facilities and schools. Though opinions were mixed, 2,600 people relocated to Bougainville Island, and others immigrated to New Zealand.[12]

Kiribati is a Pacific nation of 33 coral atolls, most of which are less than 2 meters (6 feet) above sea level. Kiribati faces drought, coral die-off, diminished fishing stocks, salinization of drinking water, high unemployment, and low access to sanitation, and if the sea level rises 2.6 feet or more, more than 80% of Kiribati land surface will be unsuitable for habitation.[25] Several years ago, Kiribati began preparing for planned relocation, which they referred to as "migration with dignity," in which people have control over whether, when, where, and how they move or stay. The government at the time purchased 6,000 acres of agricultural land in Fiji as a possible relocation site.[25] However, when a new president came to power, the focus shifted to remaining in place and emphasizing tourism, fisheries, and expanded production

Figure 11.3 The New Settlement of Vunidogoloa, Fiji, October 20, 2017. Vunidogoloa is the first village in Fiji to permanently relocate because of sea level rise.

and trade of copra, the dried flesh of coconuts used in oils, creams, and pig feed.[25] In 2020, the president announced an ambitious plan to elevate entire islands above the rising sea level.[26]

Maldives is a chain of over one thousand small coral islands and sandbanks in the Indian Ocean, many of which are barely above sea level. In 2008, its former president established a fund to purchase land for relocation, possibly in Australia, India, or Sri Lanka. However, as in Kiribati, when a new president was subsequently elected, the plan shifted to remaining in place. The country has constructed several artificial islands for human habitation, including Hulhumale Island (**Figure 11.4**), by depositing sand from surrounding atolls on shallow reefs and building a sea wall three meters above sea level.[27]

Barbuda, one of two major islands comprising the Caribbean nation of Antigua and Barbuda, took a direct hit from Hurricane Irma in 2017, which damaged or destroyed nearly every structure on the island. All of its residents were evacuated, and even though 80% have now returned, most live without improved housing or public services. Most residents make a subsistence living through fishing or low-level tourism, and they face intense pressure to give up property for major

Figure 11.4 Hulhumale Island in the Maldives.
© by sharaff/Moment/Getty Images.

tourism development pushed by banks and wealthy elites.[28]

A major concern among SIDS leaders and activists is that emphasizing migration as a climate adaptation strategy takes the focus off the primary responsibility of developed countries—which are the world's biggest GHG emitters—to commit to mitigation strategies to greatly reduce their own emissions, thus reducing displacement risk. In addition, the existential threat of the complete disappearance of some SIDS because of sea level rise opens new questions that must be addressed in regional and international negotiations about the territorial sovereignty of these disappearing countries and the human rights of islanders who could effectively become stateless.

Climate-change-related threats to SIDS are dire. "The international community needs to work hard to help small island nations build resilience as the threat of destruction expands and the cost of seasonal storms becomes increasingly devastating for SIDS," said Mr. Paul Akiwumi, the UN Conference on Trade and Development's director for Africa and LCDs.[29] This means a commitment to creating and leveraging numerous forms of climate financing for SIDS. "Future disasters could worsen humanitarian crises and impair economic recovery," said Akiwumi. "For SIDS, urgent action is needed to protect lives and livelihoods. It is in everyone's interests to support SIDS now."[29]

Vietnam

Vietnam is highly impacted by sea level rise, with 70% of its population living along its very long coast and 52% living at elevations that will be under water with 4°C (7°F) of warming. In fact, Vietnam has the largest proportion of its population living in places at risk of submersion with 4°C warming among all countries with populations of at least 25 million.[22]

In 2020, Vietnam was hit by seven consecutive tropical storms, including four cyclones, in a five-week span, which caused unprecedented destruction and displaced 375,000 people to evacuation centers.[30] Many shelters were overcrowded, lacked adequate safe water, sanitation, and healthcare services, and did not offer sufficient protections against COVID-19. In 2017, Typhoon Tembin hit Vietnam, affecting more than 7 million people and displacing over 400,000.[5] Health impacts among those displaced included wounds and infections from wading in floodwaters, severe malnutrition in children due to the food supply being destroyed, and an outbreak of diphtheria. The United Nations Children's Fund (UNICEF) disbursed emergency supplies, including ready-to-use therapeutic foods, water filters, and soap. In 2008, extreme rainfall led to destructive flooding in Vietnam, and 20–50% of people in the hardest hit areas were forced to migrate. Most evacuated to the home of a relative or friend nearby and did not receive financial or medical aid from the government. Affected populations reported increased health problems, particularly psychological disorders and dengue, which were associated primarily with exposure to flooding.[31]

Recurrent flooding in the Mekong River Delta and other areas (**Figure 11.5**) has resulted in significant internal displacement and migration to major cities.[21] The health impacts of urban migration tend to vary by age and gender. In Vietnam, women often find employment as domestic workers or in factories, where they may be away from family protection and familiar social networks and subjected to gender-based violence. Men often work in construction, where they face occupational hazards, including heat exposure. Among child migrants, girls tend to do

Figure 11.5 Flooding in Vietnam.
© robas/iStock/Getty Images Plus/Getty Images.

domestic work, and some end up in the sex industry. Boys tend to work in fishing or coal picking, and some are lured into drug trafficking.[32] Often, youth are forced to work long hours for little pay, and they may suffer stress and physical, mental, or sexual abuse. Family members left behind, usually older women, must care for children, home, and property and tend farm fields and aquaculture ponds. This creates multiple physical and mental health stressors on women and leaves them vulnerable when future storms hit.

Preventing displacement is a high priority in Vietnam. Government subsidies have been offered to coastal communities to cover the costs of flooding, but they are often inadequate and are only distributed after harm has occurred.[32] Mangroves are being restored to provide coastal protection from storm surges and saltwater intrusion, but some efforts have been met with resistance from local fisherfolk who believe it will reduce their income from aquaculture.[33] A recent resettlement program offered coastal residents loans to build houses on higher ground that are protected from storms and flooding by dykes along rivers and canals.[32] These programs have been met with mixed reactions because even though exposure to disasters lessened and access to services such as water, electricity, education, and health care expanded, debt burdens increased for many relocated families, and social networks and access to farm land and fishing grounds were disrupted.[32]

Bangladesh

Bangladesh is one of the most climate-vulnerable countries because of its low-lying, densely populated coastal region on the Bay of Bengal, its geographic susceptibility to storms, its expansive river systems, and high poverty rates. Most Bangladeshis earn a living in agriculture, fishing, aquaculture, and forestry—all vulnerable to climate change. In southern Bangladesh, the vast majority of the population is dependent on unprotected drinking water sources, such as rainwater, rivers, canals, and ponds, vulnerable to contamination and saltwater intrusion.[34] Flooding and use of unsafe water sources cause high burdens of skin diseases, diarrheal diseases, including cholera, and mosquito-borne diseases, particularly dengue.[21] Persistent and recurrent floods damage homes, crops, soils, and shrimp ponds. The most common reasons for displacement in Bangladesh are riverbank erosion, cyclones, and loss of agricultural and fishing livelihoods.

Powerful Cyclone Aila hit Bangladesh in 2009, destroying 250,000 homes along with coastal embankments that were protecting homes from being submerged.[9] Some areas were still flooded one year later, and 200,000 internally displaced people lived in temporary shelters. Humanitarian assistance in the immediate aftermath of the storm, including food and water aid, was insufficient, and even though some recovery assistance was available from rural leaders, poor and landless rural inhabitants did not have sufficient access to power to receive aid.[9] In general, Bangladeshis living in flood-prone areas tend to have lower incomes than people in non-flood-prone areas.[9] Environmental destruction in rural areas is driving unprecedented migration to Bangladesh's cities, and the vast majority of settlers in the urban slums of the capital Dhaka are considered climate change migrants.[35]

Mozambique

Mozambique and other countries along the southeastern coast of Africa are experiencing more intense tropical storms that scientists believe are linked to increased sea surface temperature in the Indian Ocean.[36] Climate change impacts are particularly devastating in Mozambique because of its long coastline and status as a least-developed country that is not well-prepared to deal with disasters. In 2019, Cyclone Idai, which turned out to be the deadliest cyclone in southern Africa up to that time, hit Mozambique, killing more than 1,000 people and displacing over 500,000.[5] Doctors in the low-lying and hard-hit port city of Beira reported treating people for bone fractures, blunt force trauma injuries, puncture wounds from storm debris, and bluish skin and chest pains caused by near drownings.[36] It took eight months after the storm for the hospital operating room to be restored.

Following the storm were many days of soaking rains and flooding. Rivers burst their banks, and many villagers had to flee as their homes became submerged. Said one woman interviewed later at a displacement camp, "People were trying to run from it but the water was coming really fast, really strong. The only thing we could do was climb a tree."[36] This woman reported clinging to a tree branch with her young daughter and watching as debris, livestock, and human bodies washed away below her. Her brother and nephew both drowned. Schools and churches were also destroyed, and grocery stores ran out of food. Many people found themselves in makeshift shelters in displacement camps and reliant on humanitarian aid. Some people returned home, but months later, tens of thousands were still living in these camps. Extreme rainfall and a heat wave destroyed two cycles of crops they had planted to survive.[36] In 2020, a locust outbreak reached Mozambique and caused major crop destruction, and a strong tropical storm led to further displacement. In addition, people faced the threat of contracting COVID-19 in crowded shelters. In early 2021, yet another disaster, powerful Cyclone Eloise, hit the same region, causing severe flooding.

Before Cyclone Idai hit, many displaced people in Mozambique earned a living by fishing, but they have not been able to afford the costs of replacing their destroyed boats and fishing nets. Some relocated away from the coast, which is less prone to flooding, but there, they had to resort to farming instead of fishing. Said one man, "I wanted to remind the children that their father used to bring back fish. Right now, he brings nothing."[36] According to the mayor of Beira, "Every day I see how the climate is changing. The sea is rising, the waves are stronger and bigger. I watch how the temperature changes. It is not like it was before."[36]

Dadaab Refugee Complex (Kenya)

The public health situation in one of the world's largest refugee camps, the Dadaab Refugee Complex in Kenya, illustrates what happens when migration occurs on a massive scale (**Figure 11.6**). Over the past several decades, hundreds of thousands of Somalis have been forced to settle at Dadaab because of recurring droughts, famines, and conflict. In **Somalia**, nomadic pastoralists make their livelihood rearing livestock, but persistent lack of rainfall and political violence

Figure 11.6 Newly Arrived Somali Refugees Carry Their Belongings Inside the IFOR Refugee Camp on July 29, 2011, outside Dadaab, Kenya.
© ymphotos/Shutterstock.

have destroyed this livelihood and displaced people across international borders to Kenya and Ethiopia. In 2011, a severe drought and famine, coupled with conflict and weak governance, led to a quarter of Somalia's population (over 2 million people) being displaced.[9] In July 2011, 1,300 Somalis arrived each day at the Dadaab complex in Kenya, and nearly 2,000 Somalis arrived each week at camps in Ethiopia.[18] In 2020, the population of Dadaab was 200,000, mostly Somalis.

In 2010, the U.S. Centers for Disease Control and Prevention's Division of Global Migration and Quarantine responded to many disease outbreaks at Dadaab, including cholera, polio, H1N1 flu, measles, meningitis, pertussis, and shigellosis.[37] In 2012, a hepatitis E outbreak occurred at the camp, sickening hundreds and killing four women who had just given birth.[38] Unsafe WASH and disruption of child immunization programs are persistent risk factors, as well as inadequate public health services generally.

A mental health crisis occurred at Dadaab in 2020 as refugees reported despair, anxiety, and fear, combined with new uncertainties brought about by the COVID-19 pandemic. The international medical humanitarian organization Médecins Sans Frontières (MSF) runs a mental health clinic at the camp that provides medical treatment for patients with mental illnesses, including depression, schizophrenia, and personality and anxiety disorders. In 2020, MSF reported a dramatic deterioration in the mental health status of camp residents and increases in mental health consultations and attempted suicides.[39] "COVID-19 has ended what little chance refugees had of escaping their degrading lives in the camps, compounding the mental distress for many who had nothing left but hope to cling to," said Jeroen Matthys, MSF project coordinator for Dadaab. "We are seeing a groundswell of desperation in the camp."[39]

United States

The U.S. state of **Alaska**, along with the entire Arctic region, is warming faster than the rest of the world and already experiencing devastating climate change impacts. In 2009, the U.S. government issued a report identifying 31 Alaska Native villages, mostly coastal, at imminent risk of destruction. These villages are facing severe flooding and erosion from sea level rise, storm surges, thawing permafrost, and delayed freeze-up and earlier breakup of sea ice, which inhibits protection from ocean storms and hinders traditional hunting and transport activities.[12,40] Twelve of these villages have begun making plans to permanently relocate, including Newtok, Shishmaref, and Kivalina, but none has been entirely successful to date. Even though most Alaska Natives anticipate that resettlement is necessary, many remain in situ because of a lack of funding and planning capacity, and, importantly, strong cultural ties and sense of attachment to place.[12]

Many residents of these villages face significant health impacts when staying in place. Infectious diseases such as pneumonia and skin infections result from floodwater exposure and deteriorating sanitation. Declining water quantity and quality from permafrost melting beneath water sources raises the risk of waterborne diseases. Food insecurity and storm- and ice-related injuries are also becoming more prevalent. Changing dietary consumption patterns are the result of lost traditional hunting, fishing, and gathering activities. The need to switch to store-bought foods, many of which are processed and less nutritious, leads to increased chronic diseases such as diabetes and heart disease.[12]

Kivalina is an Iñupiat village on a barrier island in Alaska surrounded by the Chukchi Sea. The area is prone to heavy winds and rainfall, and storm protection from sea ice has become less reliable.[41] Severe coastal land erosion due to storms and sea level rise is making it inevitable that residents will have to relocate. During floods, it is difficult for people to evacuate, so in 2020, they constructed a gravel road from their village to higher ground, and planned next steps are to construct a school there and eventually relocate the entire village.[41]

Residents of Kivalina, like many Alaska Natives, face mental health impacts caused by eroding safety and loss of land, resources, and social and cultural traditions. People in Kivalina and two other coastal villages have reported feelings of psychological stress in response to climate change.[42-45]

> "Every time the waves get high, people get anxious. Some people walk all night. My ten-year-old son was worried the house would blow away."
>
> "[The ice was] too thin. This spring it was so bad people were falling through the ice all over the place. It was very dangerous."

"My concern is the social aspect, that we can't get traditional foods, cost of food is too high to eat healthy, and it affects us, how we feel."

"Last winter was bad for mental health. The weather was so bad it was very hard to get out of the house."

"When people start talking about climate change, it really scares me."

Additional mental health impacts may include solastalgia, the deep sense of loss of an important place that provides stability, security, and personal identity. To what extent relocation may be *beneficial* for the mental health status of Alaska Natives is not known, but a safer settlement may ease some of this distress.

The Iñupiat village of **Shishmaref** on Sarichef Island, another barrier island in the Chukchi Sea, is becoming uninhabitable because of severe erosion from storm damage and melting permafrost (**Figure 11.7**). Sea ice provides storm protection and protects the shoreline for 10 months of the year, but now, the ice is thinner, freezes later in the season, and does not stay frozen. People here depend on the ice to hunt for walrus and seals, a major part of their traditional diet,

but recently, several residents have died trying to hunt or travel on thin ice.[46]

Shishmaref is planning to move to the mainland with the assistance of Alaska's government. Residents characterize this project not as a *relocation* but rather as an *expansion*.[46] However, attitudes are split between younger people who want to find a new place to live, and older people who want to stay in the place where they have spent their whole lives and to which they have a strong connection.

Newtok, a small Yup'ik village on the Ningliq River, is facing severe riverbank erosion and sinking permafrost. Residents traditionally fished on the river but now have only a short window to do so, and many families are not able to gather a sufficient fish supply to dry for winter.[47] Children's respiratory health suffers because of mold buildup from persistent flooding, and raw sewage is dumped in the river because of a lack of sanitation. As the permafrost thaws, the entire village is sinking, and people frequently became mired in mud. Boardwalks are necessary for people to move around (**Figure 11.8**).

One-third of the population of Newtok has now relocated to more stable ground about 9 miles away in Mertarvik, and the

Figure 11.7 Abandoned House in Shishmaref, Alaska, Slides Off the Land after a Storm in 2005.
© Diana Haecker/AP/Shutterstock.

Figure 11.8 Children Play along a Boardwalk That Stabilizes the Ground in Newtok, Alaska.
© Andrew Burton/Getty Images News/Getty Images.

plan is for the entire village to eventually relocate. Many positive health benefits have been reported as a result of gaining more stable ground and the ability to fish on the river any time. Residents say they have improved health and the ability to live "a more traditional Yup'ik lifestyle."[47] They also report eating more subsistence foods such as seal, halibut, salmon, and moose because there is no commercial airport in Mertarvik to bring in grocery supplies. One resident commented, "We've had more energy and improved mood."[47]

The **U.S. Gulf Coast** is the most hurricane-prone region of the country. When Hurricane Katrina struck in 2005, nearly 400,000 people in the New Orleans area and about 1 million people along the entire Gulf Coast were displaced. Both Blacks and whites were displaced, but only 38% of Blacks returned to their original county of residence, compared with 76% of whites.[9] Black people impacted by Hurricane Katrina received less relief aid and assistance than whites, and their neighborhoods in New Orleans were more likely to be abandoned than rebuilt.[9] Adverse mental health conditions were common in the aftermath of Katrina, particularly among those in temporary shelters. Similarly, when Hurricane Sandy hit the northeastern United States in 2012, people who were displaced by the storm had more than double the odds of reporting symptoms of PTSD than people who were not displaced.[48] The odds of depression, anxiety, and perceived stress were also higher.

Residents of the island of **Isle de Jean Charles**, **Louisiana**, in the Gulf of Mexico, received federal funds in 2016 for relocation to the mainland because of climate change. They are moving to former sugar plantation land 40 miles to the north. Nearly all the land on Isle de Jean Charles has been lost because of coastal erosion and sea level rise, and the only road connecting the island to the mainland is often impassable because of high water, winds, tides, and storm surge.[10] In 2021, the island suffered extensive damage from category 4 Hurricane Ida. Most of its residents are members of the Biloxi-Chitimacha-Choctaw tribe of Indigenous Americans, and it is not yet known what impacts the move will have on their ancestral and cultural traditions. In addition to these island residents, an estimated 13 million Americans will have to move away from the coast by 2100 because of sea level rise,[49] and the U.S. government will need to guide federal planning for climate migration.

Adaptations

Human displacement driven by climate change tends to impact populations that have contributed minimally to global GHG emissions yet face disproportionate consequences. Actions on the part of major GHG emitter countries to lower their emissions are critical to reduce the threat of displacement. Protecting public health in situations of migration is a global priority even without climate change, but incorporating increased healthcare capacity, infrastructure, and sustainable economic opportunities into climate adaptation plans is critical to saving lives and protecting the well-being of those displaced.[18] To support the choice to migrate and safely manage the migration process, proactive local, national, regional, and international policies, as well as guaranteed legal protections, are important.

The 2016 UN **Declaration for Refugees and Migrants** and the 2018 UN **Global Compact for Safe, Orderly and Regular Migration** are nonbinding international agreements that foster cooperation on displacement issues. The Global Compact is linked to Sustainable Development Goal 10 (*Reduce inequality within and among countries*), which set a target to "**facilitate orderly, safe, regular and responsible migration and mobility of people, including through the implementation of planned and well-managed migration policies**."[50] It aims to optimize the benefits of migration and minimize its risks, noting that migration can be "a source of prosperity, innovation and sustainable development."[10] The UN has also created **Guiding Principles on Internal Displacement** to avert conditions leading to displacement and to protect, support, and empower people who are displaced.[51] The **Kampala Convention**, a treaty adopted by the African Union in 2009, protects people in Africa who are internally displaced by natural or human-made disasters and calls for the prevention of internal displacement and the mitigation of root causes, including climate change.

The UN High Commissioner for Refugees (UNHCR) acknowledges that most of the nearly 60 million refugees at risk of human rights violations around the world live in climate change "hot spots." In the Office of the UNHCR is the **Special Rapporteur on the Human Rights of Internally Displaced Persons**, who, in 2020, made numerous recommendations to countries for action on climate-change-related displacement[7]:

- Enhance climate change mitigation to reduce GHG emissions to prevent human rights abuses and the conditions that lead to displacement.
- Incorporate climate change as a cause of displacement "into laws, policies, and programs on human mobility, ensure meaningful and effective participation of affected communities in decision-making, and address disproportionate impacts on vulnerable groups."[7]
- "Adopt and allocate resources for comprehensive climate change mitigation, adaptation, and risk reduction strategies

to be implemented with respect for human rights, including policies on urban planning, rural development, land use, sustainable livelihoods, and the provision of basic services . . ."[7]
- Enhance data collection and analysis on human mobility in the context of climate change.
- Create durable solutions to prevent, prepare for, and respond to disaster displacement using a human rights approach and incorporating best principles and practices for sustainable development.

UNHCR strongly supports the **Nansen Initiative**, a collaborative international project adopted by 108 countries in 2015 to create "a more coherent and consistent approach at the international level to meet the protection needs of people displaced externally owing to sudden-onset disasters."[52] The Nansen Initiative recommends (1) preventing displacement in the first place; (2) helping people move in a safe, regular, and planned manner before disasters make forced displacement inevitable; and (3) providing protection when forced displacement cannot be avoided.[52] The Nansen Initiative's successor initiative, the **Platform on Disaster Displacement**, offers guidance to protect people displaced across borders specifically because of climate change.[53]

Human displacement in the face of environmental hazards is a persistent threat all over the world. With increasing impacts of climate change, effective responses to protect public health are needed, including monitoring health outcomes in affected populations, increasing quality and access to health care, and generally guaranteeing fundamental economic, social, and cultural rights, including the right to health for all people.[9,18] Climate adaptation plans must build and support the resilient capacity of at-risk populations to reduce the need for movement away from climate-affected areas, respect the preferences of those at risk to move or not to move, and support those who are constrained in their ability to move.[21] Major GHG-emitting countries have a moral obligation to take the lead in the prevention of human displacement and assist people facing displacement threats, now and in the future.

Discussion Questions

1. Discuss the complexities of how we refer to people on the move because of climate change.
2. What puts populations at particular risk of displacement caused by climate change? Why are SIDS at highest risk?
3. What are the major climate stressors leading to displacement?
4. In what ways are metrics such as the WorldRiskIndex useful or harmful to our understanding of the complexities of human disaster displacement?

5. Discuss the major health impacts of displacement. How can adverse health impacts be prevented and beneficial health impacts be promoted?
6. Discuss examples of climate adaptations being taken by countries to lower the risk and health impacts of displacement in the face of climate change.
7. How can the issue of human displacement due to climate change be messaged effectively to catalyze strong climate mitigation and adaptation actions?

References

1. Gottesdiener L, Diaz L. "We lost everything": Central Americans flee north after back-to-back hurricanes. *Reuters*. December 4, 2020. Accessed February 14, 2021. https://www.reuters.com/article/us-usa-immigration-storms-feature/we-lost-everything-central-americans-flee-north-after-back-to-back-hurricanes-idUSKBN28E1BY

2. Lew L. Flooding hits southern China with 14 million affected. *South China Morning Post*. June 27, 2020. Accessed February 14, 2021. https://www.scmp.com/news/china/society/article/3090854/after-coronavirus-flooding-hits-southern-china-14-million

3. Hussain Z, Sharma G. Floods in India, Nepal displace nearly four million people, at least 189 dead. *Reuters*. July 19, 2020. Accessed February 14, 2021. https://www.reuters.com/article/us-india-floods/floods-in-india-nepal-displace-nearly-four-million-people-at-least-189-dead-idUSKCN24K06S

4. California Department of Forestry and Fire Protection. 2020 incident archive. 2021. Accessed January 5, 2020. https://www.fire.ca.gov/incidents/2020/

5. Internal Displacement Monitoring Centre. Global internal displacement database: IDMC query tool — disaster. Accessed February 14, 2021. https://www.internal-displacement.org/database/displacement-data

6. Internal Displacement Monitoring Centre. Internal displacement 2020: mid-year update. 2020. Accessed February 14, 2021. https://www.internal-displacement.org/sites/default/files/publications/documents/2020%20Mid-year%20update.pdf

7. Jimenez-Damary C. Report of the Special Rapporteur on the human rights of internally displaced persons. July 21, 2020. Accessed February 14, 2021. https://www.undocs.org/A/75/207

8. Gemenne F, Kaelin W, Warner K. Conceptualising policy—do "climate refugees" or "environmental

migrants" really exist? Platform on Disaster Displacement. October 7, 2020. Accessed February 14, 2021. https://disasterdisplacement.org/do-climate-refugees-or-environmental-migrants-really-exist

9. Jayawardhan S. Vulnerability and climate change induced human displacement. *Consilience: J Sustain Dev*. 2017;17:103–142. doi:10.7916/D8639VFH

10. International Organization for Migration. *World Migration Report 2020*. November 2019. Accessed February 14, 2021. https://www.un.org/sites/un2.un.org/files/wmr_2020.pdf

11. McAdam J. Seven reasons the UN Refugee Convention should not include "climate refugees." June 7, 2017. Accessed February 14, 2021. https://disasterdisplacement.org/staff-member/seven-reasons-the-un-refugee-convention-should-not-include-climate-refugees

12. Schwerdtle P, Bowen K, McMichael C. The health impacts of climate-related migration. *BMC Med*. 2018;16(1):1. doi:10.1186/s12916-017-0981-7

13. Intergovernmental Panel on Climate Change. *AR5 Climate Change 2014: Impacts, Adaptation, and Vulnerability*. 2014. Accessed January 30, 2021. https://www.ipcc.ch/report/ar5/wg2

14. International Organization for Migration, United Nations Office of the High Representative for the Least Developed Countries, Landlocked Developing Countries and Small Island Developing States. Climate change and migration in vulnerable countries: A snapshot of least developed countries, landlocked developing countries and small island developing States. 2019. Accessed January 30, 2021. https://publications.iom.int/books/climate-change-and-migration-vulnerable-countries

15. United Nations Department of Economic and Social Affairs. Least developed countries (LDCs). Accessed January 2, 2021. https://www.un.org/development/desa/dpad/least-developed-country-category.html

16. Robinson S-a. Climate change adaptation trends in small island developing states. *Mitig Adapt Strateg Global Change*. 2017;22:669–691. doi:10.1007/s11027-015-9693-5

17. Behlert B, Diekjobst R, Felgentreff C, et al. *WorldRiskReport 2020: Focus: Forced Displacement and Migration*. Bündnis Entwicklung Hilft; 2020. https://reliefweb.int/sites/reliefweb.int/files/resources/WorldRiskReport-2020.pdf

18. McMichael C, Barnett J, McMichael AJ. An ill wind? Climate change, migration, and health. *Environ Health Perspect*. 2012;120(5):646–654. doi:10.1289/ehp.1104375

19. World Health Organization. Tsunami wreaks mental health havoc. June 1, 2005. Accessed January 7, 2021. https://www.who.int/bulletin/volumes/83/6/infocus0605/en/

20. Siriwardhana C, Stewart R. Forced migration and mental health: prolonged internal displacement, return migration and resilience. *Int Health*. 2013;5(1):19–23. doi:10.1093/inthealth/ihs014

21. McMichael C. Climate change-related migration and infectious disease. *Virulence*. 2015;6(6):548–453. doi:10.1080/21505594.2015.1021539

22. Strauss B, Kulp S, Levermann A. Mapping choices: carbon, climate and rising seas, our global legacy. November 2015. Accessed January 2, 2021. https://sealevel.climatecentral.org/uploads/research/Global-Mapping-Choices-Report.pdf

23. Tabe T. Climate change migration and displacement: learning from past relocations in the Pacific. *Soc Sci*. 2019;8:218. doi:10.3390/socsci8070218

24. Perumal N. The place where I live is where I belong: community perspectives on climate change and climate-related migration in the Pacific island nation of Vanuatu. *Inf Syst J*. 2018;13:45–64. doi:10.24043/isj.50

25. Walker B. An island nation turns away from climate migration, despite rising seas. *Inside Climate News*. November 20, 2017. Accessed January 7, 2021. https://insideclimatenews.org/news/20112017/kiribati-climate-change-refugees-migration-pacific-islands-sea-level-rise-coconuts-tourism/

26. Pala C. Kiribati's president's plans to raise islands in fight against sea-level rise. *The Guardian*. August 9, 2020. Accessed January 7, 2021. https://www.theguardian.com/world/2020/aug/10/kiribatis-presidents-plans-to-raise-islands-in-fight-against-sea-level-rise

27. Dauenhauer NJ. On front line of climate change as Maldives fights rising seas. *New Scientist*. March 20, 2017. Accessed January 7, 2021. https://www.newscientist.com/article/2125198-on-front-line-of-climate-change-as-maldives-fights-rising-seas/

28. Sou G. Barbudans are resisting "disaster capitalism," two years after Hurricane Irma. *The Conversation*. July 17, 2019. Accessed January 7, 2021. https://theconversation.com/barbudans-are-resisting-disaster-capitalism-two-years-after-hurricane-irma-119368

29. United Nations Conference on Trade and Development. For heavily indebted small islands, resilience-building is the best antidote. January 7,

2021. Accessed January 7, 2021. https://unctad.org /news/heavily-indebted-small-islands-resilience -building-best-antidote?s=03

30. Nguyen S. Vietnamese pick up the pieces after 2020's relentless storms. *Al Jazeera*. December 31, 2020. Accessed January 8, 2021. https://www .aljazeera.com/news/2020/12/31/vietnam-storms

31. Bich TH, Quang LN, Ha LTT, et al. Impacts of flood on health: epidemiologic evidence from Hanoi, Vietnam. *Global Health Action*. 2011;4:6356–6356. doi:10.3402/gha.v4i0.6356

32. International Organization for Migration. Migration, resettlement and climate change in Viet Nam: reducing exposure and vulnerabilities to climatic extremes and stresses through spontaneous and guided migration. 2014. Accessed January 8, 2021. https://www.undp.org/content/dam/vietnam/docs /Publications/Migration%20&%20Climate%20 change%20-%20Eng.pdf

33. Vu N, Schroll H, Andersen J, Lund S. Is climate change a reality for agriculture in Quang Nam Province? In: Bruun O, Casse T, eds. *On the Frontiers of Climate and Environmental Change*. Springer; 2013:43–69. doi:10.1007/978-3-642-35804-3_3

34. Chowdhury MA, Hasan MK, Hasan MR, et al. Climate change impacts and adaptations on health of internally displaced people (IDP): an exploratory study on coastal areas of Bangladesh. *Heliyon*. 2020; 6(9):e05018. doi:10.1016/j.heliyon.2020.e05018

35. McPherson P. Dhaka: the city where climate refugees are already a reality. *The Guardian*. December 1, 2015. Accessed January 8, 2021. https://www.theguardian.com/cities/2015/dec/01 /dhaka-city-climate-refugees-reality

36. Williams S. "The sea is rising, the climate is changing": the lessons learned from Mozambique's deadly cyclone. *The Guardian*. January 2, 2021. Accessed January 8, 2021. https://www.theguardian .com/world/2021/jan/02/the-sea-is-rising-the -climate-is-changing-the-lessons-learned-from -mozambiques-deadly-cyclone

37. Centers for Disease Control and Prevention. Improving health for Kenya's refugees by building laboratory capacity. January 2, 2015. Accessed January 8, 2021. https://www.cdc.gov/globalhealth /stories/dgmq_story.htm

38. United Nations High Commissioner for Refugees. UNHCR responds to public health threats in Dadaab refugee camps. Updated September 28, 2012. Accessed January 8, 2021. https://www .unhcr.org/news/briefing/2012/9/506578369/unhcr

-responds-public-health-threats-dadaab-refugee -camps.html

39. Médecins Sans Frontières. Kenya: in the shadow of COVID-19, a growing mental health crisis in Dadaab refugee camp. Updated October 9, 2020. Accessed January 8, 2021. https://www .doctorswithoutborders.org/what-we-do/news -stories/story/kenya-shadow-covid-19-growing -mental-health-crisis-dadaab-refugee

40. Government Accountability Office. Alaska Native villages: limited progress has been made on relocating villages threatened by flooding and erosion. June 2009. Accessed January 9, 2021. https://www.gao.gov/new.items/d09551.pdf

41. Early W. Kivalina emergency access road now open for use. *Alaska Public Media*. November 30, 2020. Accessed January 9, 2021. https://www .alaskapublic.org/2020/11/30/kivalina-emergency -access-road-now-open-for-use/

42. Brubaker M, Berner J, Chavan R, et al. Climate change and health effects in Northwest Alaska. *Global Health Action*. 2011;4(1):8445. doi:10.3402 /gha.v4i0.8445

43. Brubaker M, Berner J, Bell J, Warren J. Climate change in Kivalina, Alaska: strategies for community health. January 2011. Accessed January 9, 2021. https://anthc.org/wp-content/uploads/2016/01 /CCH_AR_012011_Climate-Change-in-Kivalina.pdf

44. Brubaker M, Bell J, Dingham H, et al. Climate change in Wainwright, Alaska: strategies for community health. June 2014. Accessed January 9, 2021. https://anthc.org/wp-content/uploads/2016 /01/CCH_AR_062014_Climate-Change-in -Wainwright.pdf

45. Brubaker M, Flensburg S, Shanigan N, Skarada J. Climate change in Pilot Point, Alaska: strategies for community health. September 2013. Accessed January 9, 2021. https://anthc.org/wp-content /uploads/2016/01/CCH_AR_092013_Climate -Change-in-PilotPoint.pdf

46. Hofstaedter E. Amid an erosion crisis, Shishmaref takes small steps toward expansion. *KNOM Radio Mission*. June 28, 2019. Accessed January 9, 2021. https://www.knom.org/wp/blog/2019/06/28/amid -an-erosion-crisis-shishmaref-takes-small-steps -toward-expansion/

47. Kim G. Former residents of erosion-threatened Newtok say they're healthier after a year in new village. *Anchorage Daily News*. August 4, 2020. Accessed January 9, 2021. https://www.adn.com /alaska-news/rural-alaska/2020/08/04/former

-residents-of-erosion-threatened-newtok-say-theyre
-healthier-after-a-year-in-new-village/

48. Schwartz RM, Gillezeau CN, Liu B, et al. Longitudinal impact of Hurricane Sandy exposure on mental health symptoms. *Int J Environ Res Public Health*. 2017;14(9):957. doi:10.3390/ijerph 14090957

49. Gopal P. America's great climate exodus is starting in the Florida Keys. *Bloomberg*. September 20, 2019. Accessed January 10, 2021. https://www .bloomberg.com/news/features/2019-09-20 /america-s-great-climate-exodus-is-starting-in-the -florida-keys

50. United Nations. The 17 Goals. Accessed January 11, 2021. https://sdgs.un.org/goals

51. Global Protection Cluster. 20th anniversary of the Guiding Principles on Internal Displacement: a plan of action for advancing prevention, protection and solutions for internally displaced people 2018–2020. May 23, 2018. Accessed January 2, 2021. https://www.globalprotectioncluster.org /_assets/files/20180523-gp20-plan-of-action-final .pdf

52. United Nations High Commissioner for Refugees. UNHCR, the environment and climate change. October 2015. Accessed January 2, 2021. https:// www.unhcr.org/en-us/540854f49

53. Platform on Disaster Displacement. Accessed January 6, 2021. https://disasterdisplacement.org/

CHAPTER 12

Health Professions and the Health Sector

"Family physicians have an ethical obligation to address the health effects of climate change to safeguard the health and well-being of disenfranchised people, our children and grandchildren, and future generations."

—**Dr. Caroline Wellbery**[1]

LEARNING OBJECTIVES

- Understand the predominant attitudes and knowledge of health professionals about climate change and its impacts on their patients' health.
- Understand which health conditions physicians are seeing in their patients that they attribute to climate change.
- Understand the roles that health professionals can play in the climate movement.
- Describe the environmental impacts of the healthcare sector.

The health effects of climate change are being witnessed and experienced firsthand in front-line communities all over the world. They are also being observed by health professionals when people seek medical care for conditions that may be connected to climate change. For example, in June 2021, a staggering heat wave in the Pacific Northwest sent thousands of people to the hospital for heat-related illnesses.[2] Dr. Renee Salas, emergency medicine physician at Massachusetts General Hospital, sees the links to climate change in many of her patients' health concerns. One man came to her emergency department (ED) 30 times in a single year, suffering from persistent and debilitating symptoms of Lyme disease, a climate-sensitive tick-borne disease spreading in the United States and Canada.[3] Children regularly visit her ED suffering from asthma exacerbations, which are strongly linked to air pollutant and aeroallergen exposures that are rising with climate change. Said Dr. Salas, "My job as an emergency medicine doctor is to protect my patients and keep them healthy. Climate change increasingly threatens my ability to do that."[3]

Health professionals—including primary care and emergency medicine doctors and nurses, allergists, infectious disease specialists, respiratory therapists, dieticians, psychiatrists, public health practitioners, and social workers—have important roles to play in preventing and treating the health impacts of climate change. According to bioethicist Dr. Wendy Rogers, climate change is "squarely in the medical domain, making a case for attributing role responsibility to doctors. Doctors' relevant expertise on the health impacts of the climate crisis makes them effective actors. On my understanding of moral responsibility in medicine, we all have a duty to act."[4]

Health professionals are seeing climate-sensitive illnesses become more common, but most are not trained to make the link to climate change or to communicate these connections to their patients. According to Dr. Jay Lemery, associate professor of emergency medicine at the University of Colorado School of Medicine, "There are so many really brilliant, smart clinicians who have no clue."[5] To fulfill this critical role, health professionals need information and tools to address climate change as a complex health determinant relevant in their clinical practices, to communicate with the public, and to advocate effectively on climate policy-making. "In the end, action on climate change is a prescription for improved health and equity," said Dr. Salas.[3]

Climate action also saves on healthcare costs. A 2021 report estimated that the health-related costs of climate change in the United States may be $1 trillion per year, and are likely disproportionately borne by low-income Americans, people of color, the elderly, children and their parents, pregnant women, and Indigenous peoples.[2] These costs include over $800 billion for diseases linked to fine particulate matter exposure, $16 billion for

wildfire smoke exposure, $8 billion for ozone exposure, $1 billion for Lyme disease, and many billions for hurricanes that make landfall.[6] Included in these figures are direct costs for medical care, prescription drugs, rehabilitation, therapy and home health services, and lost wages and worker productivity due to morbidity and premature deaths.

General and family practice doctors and nurses are on the frontlines of observing climate-related health impacts in their patients. In addition, many of the health impacts of climate change are seen first in hospital EDs, and emergency medical care contributes significantly to overall costs. Extreme heat and weather result in spikes in ED visits for heat-related illnesses, asthma, and traumatic injuries, as well as more frequent use of EDs by people suffering from chronic diseases for which ongoing medical care is disrupted. After Hurricane Sandy in 2012, hospital ED visits surged for complications of heart disease and diabetes, due in part to limited access to medications, treatments, and medical providers during and immediately after the storm.[7] Climate change raises the risk of disruptions in healthcare operations, and in Hurricane Sandy, many New York City hospitals lost power or were flooded and had to close and evacuate patients to other facilities.[8]

Emergency clinicians have an important role to play to ensure that patients protect themselves from climate-related exposures and take precautions once they leave the ED. For example, the elderly and those suffering from heat-related illnesses will likely have better outcomes if they are informed about cooling measures and given access to a "buddy system" to make sure someone is checking up on them.[8] Patients with respiratory illnesses should be given information to protect themselves against exposure to polluted air or wildfire smoke. In regions where vector-borne diseases are endemic, clinicians may need to communicate risks and vector-protective strategies, such as wearing long pants or using mosquito nets.

Emergency medicine tends to provide a safety net for healthcare systems. Emergency personnel are well positioned to recognize impacts on vulnerable populations that more frequently visit the ED for health care and the ways that these groups may be burdened by widening health inequities magnified by climate change. EDs are also the primary access point for mental health care, and in the United States, at least 1 in 8 ED visits are mental health emergencies.[8] EDs are often the main access point in health systems for populations displaced by disasters and forced to relocate, even temporarily, away from their places of residence. EDs are also important sites for data collection on disease incidence, morbidity, and mortality, as well as evidence for possible links to localized climate conditions.[8]

In general, Americans place a great deal of faith in their doctors and list them first as trusted sources of information about the health impacts of climate change (**Table 12.1**).[9] Dr. Mary B. Rice, a pulmonologist in Boston, Massachusetts, recently wrote, "As physicians, we offer a clinical perspective that no climate expert can claim. With this expertise and perspective comes a certain amount of trust from the public and our peers and a responsibility to 'weigh-in' on such an important public health issue for our patients and their families."[10]

In a recent survey of family practice doctors and their patients in Wisconsin, patients reported higher *trust* in doctors and scientists than in social media for information about environmental issues, even though they chose "news outlets," "social media," and "family or friends" as their top *sources* of information.[11]

Table 12.1 Percent of Americans Who Trust Various Sources of Information about the Health Impacts of Climate Change (*n* = 1,275).

Source	Strongly/ Moderately Trust
Primary care doctor	49%
Family and friends	41%
U.S. Centers for Disease Control and Prevention (CDC)	41%
Climate scientists	40%
World Health Organization (WHO)	37%
U.S. Environmental Protection Agency (EPA)	34%
Other scientists	33%
Local public health department	33%
Environmental groups	31%
TV weather reporters	25%
Religious leaders	24%
Military leaders	22%

Data from Maibach EW, Kreslake JM, Roser-Renouf C, et al. Do Americans understand that global warming is harmful to human health? Evidence from a national survey. *Ann Glob Health*. 2015;81:396–409. doi:10.1016/j.aogh.2015.08.010

Perspectives of Health Professionals

Which climate-related health problems are people facing today? How do they differ by region or by patient population? Are these problems getting worse over time? Clinical health professionals may be best able to answer these questions because they have a key vantage point that most people lack. Several recent surveys collected information on clinical perspectives and climate change knowledge and attitudes among physicians, offering clues about how human health is currently being impacted and how health systems can be made more responsive to people's well-being in the climate crisis.

Members of the U.S.-based **National Medical Association (NMA)** and **American Thoracic Society (ATS)** were among the first physicians to be surveyed about climate change. NMA, founded in 1895, is the oldest and largest national medical organization, representing more than 30,000 Black physicians and their patients and is "a leading force for parity and justice in medicine and the elimination of disparities in health."[12] ATS was founded in 1905 as a professional organization for physicians treating people with tuberculosis. It has grown into an international society with more than 16,000 members doing clinical care and scientific research on pulmonary diseases and breathing disorders. More than half of NMA study participants practiced primary care, and most ATS respondents were pulmonary specialists.[13] ATS has both U.S.-based and international members (representing 68 countries on all continents), who were surveyed separately to distinguish differences in attitudes or knowledge based on medical practice location. These three surveys were conducted in 2014, 2015, and 2016, and even though climate change impacts, research, policy action, and media coverage have expanded since then, knowledge and attitudes were similar to those of two more recent surveys of physicians in two U.S. states, Wisconsin and Maine.[11,14]

In general, physicians recognize that climate change is impacting the health of their patients (**Table 12.2**). International members of ATS had the highest proportion (69%) responding affirmatively about climate change

affecting their patients at least a "moderate amount." This may be because these doctors specialize in treating highly climate-sensitive pulmonary diseases, and they practice in countries around the world that may have higher climate-attributable disease burdens than the United States. A majority of NMA members, who treat large numbers of people of color and low-income patients who may suffer more from chronic diseases, also responded affirmatively.[13] Fewer U.S. ATS members and Maine doctors reported being fairly certain that they are witnessing impacts in their patients, but 66% of Maine doctors reported being "extremely" or "somewhat" *concerned* about how climate change is currently affecting their patients' health.[14]

Physicians were asked which of their patients' health conditions were being affected by climate change. NMA and ATS members listed as their top concern diseases increasing in severity because of air pollution, followed by injuries from disasters, and allergies (**Table 12.3**). The most notable difference among these three physician groups was that NMA members and international ATS members were more likely than U.S.-based ATS members to identify heat-related effects and food- and waterborne diarrheal diseases as climate-sensitive. This result may be explained because NMA and international ATS members treat patients who may have higher rates of heat impacts, particularly in the absence of access to cooling or other preparedness measures, and diarrheal diseases that are due to unsafe water, sanitation, and hygiene. Surveying physicians in different countries and those who treat different populations provides an informative snapshot of regional and local climate impacts.

In a 2020 survey of Maine physicians, the most common illnesses reported to be linked to climate change were asthma, vector-borne diseases, chronic obstructive pulmonary

Table 12.2 **Percentage of Physicians Reporting That Climate Change Impacts Their Patients' Health. The relevant question or statement was phrased differently in the five surveys reported here.**

Survey Statement/Question	Survey Responses				
	NMA	ATS (U.S.)	ATS (Int'l)	Wisconsin	Maine
Climate change is affecting the health of my patients a "great deal" or a "moderate amount."	61%	44%	69%		
Have you witnessed climate change impacts on your patients' health?				Yes 64%	Yes 38% Maybe 15% Not sure 30%
How concerned are you about the current health impacts of climate change in your patients?					Extremely 13% Moderately 53% Somewhat 27%

Note. NMA = National Medical Association; ATS = American Thoracic Society.

Data from Boland TM, Temte JL. Family medicine patient and physician attitudes toward climate change and health in Wisconsin. *Wilderness Environ Med.* 2019;30(4):386–393. doi:10.1016/j.wem.2019.08.005; Sarfaty M, Mitchell M, Bloodhart B, Maibach EW. A survey of African American physicians on the health effects of climate change. *Int J Environ Res Public Health.* 2014;11(12):12473–12485. doi:10.3390/ijerph111212473; Andersen M, Carlson G. Perspectives of Maine physicians on the health impacts of climate change. 2021. Unpublished data; Sarfaty M, Bloodhart B, Ewart G, et al. American Thoracic Society member survey on climate change and health. *Ann Am Thorac Soc.* 2015;12(2):274–278. doi:10.1513/AnnalsATS.201410-460BC; Sarfaty M, Kreslake J, Ewart G, et al. Survey of international members of the American Thoracic Society on climate change and health. *Ann Am Thorac Soc.* 2016;13(10):1808–1813. https://www.ncbi.nlm.nih.gov/pmc/articles/PMC6944384

Table 12.3 **Percentage of Physicians Reporting Specific Health Conditions Linked to Climate Change In Their Patients.**

"In which of the following ways, if any, do you think your patients are currently being affected by climate change?"	Survey Responses		
	NMA	ATS (U.S.)	ATS (Int'l)
"Air pollution-related increases in severity of illness"	88%	77%	88%
"Injuries due to severe storms, floods, droughts, fires"	88%	57%	69%
"Increased care for allergic sensitization and symptoms of plant/mold exposure"	80%	58%	72%
"Heat-related effects"	75%	48%	70%
"Vector-borne infection"	58%	40%	59%
"Diarrhea from food/waterborne illnesses"	56%	26%	55%

Note. NMA = National Medical Association; ATS = American Thoracic Society.

Data from Sarfaty M, Mitchell M, Bloodhart B, Maibach EW. A survey of African American physicians on the health effects of climate change. *Int J Environ Res Public Health.* 2014;11(12):12473–12485. doi:10.3390/ijerph111212473; Sarfaty M, Bloodhart B, Ewart G, et al. American Thoracic Society member survey on climate change and health. *Ann Am Thorac Soc.* 2015;12(2):274–278. doi:10.1513/AnnalsATS.201410-460BC; Sarfaty M, Kreslake J, Ewart G, et al. Survey of international members of the American Thoracic Society on climate change and health. *Ann Am Thorac Soc.* 2016;13(10):1808–1813. https://www.ncbi.nlm.nih.gov/pmc/articles/PMC6944384

disease (COPD), allergies, and mental health (**Figure 12.1**).[14] Respondents to this survey practiced more than 20 medical specialties, with the most common being family medicine (22% of respondents), pediatrics (15%), internal medicine (11%), and emergency medicine (8%). In 2019, doctors in a family medicine practice in Madison, Wisconsin, responded with a very similar list of illnesses of highest concern: allergies and lung diseases, followed by heat stroke, vector-borne diseases, and mental health.[11] In all surveys, respondents reported that health impacts will be worse in the future than today.

Anecdotes from clinicians reveal specific details about what they are seeing in their patients today (**Table 12.4**), with conditions ranging from heat stroke and chronic respiratory illnesses to worsened allergies, vector-borne diseases, and anxiety. Responses include those from a survey of members of the **American Academy of Allergy, Asthma, and Immunology (AAAAI)**. One AAAAI member noted that everyone is seeing climate change affecting patients, and "if you have not seen it, your eyes are not open."[17]

Physicians were asked which groups were most likely to be disproportionately impacted by climate change. Most identified people with chronic diseases as the most vulnerable (75–88%), followed by children under 5 (66–83%), people living

(a)

(b)

(c)

(d)

Figure 12.1 Climate-Attributable Diseases Identified by Physicians. **(a)** Asthma; **(b)** Seasonal allergies; **(c)** Climate anxiety; **(d)** Heat stroke.

Table 12.4 **Selected Anecdotes from Clinicians about Patient Experiences with Climate Change-Related Health Problems. All entries are direct quotes.**

Survey	Anecdote
NMA members (2014)	With the aging of the population, the incidence of heat stroke has risen in my practice.
	Extreme weather (heat and dry climate) is causing heat stroke and brush fires, with subsequent smog and worsening of asthma symptoms.
	My patient experienced atrocities during Hurricane Katrina. As a result, she had PTSD and severe depression that prevented her from holding a stable job.
	The severe weather and snow have limited patient access to the doctor.
ATS members (U.S.) (2014)	Many of my patients with chronic lung diseases report increased symptoms on high pollution days, particularly when there are wildfires in close proximity to urban areas.
	Several of my patients have remarked on earlier and longer allergy seasons leading to worse asthma control.
	Tick bite in Vermont developing *erythema migrans* (rash). There were not ticks until a few years ago in Southern Vermont.
	In Florida, it has been raining and has caused the incidence of dengue.
ATS members (International) (2015)	Frequency and severity of bronchial asthma, COPD, and related cardiovascular diseases are increasing and causing increased hospital bed occupancy as well as health-related budgets. (Bangladesh)
	Change in the classic characteristic of the seasons makes winters warmer with abrupt change of humidity and temperature, which apparently increased the rate of COPD and asthma exacerbations. (Argentina)
	Peak summer temperatures have gone up noticeably and stayed up for longer periods of time. As a result, those people who have not installed air conditioners in their houses find it difficult to fall asleep and stay asleep at night in the summer months. In our practice, we have seen an increase in the frequency and intensity of complaints relating to the lack of sleep and poor sleep quality. (Northern Canada)
	More flooding, more malnutrition and ill-health. (Nigeria)
AAAAI members (2015)	The onset [of] symptoms due to ash tree pollination used to be early February; now patients are becoming symptomatic in early January. (Southern California)
	Numerous patients with fall mold allergies [have] symptoms now last well into December because the ground takes longer to freeze. (Michigan)
	A combination of high automobile pollution with heat, humidity, and high pollen produced not only nasal allergy and wheeze, but also very severe redness, itch, and eye irritation in August. (Washington, DC)
	We have all seen increasing pollen and pollution levels affecting our patients—if you have not seen it, your eyes are not open. (Illinois)

(continues)

Table 12.4 Selected Anecdotes from Clinicians about Patient Experiences with Climate Change-Related Health Problems. All entries are direct quotes.   *(continued)*

Survey	Anecdote
Maine physicians (2020)	Anxiety, depression, and breathing issues from abnormal heat and humidity.
	More Lyme disease, directly attributable to warming of the Maine climate, as the vector moves north.
	Inability to exercise/stay active in extreme heat.
	I see many children who have a great deal of anxiety and in adolescents, they frequently verbalize worry specifically about climate change.

Note. NMA = National Medical Association; PTSD = Posttraumatic stress disorder; ATS = American Thoracic Society; AAAAI = American Academy of Allergy, Asthma, and Immunology.

Data from Sarfaty M, Mitchell M, Bloodhart B, Maibach EW. A survey of African American physicians on the health effects of climate change. *Int J Environ Res Public Health.* 2014;11(12):12473–12485. doi:10.3390/ijerph111212473; Andersen M, Carlson G. Perspectives of Maine physicians on the health impacts of climate change. 2021. Unpublished data; Sarfaty M, Bloodhart B, Ewart G, et al. American Thoracic Society member survey on climate change and health. *Ann Am Thorac Soc.* 2015;12(2):274–278. doi:10.1513/AnnalsATS.201410-460BC; Sarfaty M, Kreslake J, Ewart G, et al. Survey of international members of the American Thoracic Society on climate change and health. *Ann Am Thorac Soc.* 2016;13(10):1808–1813. https://www.ncbi.nlm.nih.gov/pmc/articles/PMC6944384; Sarfaty M, Kreslake JM, Casale TB, Maibach EW. Views of AAAAI members on climate change and health. *J Allergy Clin Immunol Pract.* 2016;4(2):333–335.e26. doi:10.1016/j.jaip.2015.09.018

in poverty (64–86%), and adults over 60 (63–80%). Interestingly, 73% of NMA respondents, who are Black physicians practicing in the United States, identified people of color as another at-risk group, in contrast to only 16% and 27% of international and U.S.-based ATS respondents, respectively. Among Maine doctors, most identified people with chronic diseases, children, and older adults as most vulnerable, but some also listed Black and Indigenous communities, people of color, and workers in outdoor industries.[14] A large portion of Maine's workforce is engaged in fishing, forestry, agriculture, and recreation.

When asked if they have *personally* experienced climate change outside their role as a health professional, 61% of international ATS doctors, 48% of U.S.-based ATS doctors, and 38% of NMA doctors responded "a great deal" or "a moderate amount."[13,15,16] A higher percentage of international clinicians responded affirmatively, which may reflect that climate change impacts in countries other than the United States have been better recognized to date. For comparison, in 2014, only 35% of the American public reported directly experiencing climate change.[13] Nearly three-quarters of AAAAI survey respondents indicated that they personally experienced climate change to any extent.[17] Among Maine physicians, only 26% reported personally experiencing climate change, with an additional 16% saying maybe, 43% saying no, and 14% not sure.[14] In a study of Wisconsin family practice doctors, 89% reported that climate change is currently affecting their *community*, although they were not asked if they personally impacted.[11]

In the NMA, ATS, and Wisconsin studies, most doctors reported at least a "modest" or "moderate" level of knowledge about how climate change impacts human health (71–94%). Among members of AAAAI, only

43% felt "very" or "moderately" knowledgeable about climate change and health.[17] In the NMA and ATS surveys, 62–80% of respondents "agreed" or "strongly agreed" that physicians have a responsibility to bring the health effects of climate change to the attention of their patients. However, they also reported significant barriers to doing so, primarily "lack of time" and "lack of knowledge regarding how to approach the issue with patients."[13,16] Other barriers included "addressing these issues with patients will not make much difference in their overall health," and "patients [are] not interested or knowledgeable enough about climate impacts to discuss the issue."[15] Doctors in Wisconsin and Maine reported low frequency of addressing this issue with patients. Only 17% of Wisconsin doctors were "extremely" or "somewhat" comfortable talking to their patients, and only 25% of Maine doctors "frequently" or "sometimes" talked to their patients about climate change.[11,14] These responses suggest a gap in knowledge and practice that could be filled with better training. Most respondents in the NMA and ATS surveys (73–89%) believed that climate change and its health impacts should be taught in medical schools.

Most NMA and ATS doctors reported that they "agree" or "strongly agree" that physicians and medical societies have a responsibility to advocate for climate action (74–85%), but only 31% of doctors in the Wisconsin survey "agreed" or "strongly agreed" that physicians should actively address climate change. At least 80% of NMA and ATS doctors believed that the health sector should take action to become more sustainable, but only about 30% felt that their place of work was doing an effective job reducing its fossil fuel consumption, and fewer than half reported that their hospital was well prepared for climate-related disasters.

A recent survey of students in medical, nursing, and physician assistant programs at Yale University found that 94% agreed or strongly agreed with the statement, "I am concerned about the health impacts of climate change." Most (90%) also agreed or strongly agreed that health professionals "have a responsibility to conserve resources and prevent pollution within their professional practice" (**Table 12.5**).[18] A majority of students responded that it is important for them to understand the health impacts of climate change, and most want this subject covered in the classroom. Female students were more than twice as likely as male students to believe that climate change is "pertinent to patient care" and that they need to understand it to help their patients. Physician assistant students were much more likely than medical students to agree that they have an "important role to play in educating patients and the public about the impacts of pollution and climate change on health." These perspectives are particularly important given that today's students training for health professions are part of a younger generation that will be significantly impacted by climate change, and they have a critical role to play in climate action in society and in the health sector.[19]

If decision-makers and the public are made aware of this strong level of concern from physicians about climate change impacts on people's health, it may spur more rapid and effective climate action, including measures to mitigate greenhouse gas (GHG) emissions, build climate adaptation capacity, and strengthen resilience and equity in health systems. That health professionals are seen as trusted authorities, with specific observations of and stories from their patients, makes them critically important to help develop effective measures to protect health and the climate (**Box 12.1**).

Table 12.5 Attitudes about Climate Change from a Survey of Health Professions Students at Yale University.

Statement	Strongly Agree/Agree
I am concerned about the health impacts of climate change.	94%
It is important to understand this issue so I can help my patients.	67%*
It is important to understand this issue because [we] have an important role to play in educating patients and the public about the impacts of pollution and climate change on health.	77%†
I feel we should cover this issue in the classroom, and it should be reinforced in the clinical setting.	63%
Health professionals have a responsibility to conserve resources and prevent pollution within their professional practice.	90%

* Females more likely to agree than males
† Physician assistant students more likely to agree than medical students

Data from Ryan EC, Dubrow R, Sherman JD. Medical, nursing, and physician assistant student knowledge and attitudes toward climate change, pollution, and resource conservation in health care. *BMC Med Educat.* 2020;20:200. doi:10.1186/s12909-020-02099-0

Box 12.1 Florida Physicians Advocate for Climate Action

Courtesy of Florida Clinicians for Climate Action.

Dr. Cheryl Holder is an internal medicine specialist who has spent decades treating patients in South Florida, many of whom are poor, uninsured, or homeless.[20] She is also an associate professor of medicine at Florida International University's Herbert Wertheim College of Medicine in Miami

and Associate Dean for Diversity, Equity, Inclusivity, and Community Initiatives at the college. She cochairs **Florida Clinicians for Climate Action** (FCCA), which Dr. Holder founded after seeing patients with worsening breathing ailments no longer effectively controlled with medications. Dr. Holder and other physicians in the FCCA network make the health link to climate change with patients and the public.

Dr. Holder was also seeing patients with exacerbated allergies due to longer pollen seasons, and more heat-related illnesses, including dehydration and kidney disease, and deaths from record high heat and humidity. Patients without air conditioning (AC) reported sleeping difficulties, and those with AC often used old, moldy units, another health risk.

One of Dr. Holder's patients is Jorge, a man in his 70s from Ecuador who sells fruit on the street to earn money to send to his family back home. He also must cover his own medical expenses due to suffering from diabetes and cancer. He lives in a shed with no running water, only one window for air flow, and no screens to keep mosquitoes out. People like Jorge who work outside are at high risk of exposure to heat, humidity, and rainfall, and Jorge has noticed that it's getting hotter in Miami. "When you work in the streets, you really feel it, [but] rain or shine, cold or heat, I have to work."[20] Jorge has been able to adapt by staying in the shade as much as possible, taking breaks, and wearing protective clothing to block the sun. When he can afford it, he uses a small AC unit at night if he cannot sleep in the heat.

The city of Miami is taking steps to build climate resilience because of its extreme vulnerability to hurricanes and sea level rise. Miami-Dade County just installed a new Chief Heat Officer. One measure that may benefit outdoor workers like Jorge is that the city is planting trees to reduce the urban heat island effect. But Dr. Holder knows that much more is needed. "I hear a lot more about sea level rise and raising the sidewalks and

replenishing the beaches. But it's going to be very, very difficult for the poor populations. I don't know how they're going to survive."[20]

FCCA recognizes that clinicians are trusted messengers for patients, fellow health professionals, decision-makers, and the public, and that they therefore play an important role in protecting public health from the effects of climate change.[21] FCCA aims to recruit clinicians—physicians, nurses, and pharmacists—from all over Florida to engage with, educate people about, and advocate for climate action that protects public health. In 2018, they issued the Tampa Declaration on Climate and Health, a call to action stating that climate change is impacting the health and well-being of Florida residents today—impacts that will get worse in the future. FCCA stresses that the state must reduce its GHG emissions to tackle heat waves, sea level rise, hurricanes, and vector-borne diseases. Joining Florida are more than a dozen other states that have formed Clinicians for Climate Action groups.

Climate Change Education

Many health professionals have noted that medical school curricula lack content about the health impacts of climate change and have not kept pace with scientific advances.[19] In fact, a recent review of medical school curricular content revealed that there was little mention of climate change at all.[22] In 2019, the **American Medical Association** approved a policy to require inclusion of topics related to the health impacts of climate change in medical education at all levels.[23] In addition, many people advocate for training more diverse health professionals, especially people of color, recognizing that health outcomes for Black patients, for example, are better if they see Black doctors.[24]

The U.S. **National Institute of Environmental Health Sciences** has created climate and health learning modules at various literacy levels that target high school and college students, along with public health and clinical health students and practitioners.[25] Content is based largely on the U.S. Global Change Research Program's 2016 climate and health assessment.[26] These learning modules align with educational standards and competencies described by the American Association of Graduate Medical Education, the American Osteopathic Association, the American Association of Medical Colleges, the Association of Schools and Programs of Public Health, the United Nations Framework Convention on Climate Change, and the World Health Organization. In addition, the U.S.-based **Center for Climate Change**

and Health has created a helpful online resource for health professionals, *A Physician's Guide to Climate Change, Health and Equity.*[27]

The **American Academy of Pediatrics** (AAP) issued a policy statement on climate change in 2015 that made six recommendations for pediatricians' roles and responsibilities and the importance of educating health professionals about the impacts on children.[28]

1. Include content about climate change and its impacts on child health in medical education.
2. Actively work to reduce the carbon footprint and environmental impacts of medical facilities and transport services.
3. Discuss climate change with patients and their families, and encourage healthier and more climate-friendly choices for diet, energy use, and transportation. Importantly, pediatricians can serve as role models for their patients for healthy and sustainable practices.
4. Discuss emergency and disaster preparedness with patients, their families, and community members.
5. Educate decision-makers and the public about the impacts of climate change on child health, and advocate for climate action at local, national, and international levels.
6. Participate in disease surveillance and research on climate-sensitive impacts on child health.

The **Global Consortium on Climate and Health Education** was launched in early 2017 in response to a call at COP 21 in Paris, France, for greater investment in research and planning for the health impacts of climate change. The mission of this consortium, hosted at Columbia University in New York City, is "to secure commitments from all health professions schools around the world to educate their students on the health impacts of climate change and other planetary changes that impact human health and well-being, and to provide the curricular resources and guidance needed to implement those commitments."[29] The Consortium developed a set of Core Climate & Health Competencies for Health Professionals, with the following goals:

- "Demonstrate an understanding of the complex relationships between climate change and health"
- "Demonstrate competence in recognizing population-based hazards, and designing and implementing public health interventions"
- "Demonstrate competence in diagnosis and management of climate-sensitive and climate-induced illness and management of healthcare facilities"
- "Demonstrate familiarity with international and domestic policies relevant to climate change and health"
- "Demonstrate competence on how to communicate health and climate information to different audiences"[29]

In 2017, the University of Colorado School of Medicine, under the guidance of Dr. Jay Lemery, launched the first climate-focused medical fellowship.[30] Initially, one emergency medicine physician was awarded the fellowship each year. In their ED rotation, they paid particular attention to climate-sensitive illnesses and deaths, and they also visited frontline communities and learned about climate policymaking. In 2021, five National Climate & Health Science Policy Fellows were selected who will work in partnership with U.S. government agencies and nonprofit organizations to amplify work on climate and health.

In 2019, U.S. medical students formed a new collective, **Medical Students for a**

Sustainable Future, with chapters at dozens of medical schools around the United States. These students "recognize climate change as an urgent threat to health and social justice," and their mission is to "catalyze action to prevent and address the health harms of climate change."[31] One of their recent activities was hosting a summer climate change policy series to bring medical students and experts together to discuss climate action.

Because science and policy are constantly advancing, these curricula will need to be iterative and also focus not just on specific diseases or health outcomes but also on health inequities, climate justice principles and practices, and effective communications with a range of groups, including the public, patients, and decision-makers. Framing climate change as fundamentally about human health will invite a wide spectrum of people to get involved and take action.[22] Even children can be empowered with knowledge about how climate change is impacting them, their communities, and the world, which may help combat climate anxiety, help children make decisions about how they might respond to disasters such as hurricanes or wildfires, and create opportunities for greater youth engagement in the climate movement.

Climate Advocacy

A 2021 editorial in the medical journal *The Lancet* noted that "the health profession has an obligation to advocate for health comprehensively, and political involvement is likely to be necessary."[32] Health professionals often play a key role to advocate for policies that safeguard their patients' health. With respect to climate change, this advocacy includes lobbying decision-makers at local, state, national, and international levels on the necessity of health-protective climate action, developing consensus statements on climate change written by professional groups, and encouraging climate-friendly changes to their practices or the health sector generally.[23]

Many health professionals accept this advocacy responsibility. For example, in 2019, representatives from the **American College of Physicians (ACP)** testified before the U.S. Congress in support of the United States upholding its commitments under the Paris Agreement.[33] Joining ACP in lobbying was the U.S.-based **Medical Society Consortium on Climate and Health**, a collective created in 2016 that "organizes, empowers and amplifies the voice of America's doctors and other health professionals to advocate for equitable and effective health-focused climate solutions."[34] As of 2021, the Consortium includes 33 national medical societies representing more than 600,000 physicians (60% of all doctors in the U.S.).

A coalition of American health and medical organizations recently released a report, "U.S. Call to Action on Climate, Health, and Equity: A Policy Action Agenda," that is a "roadmap to develop coordinated strategies for simultaneously tackling climate change, health, and equity."[35] This plan, which has been endorsed by more than 170 medical associations, health and environmental organizations, and academic programs, recommends 10 actions:

1. "Meet and strengthen U.S. commitments under the Paris agreement."
2. "Transition rapidly away from the use of coal, oil and natural gas to clean, safe, and renewable energy and energy efficiency."
3. "Emphasize active transportation in the transition to zero-carbon transportation systems."

4. "Promote healthy, sustainable and resilient farms and food systems, forests, and natural lands."

5. "Ensure that everyone in the U.S. has access to safe and affordable drinking water and a sustainable water supply."

6. "Invest in policies that support a just transition for workers and communities adversely impacted by climate change and the transition to a low-carbon economy."

7. "Engage the health sector voice in the call for climate action."

8. "Incorporate climate solutions into all healthcare and public health systems."

9. "Build resilient communities in the face of climate change."

10. "Invest in climate and health."[35]

Relevant to the 10th action is the 2021 pledge by the Biden administration to provide $110 million in expanded funding to the National Institutes of Health and $110 million to the Centers for Disease Control and Prevention for climate and health research.[30]

The AAP specifically calls on pediatricians to reach out to policy makers and the public and to make climate action recommendations to government officials. The AAP emphasizes in particular healthier city planning, walkability, public transportation, and publicly accessible green space, which they argue will improve children's health.[28] In 2020, U.S. pediatricians formed a new volunteer network of **AAP Chapter Climate Advocates** to raise awareness about climate change and its effects on children's health. The program is the brainchild of pediatrician Dr. Lori Byron, who practiced on the Crow Indian reservation in Montana under the Indian Health Service and witnessed firsthand the impacts of extreme flooding in the area in 2011. The floods "took away everything. It just hits you,

the environmental injustice of the whole situation," said Dr. Byron.[36] South Carolina pediatrician Dr. Hayley Guilkey remarked, "Prior to joining this group, I didn't see how connected climate change is to health. It's not something that I learned about in residency or medical school."[36] It is not surprising that pediatricians are involved in climate action because of their frontline role to protect children's well-being. Said Dr. Aparna Bole, a network member, "Pediatricians are used to thinking about kids in context about their families, homes, and environments, and extending our purview of children beyond the 15 minutes we spend with them."[36]

Climate Impacts of the Healthcare Sector

The healthcare sector is on the front lines of protecting human health and well-being from the effects of climate change, but it also contributes significantly to GHG emissions and other environmental problems (**Figure 12.2**). In 2015, GHG emissions from the global healthcare sector were 4.4% of total GHG emissions that year, and the sector was also responsible for 2.8% of global particulate matter pollution and withdrawal of an enormous amount of the world's scarce water supply.[37] GHG and air pollutant emissions result mostly from *direct* operations of healthcare facilities and transport, and other impacts are *indirect* and result from agriculture, forestry, and industrial processes that contribute to supply chains for medical products, food, and other services in the health sector.

In the United States, approximately 10% of national GHG emissions come from health care, as well as significant releases of hazardous air pollutants and smog formation.[38] In 2018, the United States' health sector recorded the highest emissions per capita

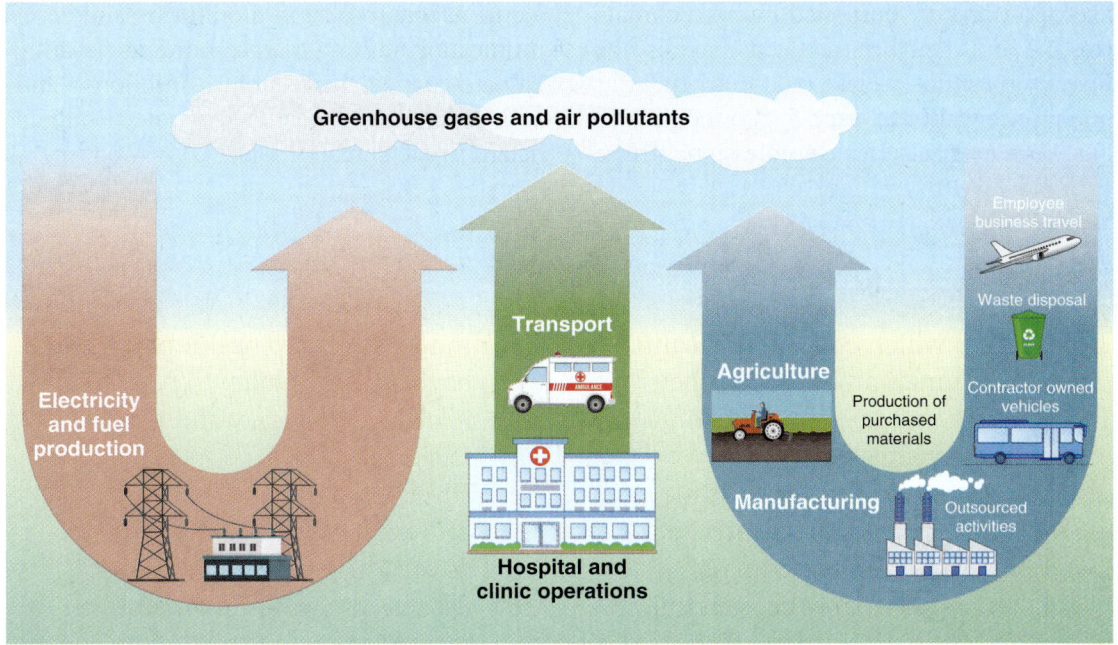

Figure 12.2 Climate Change Footprint of the Healthcare Sector.

Data from The Greenhouse Gas Protocol Initiative. Scope 3 accounting and reporting standard: supplement to the GHG protocol corporate accounting and reporting standard. World Resources Institute & World Business Council for Sustainable Development; 2009. https://www.ghgprotocol.org/sites/default/files/ghgp/ghg-protocol-scope-3-standard-draft-for-stakeholder-review-summary.pdf

of any industrialized country's health sector, which were linked to 400,000 disability-adjusted life years (DALYs) lost, mostly attributed to diseases from airborne particulate matter exposure.[39] Emissions varied by U.S. state and were not highly correlated with the quality of the state's health system, suggesting that reducing carbon footprints can be achieved without sacrificing healthcare quality.[39] The UK's National Health Service (NHS), the largest publicly funded health system in the world, contributes about 5% of its country's GHG emissions.[40] The NHS has committed to lowering its carbon footprint, and in 2019, its GHG emissions were down 26% since 1990, mostly through making low-carbon choices in its supply chains.[41]

Ultimately, climate action must improve health outcomes, quality of health care, equity in the health system, and efficacy of health expenditures.[37] Health professionals are important voices to lend support to measures that will lower the carbon and environmental footprints of hospitals, clinics, transport, and medical supply chains, in addition to advocating for climate action in policy-making. To lower GHG emissions, they can also prioritize preventive care, which may reduce the need for hospital and clinic operations, as well as telehealth services and effective online communications with patients in lieu of in-person appointments that require travel. These measures have been adopted in many places during the COVID-19 pandemic.

Because the health sector's GHG emissions contribute to fueling climate change, it is responsible for both creating and responding to the widespread health impacts of climate change. People who are disproportionately burdened by both climate stressors and their health effects may also be those who cannot access or do not benefit as much from healthcare services, so their needs must be prioritized. Health professionals have a critical role to play in monitoring and communicating shifting disease burdens, treating climate-change-attributable conditions, and advocating for decisions to protect human health in the climate crisis.

Discussion Questions

1. Discuss which conditions health professionals are seeing in their patients that they connect to climate change.
2. How important do you think it is for clinicians to attribute health problems to climate change? What are the benefits, if any? What are some of the difficulties in making this attribution?
3. Should clinicians talk to patients about the need to take climate action? Why or why not?
4. In what ways do health professionals have a responsibility to be involved in climate advocacy and awareness raising?
5. Discuss examples of health professionals organizing and advocating for change.
6. Discuss the ways that the health sector contributes to climate change and what steps can be taken to reduce its carbon footprint.
7. Generally, how can we raise public awareness that climate change is a human health issue?

References

1. Wellbery CE. Climate change health impacts: a role for the family physician. *Am Fam Physician*. 2019;100(10):602–603. https://www.aafp.org/afp/2019/1115/p602.html
2. Bella T. Historic heat wave in Pacific Northwest has killed hundreds in U.S. and Canada over the past week. *The Washington Post*. July 1, 2021. Accessed July 2, 2021. https://www.washingtonpost.com/nation/2021/07/01/heat-wave-deaths-pacific-northwest/
3. Teirstein Z. Doctors put a price tag on the annual health impacts of climate change. It's $820 billion. *Grist*. May 21, 2021. Accessed July 2, 2021. https://grist.org/health/doctors-put-a-price-tag-on-the-annual-health-impacts-of-climate-change-its-820-billion/
4. Rogers W. Moral responsibility in medicine: where are the boundaries? *The Lancet*. 2020;396(10248):373–374. doi:10.1016/s0140-6736(20)31643-3
5. Bailey M. How climate change is putting doctors in the hot seat. *KHN*. April 20, 2020. Accessed July 2, 2021. https://khn.org/news/how-climate-change-is-putting-doctors-in-the-hot-seat/
6. Medical Consortium on Climate and Health, Natural Resources Defense Council, Wisconsin Health Professionals for Climate Action. The costs of inaction: the economic burden of fossil fuels and climate change on health in the United States. May 2021. Accessed July 2, 2021. https://www.nrdc.org/sites/default/files/costs-inaction-burden-health-report.pdf
7. Lawrence WR, Lin Z, Lipton EA, et al. After the storm: short-term and long-term health effects following Superstorm Sandy among the elderly. *Disaster Med Public Health Prep*. 2019;13(1):28–32. doi:10.1017/dmp.2018.152
8. Sorensen CJ, Salas RN, Rublee C, et al. Clinical implications of climate change on U.S. emergency medicine: challenges and opportunities. *Ann Emerg Med*. 2020;76(2):168–178. doi:10.1016/j.annemergmed.2020.03.010

9. Maibach EW, Kreslake JM, Roser-Renouf C, et al. Do Americans understand that global warming is harmful to human health? Evidence from a national survey. *Ann Global Health*. 2015;81(3):396–409. doi:10.1016/j.aogh.2015.08.010

10. Rice MB. Climate change at the bedside? Observations from an ATS membership survey. *Ann Am Thorac Soc*. 2015;12(2):245–246. doi:10.1513/AnnalsATS.201412-590ED

11. Boland TM, Temte JL. Family medicine patient and physician attitudes toward climate change and health in Wisconsin. *Wilderness Environ Med*. 2019;30(4):386–393. doi:10.1016/j.wem.2019.08.005

12. National Medical Association. About Us. Accessed February 2, 2021. https://www.nmanet.org/page/About_Us

13. Sarfaty M, Mitchell M, Bloodhart B, Maibach EW. A survey of African American physicians on the health effects of climate change. *Int J Environ Res Public Health*. 2014;11(12):12473–12485. doi:10.3390/ijerph111212473

14. Andersen M, Carlson G. Perspectives of Maine physicians on the health impacts of climate change. 2021. Unpublished data.

15. Sarfaty M, Bloodhart B, Ewart G, et al. American Thoracic Society member survey on climate change and health. *Ann Am Thorac Soc*. 2015;12(2):274–278. doi:10.1513/AnnalsATS.201410-460BC

16. Sarfaty M, Kreslake J, Ewart G, et al. Survey of international members of the American Thoracic Society on climate change and health. *Ann Am Thorac Soc*. 2016;13(10):1808–1813. https://www.ncbi.nlm.nih.gov/pmc/articles/PMC6944384

17. Sarfaty M, Kreslake JM, Casale TB, Maibach EW. Views of AAAAI members on climate change and health. *J Allergy Clin Immunol Pract*. 2016;4(2):333–335.e26. doi:10.1016/j.jaip.2015.09.018

18. Ryan EC, Dubrow R, Sherman JD. Medical, nursing, and physician assistant student knowledge and attitudes toward climate change, pollution, and resource conservation in health care. *BMC Med Educ*. 2020;20(1):200. doi:10.1186/s12909-020-02099-0

19. Wellbery C, Sheffield P, Timmireddy K, et al. It's time for medical schools to introduce climate change into their curricula. *Acad Med*. 2018;93(12).

20. Stewart I, Garcia-Navaroo L. In Florida, doctors see climate change hurting their most vulnerable patients. *NPR*. March 30, 2019. Accessed February 3, 2021. https://www.npr.org/2019/03/30/706941118/in-florida-doctors-see-climate-change-hurting-their-most-vulnerable-patients

21. Florida Clinicians for Climate Action. Climate change is harming our health. Accessed February 3, 2021. https://floridaclinicians.org/

22. Limaye VS, Grabow ML, Stull VJ, et al. Developing a definition of climate and health literacy. *Health Aff (Millwood)*. 2020;39(12):2182–2188.

23. Philipsborn RP, Sheffield P, White A, et al. Climate change and the practice of medicine: essentials for resident education. *Academic Med*. 2020. doi:10.1097/acm.0000000000003719

24. Lopez G. The Black-white life expectancy gap grew in 2020—but it can be reversed. *Vox*. February 24, 2021. Accessed February 24, 2021. https://www.vox.com/22285868/black-white-life-expectancy-gap-covid-19-health

25. National Institute of Environmental Health Sciences. Climate change and human health lesson plans. March 13, 2020. Accessed January 31, 2021. https://www.niehs.nih.gov/health/scied/teachers/cchh/index.cfm

26. United States Global Change Research Program. The impacts of climate change on human health in the United States: a scientific assessment. 2016. Accessed January 29, 2021. https://health2016.globalchange.gov

27. Public Health Institute Center for Climate Change & Health. A physician's guide to climate change, health and equity. September 2016. Accessed February 2, 2021. https://climatehealthconnect.org/resources/physicians-guide-climate-change-health-equity/

28. American Academy of Pediatrics. Global climate change and children's health. *Pediatrics*. 2015;136(5):992–997. doi:10.1542/peds.2015-3232

29. Global Consortium on Climate and Health Education. Core climate & health competencies for health professionals. March 2018. Accessed January 31, 2021. https://www.publichealth.columbia.edu/sites/default/files/pdf/gcche_competencies.pdf

30. Fieseler C. The search for the Dr. Fauci of climate change. *Grist*. June 21, 2021. Accessed July 2, 2021. https://grist.org/health/doctors-climate-change-health-medicine-anthony-fauci/

31. Medical Students for a Sustainable Future. Climate change is a health crisis, and doctors aren't prepared. Accessed July 2, 2021. https://ms4sf.org/

32. Blumenthal D, Hamburg M. U.S. health and health care are a mess: now what? *The Lancet*. 2021;397(10275):647–648. doi:10.1016/S0140-6736(21)00318-4

33. American College of Physicians. Why tackle climate change? "Our health and the planet are at stake." May 17, 2019. Accessed January 30, 2021. https://www.acponline.org/advocacy/acp-advocate/archive/may-17-2019/why-tackle-climate-change-our-health-and-the-planet-are-at-stake

34. The Medical Society Consortium on Climate and Health. Consortium summary. July 2021. Accessed July 2, 2021. https://medsocietiesforclimatehealth.org/wp-content/uploads/2021/07/Consortium Summary-Jul2021.pdf

35. Health Voices for Climate Action. U.S. call to action on climate, health, and equity: a policy action agenda. 2019. Accessed January 30, 2021. https://climatehealthaction.org/cta/climate-health-equity-policy

36. Chakradhar S. New pediatrician network puts spotlight on climate change's effects on children. *STAT News*. December 19, 2020. Accessed February 3, 2021. https://www.statnews.com/2020/12/18/new-pediatrician-network-puts-spotlight-on-climate-changes-effects-on-children/

37. Lenzen M, Malik A, Li M, et al. The environmental footprint of health care: a global assessment. *Lancet Planet Health*. 2020;4(7):e271–e279. doi:10.1016/S2542-5196(20)30121-2

38. Eckelman MJ, Sherman J. Environmental impacts of the U.S. health care system and effects on public health. *PLoS One*. 2016;11(6):e0157014. doi:10.1371/journal.pone.0157014

39. Eckelman MJ, Huang K, Lagasse R, et al. Health care pollution and public health damage in the United States: an update. *Health Aff (Millwood)*. 2020;39(12):2071–2079.

40. Bawden A. The NHS produces 5.4% of the UK's greenhouse gases. How can hospitals cut their emissions? *The Guardian*. September 18, 2019. Accessed February 4, 2021. https://www.theguardian.com/society/2019/sep/18/hospitals-planet-health-anaesthetic-gases-electric-ambulances-dialysis-nhs-carbon-footprint

41. Tennison I, Roschnik S, Ashby B, et al. Health care's response to climate change: a carbon footprint assessment of the NHS in England. *Lancet Planet Health*. 2021;5(2):e84–e92. doi:10.1016/S2542-5196(20)30271-0

Glossary

A

Aedes mosquito Vector for common viral vector-borne diseases such as dengue, Zika, Chikungunya, and yellow fever, including *Aedes aegypti* (yellow fever mosquito) and *Aedes albopictus* (Asian tiger mosquito). *Aedes* mosquitoes transmit these viruses to humans. Survival, breeding, and biting behavior are sensitive to climate conditions.

Aeroallergens Airborne allergens such as plant pollen that cause allergies and asthma and are increasing with rising atmospheric CO_2 levels and temperatures, as well as changing precipitation patterns.

Aflatoxins Naturally occuring proteins produced by the molds *Aspergillus flavus* and *Aspergillus paraciticus* that frequently infect crops such as corn and groundnuts and are harmful to human health.

Air quality index (AQI) Composite air quality indicator from real-time measures of local air pollutant levels that is made available to the public and used to determine local air quality alerts.

Allergies A common health problem characterized by the immune system overreacting to exposure to a foreign substance (allergen), often certain foods, plant pollen, dust, pet dander, and insect bites.

Alveoli Tiny balloon-like sacs at the end of lung airways that have an elastic membrane across which gas exchange occurs to bring oxygen into the bloodstream and remove carbon dioxide. Pollutant gases may damage the structural and functional integrity of alveoli, leading to chronic breathing problems.

Analytic epidemiology Type of epidemiology in which studies are designed to analyze associations between a specific health outcome and risk factors or interventions (*how* and *why* people get sick). Examples are case-control and cohort studies and clinical trials.

Anopheles mosquito Vector that transmits the malaria pathogen to humans. Survival, breeding, and biting behavior are sensitive to climate conditions.

Anxiety disorders Mental health conditions characterized by a state of excessive uneasiness and apprehension that may come in many forms, including generalized anxiety (persistent and excessive worry that is difficult to control), social anxiety, separation anxiety, panic disorders, and phobias.

Arbovirus Any of a number of insect-borne viruses (**ar**thropod-**bo**rne **viruses**) that cause human diseases, including dengue.

Asthma A chronic respiratory condition affecting people of all ages in which airways are prone to inflammation and narrowing, making it hard to breathe. Asthma is often triggered by exposure to allergens, pollutants, microbes, and cold.

B

Bali Principles of Climate Justice A set of principles created by an international coalition of human rights and Indigenous groups that frames climate change as a threat to the protection of human rights and environmental justice.

Borrelia burgdorferi Bacterial pathogen that causes Lyme disease and is transmitted to humans and other hosts by the bite of infected *Ixodes* ticks.

C

Carbon sinks Reservoirs that remove CO_2 from the atmosphere via sequestration, such as forests and oceans, which has the effect of increasing the content of carbon pools on land and in water.

Cerebral malaria (CM) A severe form of malaria in which a patient has neurological complications characterized by hallucinations, seizures, lowered consciousness, coma and death.

Cholera A potentially deadly diarrheal disease caused by *Vibrio cholerae* infection. Cholera transmission is sensitive to climate conditions.

Chronic bronchitis A chronic respiratory condition in which difficulty breathing and decreased lung function result because of persistently inflamed, swollen, and irritated airways that have high levels of mucus buildup. This is one form of COPD and is often associated with prolonged exposure to tobacco smoke or air pollutants.

Chronic obstructive pulmonary disease (COPD) A potentially fatal chronic respiratory disease resulting from emphysema and/or chronic bronchitis that is characterized by lung dysfunction and difficulty breathing that becomes progressively worse over time.

Clean Air Act The major U.S. law that regulates air pollution. Clean air regulations have targeted sources of air pollution for cleanup, including coal-burning power plants and diesel- or gasoline-powered trucks and other vehicles, which also produce GHGs.

Climate Statistical description of the mean, extremes, and variability of climate features such as *temperature, precipitation*, and *wind* over a period that can range from months to thousands of years but is typically three decades; "average weather."

Climate adaptation Actions that build the capacity of populations to prevent, prepare for, and respond to the adverse impacts of climate change, including those that affect human health. Climate adaptations include investments in emergency preparedness, disaster risk reduction, climate-resilient development projects and the health sector.

Climate change Changes in the Earth's climate attributed directly or indirectly to human activities, that are altering the composition of the global atmosphere and are in addition to natural climate variability observed over the same time period.

Climate finance Mechanisms whereby high-income countries assist low- and middle-income countries in funding low-carbon and climate-resilient development projects.

Climate justice The equitable protection of all people from climate impacts, regardless of race, ethnicity, national origin, indigeneity, gender, income, or other status, and the inclusion of all people in decision-making about climate change solutions.

Climate mitigation Actions that reduce (mitigate) emissions of greenhouse gases into the atmosphere in order to limit global warming and minimize climate impacts.

Climate models Sophisticated computational representations of climate systems and future climate changes that incorporate a range of plausible scenarios about end-of-century GHG emissions and pathways of socioeconomic development that likely influence drivers of and responses to climate change.

95% confidence interval (95% CI) A statistical measure used in epidemiological studies that reflects the range of values that contains the true value for the sample population with 95% certainty (confidence). For an association between an exposure and a health outcome to be considered significant, the entire 95% CI range must be greater than 1 for a *positive association* or less than 1 for a *negative association*.

D

Dengue A highly prevalent and serious vector-borne disease caused by the dengue virus and transmitted by *Aedes* mosquitoes. Symptoms include fever, nausea, vomiting, rash, and aches and pains (behind-the-eye pain; muscle, joint, or bone pain). In some cases, particularly upon subsequent dengue virus infection, severe dengue occurs, which may be fatal.

Descriptive epidemiology Type of epidemiology in which the amount and distribution of health, including specific diseases, within a population is described (the *who, what, when, where,* and *how many* of human illness or wellness). Examples include case reports and cross-sectional and ecological studies.

Diarrheal diseases A group of potentially fatal diseases caused by a range of infectious agents that induce adverse gastrointestinal symptoms and dehydration. Diarrheal diseases cause a significant proportion of global morbidity and mortality.

Disability-adjusted life year (DALY) One year of healthy life lost in a population due to disability or premature death, usually indicated for a specific disease or risk factor. DALYs are the sum of years of life lost due to premature deaths (YLLs) and years of healthy life lost due to disability (YLDs).

Displacement The forced or voluntary movement of people away from their places of habitual residence, in particular to avoid the effects of natural or human-made disasters, armed conflict, violence, and/or human rights violations.

Drought A period of unusually prolonged dry weather that persists long enough to cause serious problems such as water supply shortages and crop losses. The severity of the drought depends upon the degree of moisture deficiency, the duration, and the size of the affected area. Drought is increasing in frequency and severity with climate change-driven warming, extreme heat, altered precipitation, and land desertification.

E

Eco-anxiety Stress or dread caused by phobic fear of extreme weather, unrelenting day-to-day despair about climate change impacts, or worrying about the future

El Niño The phase of ENSO associated with warmer than average surface waters in the tropical Pacific Ocean, which results in varied alterations in weather and climate in much of the world, including warmer temperatures and drier conditions in the northern United States and Canada, particularly in the winter months.

El Niño–Southern Oscillation (ENSO) Naturally-occurring periodic fluctuations in the temperature of surface waters in much of the tropical Pacific Ocean that affect air currents across the Pacific. The three ENSO phases are El Niño, La Niña, and the neutral phase (neither El Niño nor La Niña). ENSO phases have varying impacts on weather and climate around the world.

Emphysema A chronic respiratory condition in which difficulty breathing and decreased lung function result because of dysfunctional lung alveoli that break down and lose their elasticity. This is one form of COPD and is associated with prolonged exposure to tobacco smoke or air pollutants.

Endemic A term referring to a disease that occurs consistently in a particular geographic region or population.

Environmental justice (EJ) The fair treatment and meaningful involvement of all people regardless of race, color, national origin, or income in the development, implementation, and enforcement of environmental laws, regulations, and policies.

Environmental racism The imposition of disproportionate environmental impacts on groups or communities because of their racial or ethnic status, and exclusion of representatives of such groups from participation in environmental decision-making.

Epidemiology The study of the distribution and determinants of diseases and other health conditions in human populations.

Extreme weather Occurrences of unusually severe weather or climate conditions, including heat waves, floods, heavy rainfall, and tropical storms, that often devastate human, agricultural, and natural systems. Such events have always occurred periodically, but climate change is making these events more frequent, more intense, and more destructive.

Extrinsic incubation period (EIP) For vector-borne disease pathogens, the time interval from when the pathogen is ingested by the vector to when it can be transmitted from the vector to a host. EIP is sensitive to climate conditions.

F

Fecal–oral route A common pathway of infectious disease transmission that allows disease-causing pathogens that contaminate feces to pass from infected to uninfected people, either directly or via contaminated water, soil, food, or flies.

Flooding The overflow of water that submerges land that is usually dry, caused by extreme precipitation, hurricanes or other storms, sea level rise, coastal storm surges, rivers overflowing their banks, and sewer or storm water system backups.

Food insecurity A condition in which people lack regular access to sufficient safe and nutritious foods for adequate growth and development and the ability to lead an active and healthy life.

G

Geographic inequity The discriminatory siting, location, and configuration of environmental hazards in proximity to certain groups or populations, often based on race/ethnicity, socioeconomic condition, or gender.

Global Burden of Disease (GBD) report A comprehensive summary of the health status of people in countries all over the world compiled by the Institute for Health Metrics and Evaluation, published in the medical journal *The Lancet*, with data made publicly available on the IHME website. The GBD report summarizes at spatially-explicit scales the burdens of a wide range of communicable and noncommunicable diseases and injuries, as well as risk factors for ill health, many of which may be impacted by climate change.

Global Hunger Index (GHI) A composite measure of hunger in the world's countries that incorporates country-level prevalence of undernourishment and levels of wasting, stunting, and mortality in children under 5; GHI is updated annually.

Global warming potential (GWP) The measure of how much energy the emissions of one ton of a greenhouse gas will absorb over a given period—usually 100 years—relative to the emissions of one ton of CO_2.

Greenhouse effect The phenomenon in which gases in the earth's atmosphere trap heat radiating from the earth after being warmed by energy from the sun, which has the effect of making the earth warmer than it would be without an atmosphere. The greenhouse effect occurs naturally, allowing the earth to support life, and is also being enhanced due to the introduction of anthropogenic greenhouse gases into the atmosphere.

Greenhouse gases (GHGs) Specific gases in the earth's atmosphere that absorb and emit radiation coming from the Earth's surface, the atmosphere, and clouds after they are heated by the sun. GHGs are responsible for the greenhouse effect.

Ground-level ozone Ozone gas (O_3) that forms at the surface of the earth in a chemical reaction in the presence of heat and sunlight between volatile organic compounds and nitrogen oxides (NOx) from vehicle tailpipe emissions and industrial sources. Exposure to ground-level ozone is associated with adverse respiratory conditions, particularly asthma.

H

Harmful algal blooms (HABs) Significant growth in microalgae in oceans or freshwater systems that produce toxins that can build up in shellfish. When humans eat contaminated shellfish, they risk shellfish poisoning in which the toxins cause adverse health impacts that range from gastrointestinal to severe neurological conditions. Microalgae growth is sensitive to climate conditions, particularly ocean surface temperatures.

Health According to the World Health Organization, "a state of complete physical, mental, and social well-being and not merely the absence of disease or infirmity."

Heat acclimatization A biological phenomenon in which the body responds to frequent exposure to high temperatures by adjusting to better thermoregulate, including increased sweating and blood flow in the skin to release heat from the body.

Heat adaptation A set of human behaviors to regulate heat exposure, including seeking cool environments, that reduce the health impacts of heat.

Heat index (HI) A measure of "apparent temperature" that combines temperature in the shade and relative humidity to reflect the heat that people actually experience. Both high temperature and high humidity impair the body's ability to cool itself.

Heat stress A condition in which the body cannot get rid of excess heat upon exposure to high temperatures because regulation of its core temperature is compromised.

Heat wave According to the World Meteorological Organization, an event in which the daily maximum temperature in a specific location exceeds the average maximum temperature for that location by 9°F for more than five consecutive days.

Heatstroke The deadliest form of heat stress, in which the body stops cooling itself because dehydration prevents sweating and/or high humidity prevents sweat evaporation.

Host Human or other animal that becomes infected by a disease pathogen.

Human rights Universal entitlements that all people possess simply by virtue of being human, regardless of nationality, ethnicity, religion, sex, disability, or any other status. Everyone has a moral obligation, and sometimes also a legal obligation, to protect people's human rights.

Hunger According to the UN Food and Agriculture Organization, "an uncomfortable or painful physical sensation caused by insufficient consumption of dietary energy. It becomes chronic when the person does not consume a sufficient amount of calories (dietary energy) on a regular basis to lead a normal, active, and healthy life."

Hygiene Access to facilities for handwashing with improved water and soap.

Immobility The inability of people exposed to climate-related or other threats to move because of limited social and economic capital, creating "trapped populations," or the voluntary choice of people not to move but rather to stay in place and adapt to these threats in situ.

Incidence The number of cases of a specific disease or other health condition diagnosed in a specified population during a specified time interval (usually one year).

Incidence rate The rate at which new cases of disease occur in a specified population during a specified time interval.

Institute for Health Metrics and Evaluation (IHME) An independent global health research center at the University of Washington that provides timely, rigorous, and comparable information on the health of populations around the world and determinants of health to improve health policy and practice. IHME compiles the Global Burden of Disease report each year.

Intergovernmental Panel on Climate Change (IPCC) The United Nations body for assessing the science related to climate change. The IPCC provides regular assessments of the scientific basis of climate change, its impacts and future risks, and options for adaptation and mitigation. These Assessment Reports are made available for use by governments, researchers, organizations, and the public.

Internally displaced persons (IDPs) People who move away from their place of habitual residence but do not cross international borders. Many IDPs are forced to move due to threats of conflict, severe food insecurity, and extreme weather. IDPs are not protected by international law because they are under the legal protection of their own country.

***Ixodes* tick** Vector that transmits *Borrelia burgdorferi*, the Lyme disease pathogen, to host species, including humans. These ticks also transmit other vector-borne diseases, including anaplasmosis, babesiosis, and Powassan virus disease, and are widespread in much of the world.

K

Kwashiorkor A form of severe acute malnutrition (undernutrition) resulting from deficient intake of protein.

Kyoto Protocol Treaty negotiated under the UN Framework Convention on Climate Change that committed developed countries to reduce their GHG emissions under the principle of "common but differentiated responsibility and respective capabilities," recognizing that developed countries are largely responsible for the high levels of atmospheric GHG emissions and thus have a greater obligation to act.

L

La Niña The phase of ENSO associated with cooler than average surface waters in the tropical Pacific Ocean, which results in varied alterations in weather and climate in much of the world, including warmer and drier weather patterns across the southern United States and northern Mexico, and cooler and wetter patterns across the north.

Least developed countries (LDCs) According to the UN, "low-income countries confronting severe structural impediments to sustainable development. They are highly vulnerable to economic and environmental shocks and have low levels of human assets."

Legionellosis Also known as Legionnaires' disease, an infectious disease caused by *Legionella* bacteria, most often resulting in a serious and sometimes fatal form of pneumonia.

Leishmania Protozoan pathogen that causes leishmaniasis and is transmitted to humans via the bite of a phlebotomine sand fly.

Leishmaniasis Vector-borne disease caused by the parasitic protozoan *Leishmania* and transmitted by phlebotomine sand flies. The most

serious form of this disease, visceral leishmaniasis, causes fever, weight loss, and an enlarged spleen and liver, and may be fatal.

Leptospirosis A zoonotic disease, originating in animals and spreading to humans, that is caused by five pathogenic bacteria of the genus *Leptospira*.

Lyme disease The most common vector-borne disease affecting humans in the Northern–Hemisphere and is one of the most prevalent tick-borne diseases in the world, caused by infection by the bacterium *Borrelia burgdorferi* and transmitted by *Ixodes* ticks. Symptoms include a bull's-eye rash at the site of tick bite, fatigue, and joint pain.

M

Major depressive disorder (MDD) A serious mood disorder that causes persistent sadness, loss of interest, and interference with daily functioning.

Malaria Highly prevalent and deadly vector-borne disease caused by infection by the parasitic protozoan *Plasmodium* and transmitted through the bite of the *Anopheles* mosquito. Malaria vectors are highly sensitive to climate conditions.

Malarial anemia A serious condition in malaria patients resulting from deficiency in red blood cells (RBCs) and thus hemoglobin carried by RBCs, causing fatigue, dizziness, shortness of breath, rapid heartbeat, and neurological problems.

Malnutrition A set of adverse health conditions that result from deficiencies, excesses, or imbalances in consumption of energy, macronutrients (proteins, fats, and carbohydrates), or micronutrients (vitamins and minerals).

Marasmus A form of severe acute malnutrition (undernutrition) resulting from deficient intake of total calories.

Mental health According to the World Health Organization, "a state of well-being in which an individual realizes his or her own abilities, can cope with the normal stresses of life, can work productively and is able to make a contribution to his or her community."

Micronutrient deficiencies Lack of adequate intake of essential vitamins and minerals required in small amounts by the body for proper growth and development, mostly caused by insufficient nutritious foods in the diet. These deficiencies cause significant human illness and suffering.

Migrants People who move away from their place of habitual residence, temporarily or permanently, either within their country or across international borders, for a variety of reasons.

Migration Movement of people away from their place of habitual residence, which occurs on a continuum ranging from entirely voluntary to entirely forced, with most situations somewhere in between.

Millennium Development Goals (MDGs) A set of UN goals and targets to combat extreme poverty and improve well-being in effect from 2000 to 2015.

Minimum mortality temperature (MMT) Location-specific ambient temperature with the lowest associated human mortality. Temperatures above the MMT cause heat-related mortality, and temperatures below the MMT cause cold-related mortality.

Morbidity The condition of suffering but not dying, usually attributed to a specific disease or risk factor.

Mortality Deaths, usually attributed to a specific disease or risk factor.

Mortality displacement A phenomenon in which a period of excess deaths in a population is observed, followed by a period of fewer deaths than expected, often resulting from health threats such as heat waves, epidemics, famine, or war. These excess deaths are primarily among people who are already gravely ill and at risk of dying during the period of interest even without the harmful exposure.

Mortality rate The rate of total deaths in a specified population during a specified time interval.

N

Nationally determined contributions (NDCs) National plans mandated under the Paris Agreement in which countries prepare, communicate, and implement greenhouse gas emissions reductions targets and climate adaptation goals.

O

Odds ratio (OR) Quantitative measure from an epidemiological study of the degree of association between a disease and an exposure calculated by comparing the odds that sample populations with and without the disease have been exposed. An OR greater than one indicates an apparent *positive association* (exposure to the risk factor is correlated with more disease), and an OR less than one indicates an apparent *negative association* (exposure to the risk factor is correlated with less disease). An OR equal to one indicates no association because the odds of exposure are the same in populations with and without the disease.

Onchocerciasis Also known as river blindness, a vector-borne disease caused by the parasitic worm *Onchocerca volvulus* transmitted by blackflies that infects the skin and other tissues, including the eyes, which is a leading cause of blindness in regions where the disease is endemic.

Overnutrition Excess intake of nutrients that leads to accumulation of body fat, which may impair health.

P

Parasites Pathogens that cause disease and carry out part of their life cycle in the infected host, during which they take nutrients from the host but confer no benefit to the host in return.

Paris Agreement International climate treaty agreed to in Paris, France, in December 2015 that compels all countries to commit to ambitious climate action with the goal of limiting global warming to less than 2°C, preferably to 1.5°C compared to pre-industrial levels.

Particulate matter (PM) Major component of air pollution that includes coarse (PM_{10}), fine ($PM_{2.5}$), and ultrafine ($PM_{0.1}$) particles of complex chemical composition, with the number indicating maximum particle diameter. PM exposure causes and exacerbates numerous diseases in humans.

Pathogen Infectious agent that causes disease in humans, including bacteria, viruses, protists, and worms.

Phlebotomine sand fly Vector for the disease Leishmaniasis that becomes infected with the protist *Leishmania* and transmits it to humans.

Planned relocation A coordinated process in which people or communities are moved preemptively with the assistance of governments to protect them from the risks and impacts of environmental disasters, including climate change.

Plasmodium Pathogen that causes malaria. Several species cause disease and are transmitted to humans through the bite of *Anopheles* mosquitoes.

PM₂.₅ A particularly hazardous form of particulate matter produced from incomplete combustion of fossil fuels, wood, and vegetation and thus a major component of diesel exhaust and wildfire smoke. $PM_{2.5}$ exposure is linked to excess cardiovascular and respiratory morbidity and mortality.

Posttraumatic growth A mental health phenomenon in which people become more resilient to future traumatic events because of improved recognition of personal strength, relationships, appreciation for life, and new possibilities that arise because of the experience of living through trauma.

Posttraumatic stress disorder (PTSD) A psychiatric disorder that may occur in people who have experienced or witnessed a traumatic event such as a natural disaster or human conflict, or who have been threatened with death, violence, or serious injury.

Precautionary principle As described in the Wingspread Statement, "when an activity raises threats of harm to human health or the environment, precautionary measures should be taken even if some cause and effect relationships are not fully established scientifically."

Prevalence The total number of cases of a specific disease in a specified population, regardless of when diagnoses were made.

Prevalence rate The rate of total cases of disease in a specified population during a specified time interval.

Procedural inequity The discriminatory implementation and enforcement of environmental regulations that result in disproportionate burdens on certain groups or populations, often based on race/ethnicity, socioeconomic condition, or gender.

Project Drawdown A nonprofit organization helping to accelerate global "drawdown" (the future point when GHG emissions start to steadily decline). The organization conducts rigorous research about climate solutions and provides this information to governments, scientists, companies, community stakeholders, educators, and activists.

Protein-energy malnutrition (PEM) A form of severe acute malnutrition that results from inadequate consumption of protein-rich foods (kwashiorkor), total calories (marasmus), or both.

Public health The science and art of preventing disease and promoting health in populations, with an emphasis on controlling social and environmental determinants of health.

p-value A statistical measure of the probability, expressed as a decimal, that the observed association between a risk factor and a health outcome in an epidemiological study could have occurred by chance alone.

R

Radiative forcing A measure of the energy imbalance on Earth calculated by subtracting the energy radiating away from the earth's surface and lower atmosphere from the sun's energy flowing in. Because of anthropogenic greenhouse gas emissions, radiative forcing is increasing, indicating that the earth is warming.

Ready-to-use therapeutic food (RUTF) High-calorie, high-nutrient food supplement, usually in paste form, used to treat children and others suffering from severe acute malnutrition. RUTF does not require processing or cooking, has a long shelf life, and can be made from locally-sourced ingredients.

Refugees According to the UN, people who are forced to flee their country because of war, violence, conflict or persecution due to race, religion, nationality, political opinion, or membership in a particular social group. Refugees cross international borders seeking safety in another country, and in most cases, cannot return home or fear doing so.

Relative risk/rate ratio (RR) Quantitative measure from an epidemiological study of the degree of association between a disease and an exposure by comparing rates of disease in exposed and unexposed sample populations. An RR greater than one indicates an apparent *positive association* (exposure to the risk factor is correlated with more disease), and an RR less than one indicates an apparent *negative association* (exposure to the risk factor is correlated with less disease). An RR equal to one indicates no association because disease rates are the same in exposed and unexposed populations.

Reservoir host Animal species, including humans, that may become infected by vector-borne disease pathogens and can be a reservoir of pathogen that spreads to uninfected vectors upon vector biting, perpetuating the cycle of disease.

Respiratory system The network of organs and tissues in the human body responsible for the processes related to breathing, including absorption of oxygen from the air, removal of waste gases like carbon dioxide from the blood, and immune responses. The upper airways of the nose, mouth, and throat have mucus and cilia to capture and clear foreign substances such as large airborne particles, allergens, and some infectious microbes. The lower airways may also trap and clear foreign substances, but are also exposed to the harmful effects of fine particles, gases, aerosols, and many infectious agents that enter the lungs when inhaled and may penetrate deep into the airways.

Risk The probability that an adverse health condition will arise in an individual or a population. Risk arises when a hazard is present, when people are exposed to the hazard, and when people are susceptible to the health impacts of that hazard.

S

Sanitation Systems that physically separate humans from human excreta. A lack of sanitation causes many human diseases and significant morbidity and mortality.

Schistosomiasis Also known as bilharzia, a parasitic worm infection transmitted by certain freshwater snails. Humans become infected with parasites through skin contact with water containing these snails. Symptoms include rash, fever, cough, muscle aches, and eventually anemia, malnutrition, cognitive deficiencies, and organ damage.

Sea level rise Increase in the height of the world's oceans caused by melting ice and thermal expansion of seawater related to global warming.

Sendai Framework for Disaster Risk Reduction An international agreement to reduce the risks of disasters to lives, livelihoods, health and the broad assets of persons, businesses, communities, and countries by achieving seven specific global targets by 2030.

Severe acute malnutrition The most dire and visible form of undernutrition characterized by wasting, affecting millions of children and causing significant child mortality.

Severe dengue A potentially fatal form of dengue virus infection that requires immediate medical attention. Symptoms include fever, fatigue, restlessness, irritability, belly pain or tenderness, frequent vomiting, internal bleeding, bleeding from the nose or gums, blood in vomit or stool, and shock.

Severe malaria Infection with malaria pathogen that becomes complicated by serious organ failure or blood or metabolic abnormalities. Clinical manifestations include fever, seizures, vomiting, respiratory distress, hypoglycemia, impaired consciousness, and in the most severe cases, cerebral malaria.

Shellfish poisoning A set of human illnesses resulting from consumption of shellfish contaminated with biotoxins produced by toxic microalgae, characterized by gastrointestinal and neurological symptoms that may be severe. Risks may increase due to more frequent harmful algal blooms fueled by climate change.

Small island developing states (SIDS) A group of 58 countries located in one of three ocean regions of the world (Pacific, Caribbean, and AIS [Atlantic and Indian Oceans and South China Sea]) that face unique social, economic, and environmental vulnerabilities.

Social inequity The discriminatory impact of societal structures and practices that prevent fair participation of all people regardless of race, ethnicity, socioeconomic condition, or other status in environmental decision-making and the balance of power.

Soil-transmitted helminth infections A group of human illnesses caused by worm infections transmitted via contaminated soil that lead to significant morbidity and mortality worldwide and are often caused by a lack of sanitation.

Solastalgia An emotional response to environmental change characterized by a sense of loss of an important place that otherwise provides stability, security, or personal identity.

Stunting A condition in children characterized by low height-for-age that is due to chronic or recurrent malnutrition.

Sustainable Development Goals (SDGs) A set of 17 international goals to end poverty and promote health, education, equality, and sustainable economic growth as part of the UN's 2030 Agenda for Sustainable Development.

T

Temperature mortality response curve Graphical representation, often U-shaped, of the association between temperature and mortality in a population living in a specific location. Deaths tend to occur at high and low temperatures, and an optimal temperature in the middle range where mortality is lowest is defined by the curve.

The Global Fund to Fight AIDS, Tuberculosis and Malaria A global partnership among governments, civil society, technical agencies, the private sector, and impacted communities established in 2002 to accelerate progress to reduce the burdens of HIV/AIDS, tuberculosis, malaria, and other deadly infectious diseases.

The *Lancet* Countdown A global collaboration created in 2016 among leading climate scientists, engineers, energy specialists, economists, political scientists, public health professionals, and doctors from academic institutions and UN agencies to monitor health risks, impacts, and opportunities in the climate crisis.

Thermoregulation Physiological processes the human body uses to exchange heat with the environment in order to maintain an optimal internal body temperature.

Tropical cyclones Organized rotating systems with winds of at least 74 mph that form over warm ocean waters, which are some of the most extreme storms that occur on Earth.

U

U.S. Global Change Research Program (USGCRP) A collaboration of 13 federal government agencies in the United States to coordinate scientific and policy research on climate change.

Undernutrition Any of several human conditions caused by inadequate intake of food quantity or quality that lead to unhealthy dietary deficiencies and significant morbidity and mortality.

Underweight A condition in children characterized by low weight-for-age that is often due to a lack of adequate food intake.

United Nations Framework Convention on Climate Change (UNFCCC) The overarching international agreement on climate change adopted in 1992 at the UN Earth Summit in Rio de Janiero, Brazil. It is the parent treaty of the Paris Agreement and the Kyoto Protocol, and together, these three agreements aim to "stabilize greenhouse gas concentrations in the atmosphere at a level that will prevent dangerous human interference with the climate system, in a time frame which allows ecosystems to adapt naturally and enables sustainable development."

Urban heat island effect The increased absorption of solar radiation by human-made heat-trapping materials such as asphalt and concrete commonly used to construct roads, buildings, and other structures in cities.

V

Vector-borne diseases (VBDs) A large group of human illnesses caused by infection by pathogens, most often bacteria, viruses, protozoa, or worms, that are transmitted to humans by vector species, most often a bloodsucking insect (mosquito, tick, or fly). VBDs cause significant global morbidity and mortality.

Vector A living organism that can transmit infectious pathogens between humans or from animals to humans, most often a bloodsucking insect (mosquito, tick, or fly).

Vectorial capacity (VC) A measure of the capacity of a vector to transmit disease. VC reflects the number of new cases of disease or infected hosts per infectious vector and is based on vector and pathogen life cycles, vector density and survival, and biting rate, all of which may be temperature- and moisture-sensitive.

Vibriosis A group of infectious diarrheal diseases caused by infection by any of several species of *Vibrio* bacteria that inhabit marine and estuarine waters all over the world.

Vulnerability A measure of the degree of risk to a population of a climate impact or other hazard that accounts for the presence of the hazard, the biological and social susceptibility of the population, and a low degree of adaptive capacity to minimize the risk.

W

WASH Access to safe and adequate water, sanitation, and hygiene, a lack of which is a leading environmental determinant of ill health.

Wasting A condition in children characterized by low weight-for-height that is due to sudden or acute malnutrition, with immediate risk of death.

Water In the context of WASH, improved sources of freshwater for human use—including drinking, bathing, food preparation, and irrigation—that are protected against contamination, especially by disease pathogens and toxic chemicals.

Water-based diseases A set of human diseases caused by pathogens that spend all or most of their life cycles in water or depend on aquatic organisms for part of their life cycles.

Waterborne diseases A set of human diseases transmitted via direct exposure to water contaminated with pathogens.

Water-related diseases A set of human diseases transmitted by insects that breed in water or live near water, including mosquito-borne malaria and dengue.

Water scarcity A long-term imbalance between water demand and water availability in a specific region. Major drivers are rapid economic development, increased water consumption levels, population pressures, and climate change.

Water-washed diseases A set of human diseases that result from improper sanitation or lack of adequate handwashing facilities that prevent the removal of pathogens from the skin or from foods or other surfaces.

Weather The state of the air and atmosphere—specifically, *temperature, rain, clouds,* and *storms*—at a specific time and place.

Wet-bulb globe temperature (WBGT) An indicator of "apparent temperature" that incorporates measures of temperature in direct sunlight, humidity, wind speed, sun angle, and cloud cover.

Wildfire An unplanned, unwanted, and uncontrolled fire in an area of combustible vegetation, increasing in frequency and intensity due to climate change-fueled prolonged drought, drying vegetation, and increased human settlement at the wildland–urban interface.

Wood smoke A complex mixture of airborne pollutants emitted from wildfires and other processes that burn wood that are very hazardous to human health.

World Health Organization (WHO) The United Nations agency dedicated to global health that advises and connects UN member countries, partners, and people to promote health, keep the world safe, and serve the vulnerable in order to protect the fundamental human right to the highest attainable level of health.

Index

Note: Page numbers followed by *b*, *f*, or *t* indicate material in boxes, figures, or tables respectively.